D1216804

Squaring the Circle

SCIENCE AND ITS CONCEPTUAL FOUNDATIONS
A SERIES EDITED BY DAVID L. HULL

Squaring the Circle

The War between Hobbes and Wallis

Douglas M. Jesseph

THE UNIVERSITY OF CHICAGO PRESS
CHICAGO AND LONDON

DOUGLAS M. JESSEPH is associate professor of
philosophy at North Carolina State University. He is
author of *Berkeley's Philosophy of Mathematics* (1993),
also published by the University of Chicago Press.

The University of Chicago Press, Chicago 60637
The University of Chicago Press, Ltd., London
© 1999 by The University of Chicago
All rights reserved. Published 1999
Printed in the United States of America
08 07 06 05 04 03 02 01 00 99   1 2 3 4 5
ISBN: 0-226-39899-4 (cloth)
ISBN: 0-226-39900-1 (paper)

Library of Congress Cataloging-in-Publication Data

Jesseph, Douglas Michael.
    Squaring the circle : the war between Hobbes and Wallis / Douglas M.
Jesseph.
        p.   cm.—(Science and its conceptual foundations)
    Includes bibliographical references and index.
    ISBN 0-226-39899-4 (cloth : alk. paper).—ISBN 0-226-39900-1 (paper :
alk. paper)
    1. Hobbes, Thomas, 1588–1679—Contributions in mathematics.
2. Wallis, John, 1616–1703.   3. Mathematics—England—Philosophy—
History—17th century.   I. Title.   II. Series.
QA29.H58J47   1999
510′.942′09032—dc21   99-35819
                                                                      CIP

⊗ The paper used in this publication meets the minimum
requirements of the American National Standard for
Information Science—Permanence of Paper for Printed
Library Materials, ANSI Z39.48–1992.

For Doreen

Contents

vii

Preface

This book began as a footnote. In an earlier study of Berkeley's philosophy of mathematics I was led to consider Hobbes's views on the relationship between geometry and algebra, which at the time I deemed worthy of a passing reference. My curiosity had been aroused, however, and from there I soon found myself engaged in a study of Hobbes's mathematical writings, which inevitably turned into a study of his controversy with Wallis. I certainly did not start out planning to devote a significant part of my life to the detailed examination of a controversy that began with technical issues in mathematics and grew to include questions of theology, philology, politics, and the very nature of reason. In this, I suppose, my writing of the book mirrors the controversy itself.

While conducting the research for this work I have relied upon the assistance of a number of scholars working on seventeenth-century philosophy and mathematics, and it is a pleasure to acknowledge my debts. Margaret Wilson, whose untimely death came as the project was being completed, was a frequent source of information and support. Her absence will be keenly felt by scholars of seventeenth-century philosophy. Dan Garber and Ed Curley were instrumental in a number of ways: through their writings, their conversations with me, and not least in persuading granting agencies that the world really does need to know more about Hobbes's circle-squaring efforts. Roger Ariew, Moti Feingold, and Marjorie Grene gave helpful feedback at various stages in the development of the work. Kris Heitman deserves special thanks for having read and commented on an earlier draft, and for helping me clarify my thoughts on a number of issues. Cees Leijenhorst, Paolo Mancosu, and Siegmund Probst also deserve thanks for helpful discussions and the concrete assistance they have provided. My colleagues at North Carolina State University, particularly Randy Carter, John Carroll, Joe Levine, and Tim Hinton, all read and commented on various pieces of the book, and I have always found their advice valuable. Hal Levin's aid as technology guru and keeper of the sacred mysteries of the Mac operating system was crucial in getting the text, equations, and figures into a presentable form. I also received major assistance from Michael Mahoney, who subjected the whole typescript to a detailed and careful criticism as referee for the University of Chicago Press. Whatever flaws remain, the work has been vastly improved by his valuable suggestions. Susan Abrams, executive editor at the University of Chicago Press, was a source of encouragement throughout the project. I hope her patience has been adequately rewarded.

An undertaking like this would be impossible without support from a variety of institutions. The National Endowment for the Humanities assisted twice, first with a 1991 Summer Stipend (#FT-35703) and then with a yearlong Fellowship (#FB-3343-96) for the academic year 1996–97. The National Humanities Center provided an absolutely ideal work environment during a fellowship for the academic year 1993–94, which was generously supported by Andrew W. Mellon Foundation. The largesse of the North Carolina State University College of Humanities and Social Sciences has been considerable, amounting to no less than two summer stipends, a travel grant to support research in Paris, and research funds to enable the purchase of books and equipment. It goes without saying that my efforts have been enabled by the support and cooperation from many libraries and archives. His Grace the Duke of Devonshire and the Trustees of the Chatsworth Settlement were kind enough to grant me access to the Hobbes archives at Chatsworth House, and I must especially thank P. J. Day, the keeper of collections at Chatsworth, for his assistance. Alan Tuttle and Jean Houston, librarians at the National Humanities Center, deserve special commendation for tracking down a great deal of obscure material for me. I am also grateful to the staff at the Bibliothèque Nationale, Paris; the Bodleian Library, Oxford; the British Library, London; and the Regenstein Library, the University of Chicago. The efforts of the interlibrary loan staff of the D. H. Hill Library at North Carolina State University are especially appreciated.

My greatest debt is to my wife, Doreen, whose love and support have been essential to this project from the beginning. Anything of value here is due in large measure to her.

Abbreviations

Behemoth   *Behemoth; or, The Long Parliament* (Hobbes [1889] 1969b). References are to a dialogue number, followed by a reference to *EW* after a semicolon.

CTH   *The Correspondence of Thomas Hobbes* (Hobbes 1994).

DCo   *Elementorum Philosophiae Sectio Prima De Corpore* (Hobbes 1655). References are to part, chapter, and article numbers separated by periods, followed by a page reference to *OL;* to *EW* when I need to refer to the English version (Hobbes 1656a); or to the original (Hobbes 1655) when it differs from the version anthologized in *OL*.

DH   *Elementorum Philosophiae Sectio Secunda De Homine* (Hobbes 1658). References are to part, chapter, and article numbers separated by periods, followed by a reference to *OL* after a semicolon.

DP   *Dialogus Physicus* (Hobbes 1661b). References are to work as a whole, followed by a reference to *OL* after a semicolon.

EL   *The Elements of Law Natural and Politic* (Hobbes 1969a). References are to part, chapter, and article number separated by periods, followed by a reference to *EW* after a semicolon.

EW   *The English Works of Thomas Hobbes of Malmesbury, now First Collected and Edited by Sir William Molesworth* (Hobbes [1839–45] 1966b).

Examinatio   *Examinatio et Emendatio Mathematicae Hodiernae* (Hobbes 1660). References are to dialogue number, followed by a reference to *OL* after a semicolon.

L   *Leviathan* (Hobbes 1651). References are to part and chapter number separated by a period, followed by a page number from the 1651 edition; a reference to *EW* follows after a semicolon.

Lux	*Lux Mathematica* ([Hobbes] 1672). References are to chapter number, followed by a reference to *OL* after a semicolon.
MHC	*Mr. Hobbes Considered in his Loyalty, Religion, Reputation and Manners* ([Hobbes] 1662b). References are to the work as a whole, followed by a reference to *EW* after a semicolon.
OL	*Thomae Hobbes Malmesburiensis Opera Philosophica Quae Latine Scripsit Omnia in Unum Corpus Nunc Primum Collecta* (Hobbes [1839–45] 1966a).
PP	*Problemata Physica* (Hobbes 1662a). References are to the work as a whole, followed by a reference to *OL* after a semicolon.
PPAG	*Principia et Problemata Aliquot Geometrica* (Hobbes 1674). References are to chapter number, followed by a reference to *OL* after a semicolon.
PRG	*De Principiis et Ratiocinatione Geometrarum* (Hobbes 1666). References are to chapter, followed by a reference to *OL* after a semicolon.
SL	*Six Lessons to the Professors of the Mathematiques* (Hobbes 1656b). References are to lesson number, followed by a reference to *EW* after a semicolon.
SPP	*Seven Philosophical Problems* (Hobbes 1682). References are to chapter number, followed by a reference to *EW* after a semicolon.
Στίγμαι	Στίγμαι ... *or Markes of the Absurd Geometry, Rural Language, Scottish Church-Politicks, And Barbarismes of John Wallis* (Hobbes 1657). References are to section number, followed by a reference to *EW* after a semicolon.

WORKS BY WALLIS

AI	*Arithmetica Infinitorum* (Wallis 1656b, part 3). References are to proposition number, followed (where possible) by a reference to *OM* after a semicolon.
Angle of Contact	*De Angulo Contactus et Semicirculi Disquisitio Geometrica* (Wallis 1656b, part 1). References are to chapter number, followed by a reference to *OM* after a semicolon.

Conic Sections	*De Sectionibus Conicis Tractatus* (Wallis 1656b, part 2). References are to chapter number, followed by a reference to *OM* after a semicolon.
Dispunctio	*Hobbiani Puncti Dispunctio; or The Undoing of Mr Hobs's Points* (Wallis 1657b).
Due Correction	*Due Correction for Mr. Hobbes; or Schoole Discipline, for not saying his Lessons right* (Wallis 1656a).
Elenchus	*Elenchus Geometriae Hobbianae* (Wallis 1655).
HHT	*Hobbius Heauton-timorumenos; or A Consideration of Mr. Hobbes his Dialogues in an Epistolary Discourse Addressed to the Honourable Robert Boyle, Esq.* (Wallis 1662).
Mechanica	*Mechanica; sive, De Motu, Tractatus Geometricus* (Wallis 1670). References are to chapter number, followed by a reference to *OM* after a semicolon.
Mens Sobria	*Mens Sobria Seriò Commendata* (Wallis 1657a).
MU	*Mathesis Universalis* (Wallis 1657c, part 2). References are to chapter number, with a citation to *OM* following after a semicolon.
OM	*Opera Mathematica* (Wallis 1693–99).
Treatise of Algebra	*A Treatise of Algebra, Both Historical and Practical* (Wallis 1685). References are to chapter number and page.

WORKS BY OTHER AUTHORS

AT	*Oeuvres de Descartes* (Descartes 1964–76).
Elements	*The Thirteen Books of Euclid's "Elements"* (Euclid [1925] 1956). References are to Book number and proposition, definition, or axiom number.
Exercitatio	*In Thomae Hobbii Philosophiam Exercitatio Epistolica* (Ward 1656).
HOC	*Les Oeuvres Complètes de Christiaan Huygens* (Huygens 1888–1950).
LM	Barrow, *Lectiones Mathematicae* (Barrow 1685). References are to lecture number, with a citation to page number in volume 1 of Barrow (1860) following after a comma.

References to Aristotle use title of work, followed by book and chapter number separated by periods; reference to the Bekker page numbers given marginally in the Revised Oxford Translation of Aristotle (1984) follow after a semicolon.

All translations are by the author unless otherwise indicated in the references section.

CHAPTER ONE

The Mathematical Career of the Monster of Malmesbury

> [T]he doctrine of Right and Wrong, is perpetually disputed, both by the Pen and the Sword: Whereas the doctrine of Lines, and Figures, is not so; because men care not, in that subject what be truth, as a thing that crosses no mans ambition, profit, or lust.
>
> —Hobbes, *Leviathan*

In June of 1645 the English mathematician John Pell wrote to his friend Sir Charles Cavendish seeking assistance in an ongoing controversy with the Danish astronomer-mathematician Christian Severin Longborg, who is better known by his Latinized name Longomontanus. Pell had been appointed professor of mathematics in Amsterdam in 1643 and was known for his work in algebra, and particularly for his lectures on the work of the third-century Greek mathematician Diophantus of Alexandria. The aged Longomontanus, who had once served as an assistant to the venerable Tycho Brahe, had held the professorship of mathematics at Copenhagen since 1607 and over the course of several decades had published a number of supposed solutions to the ancient problem of squaring the circle (Longomontanus 1612, 1627, 1634, 1643, 1644a). Pell saw the last of these, which bore the ponderous title *Rotundi in Plano, seu circuli absoluta mensura, duobus libellis comprehensa,* and promptly penned a two-page refutation (Pell 1644). In a move guaranteed to infuriate Longomontanus, his publisher Johannes Blaeu had the refutation printed and appended to the copies of *Rotundi in Plano* that remained in stock. Longomontanus responded with Ἐλέγξεος *Joannis Pellii contra Christianum S. Longomontanum De Mensura Circuli* Ἀνασκευὴ in the same year (1644). In this work he rejected the trigonometric principles employed by Pell and reasserted the correctness of his result.[1] Pell was determined

---

1. The theorem used by Pell asserts that if *r* is the radius of a circle, *a* the tangent of an arc less than 45°, and *x* the tangent of an arc twice as great as arctangent *a,* then

to pursue the battle and decided to solicit alternative proofs from European mathematicians of the key lemma he had used in his earlier refutation. His intent was to "appeale to ye judgements of all those that by demonstrating my fundamental Theorem . . . can shew themselves able to judge of such a controversy," and he was convinced that "yose ignorant Danes may be so much ye more confounded to see a thing demonstrated so many severall wayes, which Longomontanus sayd was *indemonstrabile.*"[2]

Sir Charles Cavendish was the brother of the first earl (and later duke) of Newcastle and a steadfast supporter of King Charles I, but he found himself in France at the time as a result of the English Civil War. While there, he pursued his scientific and mathematical interests by maintaining contacts with European scientific figures, including other Royalist expatriates.[3] Among them was Thomas Hobbes, whose ties to the Cavendish family were long-standing, and whom Sir Charles approached for a proof of the relevant result. Writing in reply to Pell in June of 1645, Cavendish remarked that "I have as you desire procured not onelie the approbation but demonstration of your fundamentall proposition by Mr. Hobbes his meanes. Fermat is not in this towne, and Mersennus is on his waye hither, so that I knowe not whither to write him. But I doute not but more handes with demonstrations might be procured if you desire it."[4] Hobbes's demonstration was eventually published in Pell's *Controversiae de verâ circuli mensurâ . . . Prima pars* (1647), along with contributions by such mathematical luminaries as René Descartes, Bonaventura Cavalieri, and Gil-

---

$xr^2 - xa^2 = 2r^2a$. As Isaac Barrow noted in commenting on this exchange, Pell's "most excellent theorem" comes to the same thing as the claim that "the difference between the square of the radius and the square of the given tangent will be in the same ratio to twice the square of the radius, as the given tangent is to the tangent sought" (*LM* 15, 243–44). Longomontanus simply denied this theorem, although his objections are not worth examining here. For a full account of this controversy see Maanen 1986 and Dijksterhuis 1931–32.

2. John Pell to John Leake, 7/17 August 1645; British Library MS. Add. 4280, f. 94^r.

3. See Jacquot 1949 and 1952c for an account of Sir Charles Cavendish and his connections to the scientific and mathematical work of the seventeenth century. For more on the Cavendish family and its history, see Bickley 1911, especially the fourth chapter, which deals with Newcastle and the English Civil War. For a general study of Royalist expatriates in France during the Civil War see Guitian 1996.

4. Cavendish to Pell, 27 June/6 July 1645; British Library MS. Add. 4278, f. 170^r. Mersennus is Marin Mersenne, the Minim friar and scientific writer who maintained a voluminous correspondence with nearly all of the important European philosophical and scientific figures of the seventeenth century. Fermat is the great French mathematician Pierre de Fermat.

les Personne de Roberval. These "notable mathematicians" were to serve rather like a jury in hearing the claims of Pell and Longomontanus, and they naturally returned a verdict in Pell's favor.[5] The irony here is quite remarkable: Hobbes, who would later spend years publishing and defending numerous failed attempts to square the circle, published his first mathematical work as part of a campaign to silence an old circle squarer. Indeed, less than a decade after his participation in Pell's battle with Longomontanus, Hobbes would find himself involved in a prolonged and bitter controversy that centered on his claims to have squared the circle, and he would go to his grave insisting that he had solved this ancient geometrical problem.[6]

Near the end of his extraordinarily long and full life, the octogenarian Hobbes penned an account of his accomplishments in the form of a Latin prose vita. In it he recounted a list of his great mathematical achievements and confidently asserted numerous claims to everlasting mathematical glory. This account of his supposed mathematical triumphs is worth considering briefly, as it outlines much that will be of concern in the present book:

> In mathematics, he corrected some principles of geometry. He solved some most difficult problems, which had been sought in vain by the diligent scrutiny of the greatest geometers since the very beginnings of geometry; namely these:

5. Richard Blackbourne's *Vitae Hobbianae Auctarium* reports the episode as follows: "And when the learned Pell publicly attacked the paralogisms of Longomontanus and readily discerned that the whole controversy turned on one theorem that was to be demonstrated, he both demonstrated it after his own style and sought to defend his opinion by inviting demonstrations from the most famous mathematicians in Europe." In evaluation of Hobbes's contributions, he adds "But of Hobbes's reasoning, in order that I not seem to offer immoderate praise through my admiration, I shall only add that it is inferior to none of the others in the brevity of its elegance or the perspicacity of its evidence" (*OL* 1:xxxi–xxxii). The successful outcome of his dispute with Longomontanus seems to have won Pell some recognition; at any rate, Aubrey (repeating the evaluation of Theodore Haak) reports that "his fame was much augmented by his refuting a large book of Longomontanus Quadratura, which caused the Prince of Orange (Henry Fredrick) being about to erect an Academie at Breda, borrowed Mr. Pell from the magistrate of Amsterdam, to grace his new Academy with a man of that fame for a few years" (Aubrey 1898, 2:130).

6. In fact, Hobbes left an incomplete tract on circle quadrature that was apparently his last written work. The fragment dates from the last year of Hobbes's life and was his last attempt to assert his claims to mathematical glory. The dedication contains a revealing statement of Hobbes's attitude toward the rest of the mathematical world at this point. He writes: "And so, after I had given sufficient attention to the problem by different methods, which were not understood by the professors of geometry, I added this newest one" (Chatsworth Library, Hobbes MS. A.9, f.1^r).

1. To exhibit a right line equal to the arc of a circle, and a square equal to the area of a circle, and this by various methods. In several books.

2. To divide an angle in a given ratio.

3. To find the ratio of a cube to a sphere. In the *Geometrical Problems*.

4. To find any number of continual mean proportionals between two given lines. In the *Geometrical Problems*.

5. To describe a regular polygon with any number of sides. In the *Geometrical Rose Garden*.

6. To find the center of gravity of the quadrant of a circle, and the bilinear figure contained in the arc of a quadrant and its subtense. In the *Geometrial Rose Garden*.

7. To find the centers of gravity of all types of parabolas *[paraboliformium]*. In the book *De Corpore*.

He was the first to construct and demonstrate these, and many other things besides, which (because they will appear in his writings *[legentibus]* and are less important) I pass over. (OL 1:xix)

This list is essentially a catalog of Hobbes's mathematical publications and, not coincidentally, a summary of the most eagerly sought mathematical results from antiquity to the early modern era. It is important to observe that Hobbes claims credit for "correcting" some geometric principles as well as solving great outstanding problems of geometry. These claims are not unrelated, for Hobbes was convinced that his reformation of geometry would render essentially any problem solvable. In his view the failures of previous generations of geometers to solve them did not stem from an intrinsic intractability of the problems or from a lack of industry and intelligence on the part of earlier mathematicians. Rather, their failures were the result of a lack of proper method, and Hobbes was convinced that his understanding of the true mathematical method would put him (or anyone else who adopted his principles) in a position to find long-sought solutions to the great problems of classical geometry.

This emphasis upon the importance of proper method recurs throughout Hobbes's writings, whether they concern politics, natural philosophy, or mathematics. The Hobbesian obsession with methodological concerns is also clearly inspired by his appreciation of the deductive structure of geometry. Indeed, it was supposedly the discovery of mathematics at the age of forty that led Hobbes to attempt to cast all of philosophy on the model of geometry. We owe to Aubrey the famous and revealing account of Hobbes's first appreciation of mathematical method and his subsequent infatuation with the subject:

> He was . . . 40 yeares old before he looked upon geometry; which happened accidentally. Being in a gentleman's library in ———, Euclid's Elements lay open, and 'twas the 47 El. libri I. He read the proposition. 'By G———,' sayd he, 'this is impossible!' So he reads the demonstration of it, which referred him back to such a proposition; which proposition he read. That referred him back to another, which he also read. *Et sic deinceps,* that at last he was convinced of that trueth. This made him in love with geometry. (Aubrey 1898, 1:332)

However embellished this account may be (and it is almost certainly embellished), there is no reason to doubt that Hobbes was deeply impressed with the deductive structure of geometry and especially with the idea that simple and indubitable first principles can yield recondite theorems.[7]

During his prolonged stay in Paris in the 1640s Hobbes was active in the intellectual circle around Marin Mersenne and came in contact with many of Europe's leading scientists, philosophers, and mathematicians. We can gain some idea of the extent of Hobbes's contacts

---

7. The principal reason for suspecting the complete accuracy of Aubrey's account is the remarkable coincidence that the proposition that supposedly initiated Hobbes into geometry is the Pythagorean theorem: proposition 47 of the first book of Euclid. Moreover, Euclid's proof of the theorem is sufficiently complex to be largely incomprehensible to a complete novice (as Hobbes allegedly was). Pacchi 1968 has drawn attention to a reading list in the library at Chatsworth that lists several hundred books and was marked up in Hobbes's hand sometime between 1625 and 1628. This list seems to have served as a guide to Hobbes's studies in this period and it contains references to a remarkably wide collection of books in mathematics, so it seems unlikely that Hobbes had no exposure to geometry in 1629. Still, there is little doubt that Hobbes undertook intensive mathematical study during his journey to the Continent in that year. In his autobiographical vita Hobbes reports that in that year: "In that journey [to the Continent in the company of Sir Gervase Clifton] he began to read Euclid's *Elements;* and, well-pleased by its method, not because of the theorems, but rather because of its art of reasoning, he read through it most carefully" (*OL* 1:xiv). Although we may be suspicious of points in Aubrey's account, it is not without a ring of truth. For example, his report that "I have heard Mr. Hobbes say, that he was wont to draw lines on his thighs, and on the sheets a-bed" (Aubrey 1898, 1:333) finds an echo in the curiously autobiographical remark in *Leviathan* that "from being long and vehemently attent upon Geometricall Figures, a man shall in the dark, (though awake) have the Images of Lines, and Angles before his eyes" (*L* 1.2, p. 6; *EW* 3:6). Bernhardt (1986) has drawn attention to a passage from the first edition of Hobbes's *Examinatio* (Hobbes 1660, 154) that is strikingly similar to Aubrey's account, but I think that the story remains somewhat exaggerated.

with the Parisian mathematical establishment from Sir William Petty's remark to Pell that "Mr. Hobbes served you in procuring the demonstrations of other French Mathematicians" in the battle against Longomontanus.[8] The intellectual stimulation provided by his extensive contacts with Parisian savants led Hobbes to devote himself to the study of mathematics and natural philosophy, and it is at this time that he set to work drafting a comprehensive summary of the grand philosophical system that he intended to encompass all of mathematics, physics, and political philosophy. These efforts were interrupted by his concern over the course of political events in England and the consequent writing of *Leviathan,* so the grand scheme of a complete system of philosophy remained unfinished by the time of his return to England in 1651. Nevertheless, as the incident with Pell suggests, Hobbes had acquired a considerable reputation in mathematics by the time his masterpiece *Leviathan* was published. On the death of Descartes in 1650, Hobbes's friend Samuel Sorbière could say of the deceased French savant that he was "one of the world's foremost men in algebra and geometry," an opinion which he supported by reference to "Roberval, Bonnel, Hobbes, and Fermat, who are the greatest masters."[9] Word of Hobbes's mathematical prowess no doubt contributed to his being engaged as tutor in mathematics to the prince of Wales (the future Charles II) in 1646, although he reminded Sorbière to "beware of thinking it more important than it is" because "I am only teaching mathematics, not politics" (Hobbes to Sorbière, 24 September/4 October 1646; *CTH* 1:141). In June of 1655 Henry Oldenburg could request his assistance on behalf of an unnamed friend (who may have been Robert Boyle) for advice on "ye best authors, y^t haue specified those uses [of mathematics]," and ask that Hobbes "would fauour us wth ye culling out of such authors, and send their names" (Oldenburg to Hobbes, 6/16 June 1655; *CTH* 1:211–12).

Hobbes remarked that the learned community's "expectation of that which should be written by me, was raised partly by the *Cogitata physico-Mathematica* of *Mersennus,* wherein I am often named with honour; and partly by others with whom I then conversed in *Paris*" (*SL* 6, 56; *EW* 7:334). Still, the lofty mathematical reputation he enjoyed in the early 1650s was not based on his publications, which up

8. Petty to Pell, 8/18 November 1645; British Library MS. Add. 4279, f. 172^r.
9. Sorbière to Claude de Saumaise, 28 February/10 March 10 1650. Extract printed in Tönnies 1975, 68. Bonnel was a mathematician and doctor from Montpellier who corresponded with Mersenne (*AT* 3:332).

to that point contained essentially no mathematical material.[10] Hobbes intended to cement his reputation as an important mathematician by putting a collection of mathematical results into his treatise *De Corpore,* a volume of first philosophy he had worked on intermittently for years.[11] When it finally appeared in 1655, *De Corpore* was supposed to guarantee Hobbes's place in the mathematical pantheon by (among other things) squaring the circle.

Hobbes's claims to mathematical glory did not go unchallenged, and they form the core of the long and bitter dispute that raged between him and John Wallis, Oxford's Savilian professor of geometry. Wallis was by no means the only critic of Hobbes; his colleague Seth Ward (the Savilian professor of astronomy) published a detailed critique of the Hobbesian philosophy that attacked everything in *De Corpore* that Wallis had passed over (Ward 1656), while many other authors of the period also weighed in with denunciations of Hobbes and his philosophy.[12] Nevertheless, for its length and intensity, the fight between Wallis and Hobbes was the outstanding dispute in Hobbes's sometimes quarrelsome career, and it is not without justice that those who have commented on it characterize the dispute as a war.[13]

10. It is worth observing that mathematical reputations were frequently not linked to published output in Hobbes's day. Such luminaries as Roberval or Viscount Brouncker could establish substantial reputations through correspondence and with little or no published output.

11. The title of the work is *Elementorum Philosophiae Sectio Prima De Corpore,* or in the English translation of 1656 *Elements of Philosophy; The First Section Concerning Body.* Both versions are universally known as *De Corpore,* however, and I will adopt that usage here.

12. See Mintz 1962 and Bowle [1951] 1969 for more on the anti-Hobbes literature. Unfortunately, Mintz ignores the great majority of the Wallis material.

13. In his prose *Vitae Hobbianae Auctarum,* Richard Blackbourne characterizes the course of events this way: "At that time the illustrious Wallis, Savilian professor of geometry at the University of Oxford, sounded the call for this long-running mathematical war by publishing his *Elenchus of Hobbesian Geometry.* This most bitter struggle, fought with compass and rule, and sometimes the exchanged of the most sharp abuse, was carried on by both men and lasted more than twenty years, nor did it finally end until the death of Hobbes" (*OL* 1:xxxviii). Isaac D'Israeli portrayed the controversy as "The Mathematical War between HOBBES and the celebrated Dr WALLIS . . . A series of battles, the renewed campaigns of more than twenty years, can be described by no term less eventful. HOBBES himself considered it as a war, in which he took too much delight. His 'Amata Mathemata' was a war of idle ambition; it became his pride, his pleasure, and his shame. He attempted to maintain his irruption into a province he ought never to have entered in defiance, by a 'new method;' but having invaded the powerful natives, he seems to have almost repented the folly, and retires, leaving 'the unmanageable brutes' to themselves!" ([1814] 1970, 90).

For all the significance that this conflict had for Hobbes personally, it has been little investigated by historians of philosophy or science.[14] To whatever extent there is a received view on the controversy, it is that there is no great point in studying it. Hobbes scholarship generally treats the whole affair as an embarrassment, and historians of mathematics generally disregard the incident, apparently on the grounds that it failed to lead to any interesting mathematical advances. On the few occasions when the controversy is mentioned, it is with an air of puzzlement at the fact that Wallis should have wasted so much time and effort on an opponent as unworthy as Hobbes.[15] In his study of Wallis's mathematical career, J. F. Scott summed up what remains the prevailing attitude of bewilderment at the duration and intensity of this quarrel. After remarking upon the "strange lack of restraint on the part of each of the disputants," Scott declares:

> For nearly a quarter of a century the two disputants had waged a contest which shed a lustre round neither of them, and one cannot help wondering why Wallis should have been so eager to expose the mathematical short-comings of one so ill-equipped as Hobbes. For apart from his published exposures, Wallis disparages, and in no unmeasured language, the claims of his antagonist in more than a score of lengthy letters to different members of the Royal Society. Hobbes' claims were so preposterous that a wiser than Wallis would have left him to work out his own destruction. (Scott 1938, 170)

This received view is plausible if we assign importance to Hobbes's mathematical work in proportion to his contributions to the advancement of seventeenth-century mathematics. But it is not obvious that

14. The dispute, and Hobbes's mathematical writings generally, have not been completely ignored. Older works that summarize the dispute without going into the technical details include Cajori 1929 and Scott 1938, chap. 10. A recent important study of the dispute is Probst 1997. Studies of Hobbes's philosophy of mathematics that do not concentrate on the dispute with Wallis include Bird 1996; Breidert 1979; Giorello 1990; Grant 1990; Grant 1996; Jesseph 1993a; Jesseph 1993b; Keller 1992; Pycior 1987; Pycior 1997, chap. 6; Sacksteder 1980; Sacksteder 1981a; Schumann 1985; and Weinreich 1911, pt. 2.

15. Thus, Richard Peters concludes his brief summary of the dispute with the remark that "[e]ven when allowance has been made for the fact that squaring the circle did not then seem quite such a preposterous project as it does now, it remains evident that Hobbes' sublime confidence in his own ability led him to make rather a pathetic exhibition of himself. Wallis had probably forgotten more mathematics than Hobbes ever knew. . . . Nevertheless Wallis was not an attractive man and his brilliant demolition of Hobbes' argument was interspersed with pontifical and boorish invective. A greater

this criterion of importance is correct. We can grant that Hobbes failed to prove anything of significance without thereby being committed to the view that his mathematical writings do not merit scholarly attention. Whatever the caliber of his mathematical work, Hobbes assigned great importance to mathematics, and an understanding of his views on the subject must be part of our understanding of his philosophy as a whole. Indeed, Hobbes's insistence on the importance of proper method and his lavish praise of mathematical method invite scrutiny of his mathematical work to see how closely his geometric practice follows his methodological theory.

More importantly, the Hobbesian mathematical corpus does not consist exclusively of attempts to square the circle. Much of what Hobbes wrote on mathematics concerns philosophical and methodological issues independent of his failed quadratures, and these writings can provide a valuable insight into seventeenth-century philosophy and mathematics. To understand Hobbes's philosophy in any depth, it is necessary to understand the place of mathematics in his grand system. Furthermore, to appreciate the groundbreaking mathematical developments of the seventeenth century, we need the perspective of those, like Hobbes, who resisted them. My purpose here is therefore not to rehabilitate Hobbes's mathematical reputation, but rather to focus on his mathematical work as a way of improving our understanding of his philosophy and the context in which it developed. Such an inquiry will necessarily lead into territory not generally covered in a study of Hobbes and his philosophy, but it is an inquiry worth undertaking for anyone interested in the Hobbesian philosophical enterprise and its reception. It also provides a chance to test some historiographic theses, and particularly the claim that the history of philosophy, science, or mathematics is driven exclusively by social and political factors.

Although I will be primarily concerned with the exchanges between Hobbes and Wallis, there were many other figures who contested Hobbes's mathematical claims either in print or in correspondence, and it will be necessary to bring some of this additional material into consideration at various points in this study. Hobbes's reputation was great enough in 1655 that his mathematical works were taken seriously in English and Continental circles, and in its early stages his dispute with Wallis seems to have attracted a good deal of attention from

---

man could have afforded to be kinder to the old gentleman in spite of such pretentious provocation" (Peters 1956, 40).

the learned world.[16] By the mid 1660s, however, Hobbes had lost essentially all credibility with the mathematical public and his works were largely disregarded. In any case, although this point is occasionally overlooked in the literature, it is worth remembering that Wallis was not the only person who read and critiqued Hobbes's mathematical work.[17]

In the course of their dispute Hobbes and Wallis covered issues that went well beyond mathematical and methodological concerns. Questions of political loyalty, church government, theology, and classical philology were all raised and debated as the two traded vituperative pamphlets. The length of the dispute and the variety of topics concerned in it make an exhaustive treatment of the issues impossible, but I intend to cover as much of the relevant territory as is possible within the confines of a single book. The rather tangled history of this protracted conflict can best begin with an overview of the exchanges and an introduction to the mathematical background, both of which will be my concern in this chapter. The first section sketches an overview of the Hobbes-Wallis controversy, while the second introduces the necessary mathematical history.

1.1 THE DISPUTE IN OVERVIEW

The roots of the conflict are complex and will be examined in detail in chapter 2, but the controversy between Hobbes and Wallis began in earnest in 1655 with the publication of Hobbes's treatise *De Corpore,* in which he claimed that his methodological principles had enabled

16. Not all observers quite understood the nature of the dispute at the outset, however. The Parisian mathematician Claude Mylon, commenting to Hobbes's friend François du Verdus, remarked that "I know that the universities do not approve too highly of his work, and I believe that the geometer Dr John Wallis, Savilian Professor in the University of Oxford, has used Aristotle to attack it" (du Verdus to Hobbes 20 February/ 1 March 1656; *CTH* 1:239). Mylon was one of many who critiqued the mathematics of *De Corpore,* and we will be concerned with his efforts in chapter 6.

17. Bird (1996, 218) mistakenly represents Wallis as nearly the only person to have taken notice of Hobbes's mathematics when he claims that "Wallis distinguished himself as the only mathematician to have taken Hobbes's geometry seriously. Only three others—the Belgian philosopher Moranus, John Pell, sometime professor at Amsterdam and Breda, and Viscount Brouncker—are known to have given it more than passing thought." This is seriously in error, first because the mathematical sections of Moranus's critique of *De Corpore* were written by André Tacquet, and more importantly because Bird fails to mention Christiaan Huygens, François de Sluse, Claude Mylon, Seth Ward, Laurence Rooke, Roberval, and Pierre de Carcavi, all of whom gave Hobbes's geometry at least as much attention as Brouncker or Pell. This will become evident in chapter 6.

him to square the circle, rectify curvilinear arcs, and solve other out-standing geometrical problems. These claims were accompanied by what appear to be proofs of some of the most eagerly sought geometrical results of the seventeenth century. Within the year Wallis responded with his *Elenchus Geometriae Hobbianae,* in which he attacked the entire account of geometry contained in *De Corpore* and pointed out numerous technical errors in Hobbes's attempted solution to such problems as the squaring the circle.[18] Hobbes replied in an appendix to the English version of *De Corpore* entitled *Six Lessons to the Professors of the Mathematiques* (1656). The professors concerned were Wallis and Ward—holders of the Savilian Chairs of Geometry and Astronomy, respectively. Ward had attacked Hobbes in his 1656 *In Thomae Hobii Philosophiam Exercitatio Epistolica,* which undertook the refutation of Hobbes's natural, moral, and political philosophy, while adding scattered references to inadequacies in his mathematics. Although the *Six Lessons* were addressed to both Ward and Wallis, they are almost exclusively concerned with replying to Wallis's *Elenchus* and going over to the attack against other of his writings. The rhetorical demands of this undertaking led Hobbes to criticize some of Wallis's Latin usage, as well as to urge purely mathematical objections to his *Arithmetica Infinitorum* of 1656, which was an influential and highly regarded text that employed a mathematical technique known as the "method of indivisibles."

Wallis did not wait long to answer and kept up the high level of invective with his *Due Correction for Mr. Hobbes; or Schoole Discipline, for not saying his Lessons right* (1656). In the following year, Hobbes responded with his Στίγμαι Αγεομετρίας, Αγροικίας, Αντιπολιτείας, Αμαθείας; or, Markes of the Absurd Geometry, Rural Language, Scottish Church-Politicks, And Barbarismes of John Wallis Professor of Geometry and Doctor of Divinity (1657). As is evident from the title, this work launched an offensive against Wallis on all fronts—mathematical, grammatical, political, and personal. While preparing his response, Hobbes entered into a correspondence with Henry Stubbe, a fellow of Oxford's Christ Church College whose classical learning proved useful in addressing the philological points that had come into dispute. Stubbe had undertaken the translation of Hobbes's *Leviathan* into Latin (a task that remained unfinished) and

18. Among other things, Wallis noted that *De Corpore* had apparently been snatched from the press and undergone (unsuccessful) correction of some key mathematical sections. The tangled publishing history of *De Corpore* will be taken up at the end of chapter 3.

was deeply involved in theological and political struggles against Wallis at Oxford, but he also found time to act as an agent for Hobbes—reporting Wallis's plans and attempting to rally support for Hobbes within the university community.[19] Hobbes appended an extract from a letter from Stubbe to Στίγμαι in which Stubbe defended Hobbes on certain finer points of Latin usage and Greek etymology while ridiculing Wallis's efforts at philological criticism. Wallis did not wait long to answer, and his reply took the form of a book entitled *Hobbiani Puncti Dispunctio; or the Undoing of Mr. Hobs's Points* (1657). This drew no direct rebuttal from Hobbes, but Stubbe responded with *Clamor, Rixa, Joci, Mendacia, Furta, Cachiny; or, a Severe Enquiry into the late Oneirocritica Published by John Wallis* (1657).

This began a three-year period of calm, but an ominous chord was struck with Wallis's publication of *Mathesis Universalis* (1657). This specimen of "universal mathematics" is a wide-ranging work that purports to expound the true nature of all mathematics (including its historical development) while also presenting the essentials of the major branches of the subject. Hobbes is briefly and contemptuously mentioned in chapter 24 as one of those who has tried and failed to square the circle, but Wallis undertakes no extended critique of his work, except to mention that the publication of *Mathesis Universalis* was delayed by the necessity of attacking Hobbes's mathematical publications.[20]

19. See Jacob 1983 (chap. 1) and Malcolm's biographical register (*CTH* 2:899–902) for accounts of Stubbe and his activities. I will investigate the Stubbe-Hobbes connection more closely in chapters 2 and 7. It is worth mentioning that Stubbe attacked Wallis and his associates on other occasions that had nothing to do with Hobbes. He published *The Savilian Professours Case Stated* (1658) as a censure of Wallis's appointment as Oxford's keeper of the archives, and later went on to write a number of denunciations of the Royal Society (Stubbe 1670a, 1670b, 1670c, 1671). These latter were evidently written at the behest of Dr. Baldwin Hamey of the Royal College of Physicians, who feared that the Royal Society would encroach upon the claims of the College of Physicians (Westfall 1958, 23). This last attack seems to have displeased Hobbes, as Aubrey reports that he "much esteemed [Stubbe] for his great learning and parts, but at latter end Mr. Hobbs differ'd with him for that he wrote against the lord chancellor Bacon, and the Royall Societie" (Aubrey 1898, 1:371).

20. Wallis notes that the search for an exact value of π (upon which the quadrature of the circle depends) has remained unsuccessful. "Which [exact value] when they imagine themselves (after all the others) to have found it, *Joseph Scaliger, Severinus Longomontanus,* and most recently *Thomas Hobbes,* in dreaming the immortal praises thence due to themselves alone, are merely hallucinating" (*MU* 24; *OM* 1:132). Of Hobbes's role in delaying the publication of the *Mathesis Universalis,* Wallis claims that "But in the meantime (aside from publishing some theological writings) it happened that *Hobbes* was twice to be castigated, now for his geometrical incompetence [ἀγεομετρη-

Hobbes's reaction to the *Mathesis Universalis* was not immediate, but in 1660 he published six dialogues under the title *Examinatio et Emendatio Mathematicae Hodiernae*. The first four dialogues are an extended commentary on and rebuttal to Wallis's *Mathesis Universalis*, while the fifth attacks other works (principally the *Arithmetica Infinitorum*), and the sixth presents sixty-eight propositions concerning the quadrature of the circle and numerous other geometrical problems. Wallis exercised uncharacteristic restraint by not publishing a reply, although he was certainly not impressed by Hobbes's efforts.[21]

Hobbes renewed his claims to preeminence in the mathematical world the next year (1661) with an anonymous solution to the ancient problem of doubling the cube, published in Paris under the title *La Duplication du Cube par V.A.Q.R.* Wallis recognized the author of the work and responded immediately with a letter to an unnamed gentleman (possibly Viscount Brouncker) exposing the error in the alleged solution.[22] Not to be deterred, Hobbes appended a Latin version of this cube duplication to the 1661 criticism of Robert Boyle and the Royal Society that he entitled *Dialogus Physicus, sive De Natura Aeris*. Wallis responded with a severe critique of all things Hobbesian in his *Hobbius Heauton-timorumenos; or A Consideration of Mr. Hobbes his Dialogues in An Epistolary Discourse Addressed to the Honourable Robert Boyle, Esq.* (1662). This attack on Hobbes takes principal

---

σίαν] in *De Corpore*, and now for the railing and abuse [λοιδορήματα] with which he has attacked all of the schools and academies, both ancient and recent. He is puffed up with nothing, and offering for sale his ignorance with this open contempt of everything" (*MU Dedicatio; OM* 1:15–16). The "theological writings" Wallis mentions here were published in the 1657 collection *Mens Sobria*, which will concern us in chapter 7.

21. In a summary review of Hobbes's mathematical publications from 1669, Wallis remarks that "I judged that these dialogues should be ignored, considering that so far as they concerned me, they barely contained anything other than the *Six Lessons,* set forth in a different form, to which I had already responded. Nor then did he seem a worthy adversary, who gravely opposed himself to all mathematicians" (Wallis 1669a, 21). It is worth noting that the version of the *Examinatio* in the 1668 Amsterdam edition of Hobbes's *Opera* differs significantly from the original by altering much of the material in the sixth dialogue.

22. The letter survives as no. 51 in the Chatsworth collection of Hobbes's correspondence and is dated 23 June/3 July 1661. Malcolm (*CTH* 1:xlvi n. 5) suggests that "the most likely recipient was Lord Brouncker" because Wallis ends the letter with the closing "Dominationis vestrae observantissimus" ("your Lordship's most dutiful servant"), which would be a very odd way to address Hobbes. In the letter, Wallis reports that the sheet containing the duplication of the cube had been brought to him the previous evening and then embarks on a close examination of Hobbes's argument and reveals the false supposition upon which it is based. This episode will be examined in more detail in chapter 6.

aim at the *Examinatio*, but it also contains an encyclopedic summary of his various attempted quadratures and showed the tortured history of the Hobbesian mathematical enterprise.[23]

Hobbes soldiered on in pursuit of mathematical glory and continued to circulate cube duplications, including one that he submitted to the Royal Society in September of 1662 (with the approval of Charles II), evidently with the purpose of gaining admission to the society by having his work approved by the fellows at the request of their royal patron. Hobbes also appended sixteen propositions on the quadrature of the circle and a revised cube duplication to the exposition of his physics in the *Problemata Physica* of 1662; this work also contained an appeal to Continental mathematicians to examine the algebraic techniques used by Wallis and accepted by English mathematicians as refutations of Hobbes's efforts. In the same year, Hobbes replied to *Hobbius Heauton-timorumenos* with *Mr. Hobbes Considered in his Loyalty, Religion, Reputation, and Manners*. This publication attempted to repair some of the damage done to Hobbes's reputation and abandoned the mathematical issues in dispute for questions of political loyalty and religious orthodoxy. Its appearance marked the end of the second phase of the dispute, but it was not the final chapter.

In 1666 Hobbes weighed in with *De Principiis et Ratiocinatione Geometrarum . . . Contra fastum Professorum Geometriae*. The aim of this work was to restate the Hobbesian principles of geometry and critique the traditional understanding of the subject. Wallis responded with a scathing review of the piece in the *Philosophical Transactions of the Royal Society* in August of 1666. Hobbes's *Opera Philosophica* of 1668 contained some revised versions of his mathematical works, which were followed in 1669 by the publication of *Quadratura Circuli, Cubatio Sphaerae, Duplicatio Cubi, Breviter demonstrata*. Wallis immediately refuted the entire work in his *Thomae Hobbes Quadratura Circuli, Cubatio Sphaerae, Duplicatio Cubi; Confutata* (1669). With lightning speed, Hobbes produced an amended second edition of the work in 1669, in which he responded to Wallis's criticisms. Wallis obliged with a second edition of his refutation, still in 1669, to which Hobbes made no direct response.

In 1671 Hobbes petitioned the Royal Society to hear his claims against Wallis and presented three papers attacking Wallis's mathematical work. These were answered by Wallis in the Royal Society's *Philo-*

---

23. See *HHT* (104–20) for a remarkable summary of the dozen Hobbesian quadratures that preceded the publication of the *Examinatio*.

sophical Transactions of 8/18 September 1671, even though Hobbes's original papers were not published in the *Transactions.* Hobbes reacted in the same year by publishing *Three Papers Presented to the Royal Society Against Dr. Wallis; Together with Considerations on Dr. Wallis his Answer to them.* In the meantime Hobbes published an anonymous collection of his alleged results and a further attack on Wallis under the title *Rosetum Geometricum; Sive Propositiones Aliquot Frustra antehac tentatae; Cum Censura brevi Doctrinae Wallisianae de Motu.* The "Doctrine of Motion" attacked in this piece was Wallis's *Mechanica; sive, De Motu, Tractatus Geometricus* (1670), and much of Hobbes's critique concerns Wallis's use of infinitesimal methods in his presentation of mechanics along with a restatement of other points from the controversy. Wallis replied in the *Philosophical Transactions of the Royal Society* for 19/29 June and 17/27 July of 1671.

Not to be silenced, Hobbes published a review of the various controversial issues debated between himself and Wallis under the title *Lux Mathematica, Excussa Collisionibus Johannis Wallisii . . . et thomae Hobbesii.* This anonymous work was addressed to the Royal Society and pretended to be an impartial review of the major points of contention between the two disputants from 1655 to 1671. Not surprisingly, *Lux Mathematica* declares Hobbes the winner of every contested point. Wallis again confined his reply to denunciation of the piece in the *Philosophical Transactions* for 19/29 August and 14/24 October 1672.

Two years elapsed before Hobbes ventured to publish anything further regarding mathematics or Wallis, but in 1674 (at the age of eighty-six) he brought out a collection of mathematical results in *Principia et Problemata Aliquot Geometrica Antè Desperata, Nunc breviter Explicata et Demonstrata.* This was not directly attacked by Wallis, and its publication marks the beginning of the end of the Hobbes-Wallis dispute. The final episode occurred in 1678 when Hobbes published his last work, *Decameron Physiologicum: Or, Ten Dialogues of Natural Philosophy . . . To which is added the Proportion of a straight Line to half the Arc of a Quadrant.* These ten dialogues contain an exposition of Hobbesian physics and a critique of Wallis's *Mechanica,* while the appended mathematical paper is Hobbes's attempt to rectify circular arcs. Wallis did not bother to reply to this work, and Hobbes's death in the following year closed the dispute. But even as he went to his grave, Hobbes refused to abandon the fight. Among Hobbes's manuscripts is an unfinished tract on circle quadrature dating from the last year of Hobbes's life (Library, Chatsworth House, Hobbes Manuscripts, MS A.9), and it is easy to imagine that his last intellectual efforts were devoted to the unrepentant assertion of his geometrical claims.

I.2 THE MATHEMATICAL BACKGROUND

The problems Hobbes claimed to have solved derive from the tradition of classical geometry and were standard fare in the mathematical literature of the early modern era. Although these were familiar in Hobbes's day, they are less likely to be well understood by a contemporary reader, and it is therefore necessary to spend some time exploring the mathematical context in which the Hobbes-Wallis controversy took place. This task requires a brief account of the classical conception of mathematics and an introduction to the great geometric problems from antiquity: the quadrature of the circle, the duplication of the cube, and the trisection of the angle. After setting out this part of the relevant background, I turn to a short exposition of the two most important mathematical methods of the seventeenth century: the analytic geometry of Descartes and the method of indivisibles pioneered by the Italian mathematician Bonaventura Cavalieri.

1.2.1 *The Classical Conception of Mathematics*

Philosophical discussions of mathematics in the seventeenth century were largely informed by the doctrines inherited from ancient Greek sources. Although there was by no means unanimity of opinion among classical authors on all points, a broadly shared classical approach to the philosophy of mathematics was still influential in the early modern period. According to the classical point of view there are two fundamentally distinct branches of mathematics: arithmetic and geometry. The former takes numbers as its object and is concerned with discrete quantities that can be expressed as collections of units. The latter deals with continuous and infinitely divisible magnitudes such as lines, figures, and angles.

Further distinctions and refinements are possible within this general classification of the mathematical sciences. Proclus, the fifth-century neo-Platonist whose commentary on the first book of Euclid's *Elements* is an essential source for the traditional philosophy of mathematics,[24] reports that

---

24. On Proclus and his relationship to philosophy and science, see Siorvanes 1996.

[t]he Pythagoreans considered all mathematical science to be divided into four parts: one half they marked off as concerned with
quantity *(ποσόν)*, the other half with magnitude *(πελίκον)*; and
each of these they posited as twofold. A quantity can be considered in regard to its character by itself or in its relation to another
quantity, magnitudes as either stationary or in motion. Arithmetic, then, studies quantity as such, music the relations between
quantities, geometry magnitude at rest, spherics [i.e., astronomy]
magnitude inherently moving. (Proclus 1970, 29–30)

Others, such as the first-century mathematical encyclopedist Geminus,
opted for a slightly more complex and exhaustive division of the mathematical sciences: Geminus first distinguished arithmetic and geometry as the pure sciences concerned with intelligible objects not apprehended by the senses, and then distinguished six applied mathematical
sciences (mechanics, astronomy, optics, geodesy, canonics, and calculation) that treat of sensible objects in accordance with the principles
of pure mathematics.[25] Although such classificatory schemes were not
universally accepted in their entirety, the fundamental distinction between arithmetic and geometry was not seriously challenged in the
classical period.

Because Hobbes's mathematical work is devoted almost exclusively
to geometry, we can confine our investigation to the classical philosophy of geometry, especially as it is expressed in the *Elements* of Euclid.
The Euclidean presentation begins with three kinds of elementary principles or starting points: axioms, definitions, and postulates. In accordance with Aristotle's discussion of demonstrative science in the *Posterior Analytics,* axioms (or "common notions") apply to any science
and include such principles as "Things which are equal to the same
thing are also equal to one another" (*Elements* 1, ax. 1) or "If equals
be subtracted from equals, the remainders are equal" (*Elements* 1,
ax. 3). Definitions are principles specific to a given science and state
the fundamental or essential properties of its objects, such as the definition of a plane angle as "the inclination to one another of two lines

25. As Proclus reports: "[O]thers, like Geminus, think that mathematics should be
divided differently [than the Pythagoreans]; they think of one part as concerned with
intelligibles only and of another as working with perceptibles and in contact with them.
By intelligibles, of course, they mean those objects that the soul arouses by herself and
contemplates in separation from embodied forms. Of the mathematics that deals with
intelligibles they posit arithmetic and geometry as the two primary and most authentic
parts, while the mathematics that attends to sensibles contains six sciences: mechanics,
astronomy, optics, geodesy, canonics, and calculation" (Proclus 1970, 31).

in a plane which meet one another and do not lie in a straight line" (*Elements* 1, def. 8). Finally, postulates are demands that a certain construction be admitted or effected, such as the postulate "To describe a circle with any centre and distance" (*Elements* 1, postulate 3), which permits the construction of a circle of any radius from any point in a plane.

According to the Aristotelian methodology, all sciences must take certain fundamental principles for granted, and the development of the science involves deduction from these basic principles. No science, geometry included, can prove the existence of its proper object, but instead proceeds from the assumption that its object exists. As Aristotle explains:

> I call the principles in each genus those which it is not possible to prove to be. Now both what the primitives and what the things dependent on them signify is assumed; but that they are must be assumed for the principles and proved for the rest—e.g. we must assume what a unit or what straight and triangle signify, and that the unit and the magnitude are; but we must prove that the others are. (*Posterior Analytics* 1.10; 76a 31–36)

Thus, geometry assumes the existence of points, lines, and surfaces, as well as the meaning of its fundamental terms, but it does not attempt to show that things answering to its basic definitions exist. Similarly, the geometric postulates are unchallenged "licenses" to effect constructions answering to specified conditions. But within the framework of these principles, all other geometric propositions are to be derived by a chain of rigorous logical deduction, which establishes the remaining attributes of geometric objects as firmly as the first principles themselves.

The Euclidean *Elements* were generally taken as the prime example of a properly developed science conforming to the Aristotelian canons of demonstration. Indeed, this is largely the reason that medieval thinkers were interested in geometry at all, since the geometry of Euclid was the best-developed example of a properly Aristotelian science.[26] There are notorious shortcomings in the Euclidean presentation

26. As Mahoney observes, "[T]he primary purpose for learning geometry in the Middle Ages was not to carry out further research in the area, but, rather, to understand the geometrical references of Aristotle and the Church Fathers or to be able to do the mathematics demanded by astronomy or optics or to improve mensurational practice and the instruments designed for it" (Mahoney 1978, 153). In commenting on the edi-

of geometry: key principles go unstated, some proofs contain logical gaps, and the famous "parallel postulate" (*Elements* 1, postulate 5) fails to have the simplicity and self-evidence commonly demanded of geometric first principles. The "official" distinction between postulates, axioms, and definitions is also only poorly observed in Euclid: not all of the postulates are really licenses to effect constructions; the fourth postulate of the first book, for example, simply declares that "[a]ll right angles are equal to one another" (*Elements* 1, postulate 4). In fact, much of the history of geometry after Euclid can be read as the search for improvements designed to bring geometry up to the Aristotelian standard by reorganizing its first principles and filling in apparent gaps in the deductive structure.[27] Sir Henry Savile, whose bequest endowed the Savilian Professorships of Mathematics at Oxford, remarked in his commentary on the first book of the *Elements* that "postulates and axioms have this in common, that they require no demonstration, and need be proved by no argument but are taken as manifest, and they are thereafter the principles of all the other things that follow" (1621, 131). But because the fifth postulate of book 1 and the last definition of book 6 seem to lack this kind of self-evidence, Savile regarded them as the only two "flaws or blemishes on the most beautiful body of geometry," and he left to his successors the task of proving them from the remaining principles (Savile 1621, 140–41).[28] Hobbes was convinced that his own work provided the basis for a genuine science of geometry, and he plainly regarded himself as the man who

---

tions of Euclid developed by Adelard of Bath and Campanus of Novara, Mahoney notes that "[i]ndeed, it was the argument rather than the mathematical content that seems most to have interested Adelard and Campanus. For the *Elements* was accompanied into scholastic thought by Aristotle's *Posterior Analytics,* which held it up as a model of scientifically demonstrated knowledge" (1978, 154).

27. See Whiteside 1960–62, pt. 1, and Mancosu 1996, chap. 1, for more extensive accounts of seventeenth-century investigations into geometry and Aristotelian methodological principles. Mueller (1981) investigates Euclidean proof structures with modern logical techniques. Dear (1995b) gives an account of the mathematical model of knowledge in Hobbes's day, especially as understood in philosophy.

28. As I mentioned, the fifth postulate of book 1 is the notorious parallel postulate. It asserts "if a straight line falling on two straight lines make the interior angles on the same side less than two right angles, the two straight lines, if produced indefinitely, meet on that side on which are the angles less than the two right angles." Aside from not actually being a license to construct anything, the postulate seems too convoluted to pass the test of intuitive self-evidence required of postulates. The last definition of book 6 is actually spurious. It declares that "[a] ratio is said to be compounded of ratios when the sizes of the ratios multiplied together make some (?ratio, or size)" (*Elements* 1, def. 5). We will be concerned with this definition and its interpretations in chapter 4.

could raise geometry to the standard recognized by Aristotle, Savile, and the tradition.[29]

Propositions proved in Euclidean fashion were commonly distinguished into two groups: theorems and problems. A theorem is a claim to the effect that a certain class of geometric objects has a specific property, such as "[i]n any triangle, if one of the sides be produced, the exterior angle is equal to the two interior and opposite angles, and the three interior angles of the triangle are equal to two right angles" (*Elements* 1, prop. 32). A problem, in contrast, shows how to construct a geometric object answering a certain description. Thus, "[t]o draw a straight line at right angles to a given straight line from a given point on it" (*Elements* 1, prop. 11) is a problem that begins with certain stated conditions (a given straight line and a point on the line), and then shows how to construct a perpendicular from the given point. Proclus marks the distinction between theorems and problems in essentially this way, and claims that Euclid himself observed the distinction:

> Again the propositions that follow from the first principles [Euclid] divides into problems and theorems, the former including the construction of figures, the division of them into sections, subtractions from and additions to them, and in general the characters that result from such procedures, and the latter concerned with demonstrating inherent properties belonging to each figure. (Proclus 1970, 63)

Christopher Clavius, the great Jesuit mathematician of the late sixteenth and early seventeenth centuries,[30] illustrated this classical distinction by comparing mathematics and dialectics:

> All demonstrations of mathematicians are divided by ancient writers into problems and theorems. A demonstration that demands that something be constructed and teaches how to construct it they call a *problem*. For if someone desires to show that upon a finite right line an equilateral triangle can be constructed, a demonstration of this sort will be called a problem, because it teaches how an equilateral triangle may be constructed upon a

29. In fact, he claims in the epistle to the *Six Lessons* that in chapters 7 through 13 of *De Corpore* that "I have rectified and explained the Principles of the Science [of geometry], *id est,* I have done that business for which Dr. *Wallis* receives the wages" (*SL* epistle; *EW* 7:185).

30. On Clavius and his treatment of mathematics see Dear 1995a, chap. 2; Lattis 1994, chap. 2; and Mancosu 1996, chap. 1.

finite right line. This type of demonstration is called a problem for its similitude to the problems of dialectics. Just as dialecticians call that question, each of whose contradictory parts is credible *[probabile]* (of which sort is the question whether the whole is really distinguished from its parts taken together) a problem, so also that which is sought by mathematicians when they demand that something be constructed, the contrary of which can also be effected, is called a problem. For example, if someone proposes to himself to demonstrate that upon a finite right line an equilateral triangle can be constructed, he poses a problem because both a nonequilateral triangle (namely isosceles or scalene) can be constructed upon the same line. . . . But they call that demonstration that examines only some aspect *[passio]* or property of one or several quantities at once a *theorem*. So if someone desires to demonstrate that in every triangle the sum of the three angles is equal to two right angles, they will call such a demonstration a theorem, because it does not demand or teach to construct a triangle or anything else, but solely contemplates this aspect of any constructed triangle, namely that the sum of its angles is equal to two right angles. Whence from this contemplation the demonstration is called a theorem. In a theorem it can by no means be made to happen that both parts of a contradiction are true. For if someone demonstrates that the sum of all angles of any triangle is equal to two right angles, it could by no means happen that they are also unequal to two right angles. And the same is to be understood in other theorems. And so to sum it up, that which teaches to construct something mathematical that is sought, but whose opposite can be effected, is a problem; but that which teaches to construct nothing, and whose opposite part remains perpetually false, is called a theorem. (Clavius 1612, 1:9)

This somewhat muddled distinction between problems and theorems was not always taken too seriously, since any problem can be rephrased as a theorem stating the possibility of constructing a certain kind of geometrical object.[31] Nevertheless, the general nature of geometric problems should be clear: they are constructions to be effected on the basis of first principles. Three problems in particular captured the

31. Isaac Barrow, for instance, draws attention to this fact when he notes that "every possible problem can be taken for a theorem, in so far as this possibility is demonstrable" (*LM* 6, 99). He consequently places little weight on the distinction.

imagination of geometers from antiquity, and it is to a consideration of them that I now turn.

1.2.2 The Three Great Classical Problems

To construct a square equal in area to a given circle, to divide any given angle into three equal sections, and to construct a cube double in area to a given cube: these are the three great geometric problems from classical antiquity. The long search for solutions to them led to the development of powerful new mathematical tools as well as innumerable wasted hours of intellectual effort. It is now widely known that the problems are unsolvable, but it is important to clarify the nature of the problems and to explain in a general way why they are unsolvable. In Hobbes's day it was an open question whether these problems could be solved, although the repeated failure to find appropriate solutions did provide a certain amount of "inductive" evidence that they could not be conquered. In fact, by the middle of the seventeenth century many had concluded that the problems (and particularly that of squaring the circle) could not be solved with the resources of classical geometry.

1.2.2.1 THE QUADRATURE OF THE CIRCLE The origins of the problem of squaring the circle are obscure, but it seems to have developed naturally out of simpler problems that involve finding the area of rectilinear figures. The resources of Euclidean geometry easily allow the construction of a square equal in area to any rectilinear figure. Indeed one of Euclid's most elementary results is the solution to the problem "[t]o construct a square equal in area to a given rectilinear figure" (*Elements* 2, prop. 14), and it is natural to think of the quadrature of the circle as an attempt to extend this result to the simplest of the curvilinear figures. One of the key results needed to prove *Elements* 2, prop. 14, is *Elements* 1, prop. 45, the problem "[t]o construct, in a given rectilineal angle, a parallelogram equal to a given rectilineal figure." Proclus, in commenting upon this proposition, reports that

> [i]t is my opinion that this problem is what led the ancients to attempt the squaring of the circle. For if a parallelogram can be found equal to any rectilinear figure, it is worth inquiring whether it is not possible to prove that a rectilinear figure is equal to a circular area. Indeed Archimedes proved that a circle is equal to a right-angled triangle when its radius is equal to one of the

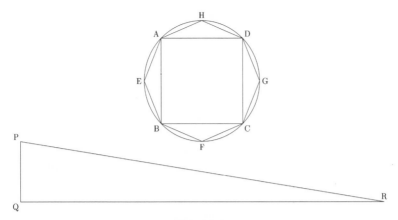

Figure 1.1

sides about the right angle and its perimeter is equal to the base. But of this elsewhere. (Proclus 1970, 335)

Whatever the truth about the origins of the problem, it is clear what it involves and what kind of solution would be required: given a circle of radius r, we need to construct a square equal in area to it, using only the means available in Euclid's geometry. The general problem of quadrature is that of finding a square equal in area to any given figure, but it is the quadrature of the circle that is the most important case of this general problem.[32]

Proclus's allusion to the Archimedean result concerning the equality of a circle and a right triangle is instructive for understanding the problem of squaring the circle, because it underscores what would be required for a complete solution to the problem. Proposition 1 of Archimedes' treatise *On the Measurement of the Circle* asserts that "[t]he area of any circle is equal to a right-angled triangle in which one of the sides about the right angle is equal to the radius, and the other to the circumference, of the circle" (Archimedes 1912, 91). The proof of this theorem employs a classical technique known as the "method of

32. Grattan-Guinness has pointed out that Euclid actually observed a strict division between curvilinear and rectilinear regions, and does not even raise the question of whether the area of a circle and a square can be compared. "[Euclid] does not equate a rectilinear region with a curvilinear one; indeed, in connection with the famous problem of squaring the circle, his commentator Proclus . . . explicitly mentioned this possibility as a worthwhile research topic. Euclid may well have deemed this problem, and similar ones such as squaring lunes, as not Element-ary—and with good justice!" (Grattan-Guinness 1996, 365).

exhaustion," which uses sequences of approximations to determine the value of a sought geometric quantity. In this particular case, Archimedes begins with the given circle *ABCD* and the right triangle *PQR*, such that the side *PQ* is equal to the radius of the circle and the side *QR* is equal to its circumference (figure 1.1). He then considers squares inscribed within and circumscribed about the circle; these provide upper and lower bounds for the area of the circle, which are then systematically improved by doubling the number of sides in the inscribed and circumscribed figures. Thus, for example, the inscribed square *ABCD* can be replaced by an octagon constructed by bisecting $\overset{\frown}{AB}$, $\overset{\frown}{BC}$, $\overset{\frown}{CD}$, $\overset{\frown}{DA}$ and connecting the points *E, F, G,* and *H.* Similarly, a circumscribed square could be replaced by an octagon using the same sort of construction. By continuing the process of bisection, Archimedes constructs a sequence of approximating regular polygons such that the difference between the area of each polygon and the area of the circle is reduced by more than half with each successive term in the sequence. If the same procedure is followed for the circumscribing figures, two sequences will be generated that "compress" the area of the circle as they converge to a common limit. The proof then employs two reductio ad absurdum arguments showing that the area of the circle can be neither greater nor less than the area of the right triangle.[33]

It might initially seem that this result suffices to square the circle: after all, it gives a precise value for the circle's area, and although the right triangle used in the proof is not itself a square, it is a trivial matter to find a square equal in area to any triangle. However, the Archimedean theorem does not actually *construct* a triangle equal in area to the circle, but only shows that a right triangle with sides the length of the circle's radius and circumference is equal in area to the circle. To solve the problem of squaring the circle it would be necessary to devise a means of constructing "from scratch" a line equal in length to the circumference of a given circle, and it is this further task that cannot be solved within the framework of Euclidean geometry.

Consider, for example, the circle *BCD* with radius *AB*, as in figure 1.2. We achieve the quadrature of *BCD* only if we can construct a

33. The exhaustion generates an exact result by assuming (for purposes of reductio) that there is some finite difference, δ, between the circle and triangle. Then it can be shown that there is a polygon inscribed within or circumscribed about the circle that differs from the area of the circle by an amount less than δ. Either case leads to an absurdity, i.e., the circle being greater than a circumscribed polygon or less than an inscribed polygon. For a detailed discussion of the theorem and its proof, see Dijksterhuis [1956] 1987, chap. 6.

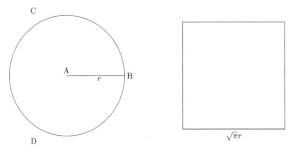

Figure 1.2

square with an area equal to it. Using the customary symbol π to designate the ratio of circumference to diameter in the circle and r for the magnitude of the radius *AB,* we know (on the basis of other results such as Archimedes' proposition above) that the area of the circle is πr^2. Thus, the square equal in area to the circle must have a side of length $\sqrt{\pi}r$. A line of length r is given in the statement of the problem, and it is a simple matter to construct a line equal to the square root of any given line,[34] so the quadrature of the circle is achieved if (and only if) we can construct a line of length π. It was not until the nineteenth century that the search for a construction of π was shown to be impossible,[35] and in Hobbes's day the problem still attracted the attention of leading mathematicians. The Dutch mathematician Willebrord Snell voiced the optimistic opinion that "such is the mechanical revolution of every circle, until it returns to the same point of the periphery, from whence the circumduction was begun; which indeed argues, and as it were sets before the eyes, that a right line may really be exhibited equal to the periphery" (Snell 1621, preface signature **2). His optimism was not shared by all writers on the subject. Wallis, for one, held that there could be no "geometrical" solution to the problem with classical

34. The extraction of the square root is essentially the problem of finding a mean proportional between two lines, one of which is taken as a unit. The construction in Euclid's *Elements* (6, prop. 13) suffices to extract the square root of any magnitude.

35. The impossibility of the construction of π was shown by the German mathematician C. L. F. Lindeman in 1882, with his proof that π is not an algebraic number, i.e., cannot be the root of an algebraic equation with integer coefficients. The difficulty is that the means of construction in classical geometry can only generate magnitudes corresponding to a subset of the algebraic numbers, but π is "transcendental"—which in effect means that it lies outside of a class of numbers that contains the classically constructable magnitudes as a subset. Thus, there is no means of constructing a line of length π.

methods, which is a view to which Descartes and Christiaan Huygens subscribed.[36]

It is important to stress here that the problem is unsolvable only with respect to a set of principles governing the construction of geometric magnitudes. The framework of Euclidean geometry permits only "compass and rule" constructions, which is to say constructions involving only the description of circles and the drawing of straight lines. The familiar compass and straight rule are the only geometric implements permitted in the Euclidean solution of problems, and the impossibility of squaring the circle is to be understood as the inability of compass-and-rule constructions to deliver a true quadrature. The Greeks were familiar with more powerful problem-solving methods, and they distinguished between "plane" solutions, which relied only upon compass-and-rule constructions, and "solid" solutions, which used curves (such as conic sections) generated by the intersection of planes and solids, and "linear" solutions, which employed curves (such as the spiral) whose definition is still more complex.[37]

Methods that extend beyond the austere Euclidean means can, in fact, square the circle, and it is instructive to consider one such method based on a special curve known as the "quadratrix of Hippias." It is unclear when the curve was discovered, or exactly who Hippias was (Knorr 1986, 80–82). No original works of Hippias survive and we know of the curve and its applications only through the report of later mathematical writers such as Pappus of Alexandria, who provides the best account of this curve in his *Mathematical Collection*. Let the

36. Wallis's doubts about the possibility of finding an exact value for π are voiced in chapter 24 of *MU* (*OM* 1:132). Mancosu (1996, 77–79) has shown that Descartes's pessimism is rooted in his mathematical epistemology. Huygens's opinion is explored in Breger 1986, with reference to the work of Leibniz and James Gregory.

37. Pappus reports in book 3 of his *Mathematical Collection* that "[t]he ancients maintained that there are three types of problems, one of which is called plane, another solid, and the third linear. Therefore those problems which can be solved by right lines and the circumferences of circles are justly called plane, since the lines by which they are solved have their origin in the plane. But those problems whose resolution is found by one or more sections of the cone are called solid problems, for it is necessary in the construction to use the surfaces of solids, namely cones. There remains the third type of problem, which is called linear; for in these cases the construction requires curves beyond those already described, which curves have a more varied and forced origin. Among these curves are the spirals, quadratrices, conchoids, and cissoids, all of which have many wonderful properties" (Pappus of Alexandria [1875] 1965, 1:55). For more on the classification of problems see Heath [1921] 1981, chap. 7. This issue will be of interest in chapter 5 when we investigate Hobbes's views on his "method of motions" and its comparison with the techniques of analytic geometry.

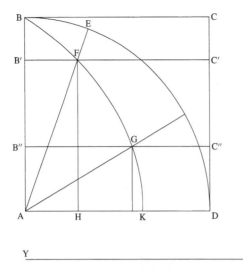

Figure 1.3

square *ABCD* be given as in figure 1.3, and describe within it the circle quadrant *BED* with center *A*. Suppose that the radius *AB* describes $\overset{\frown}{BED}$ with uniform motion, and that at the same time the line BC moves uniformly and always parallel with itself into the position *AD*. Let *B'C'* and *B"C"* be two positions of the line *BC* as it transverses the distance *BA*. Throughout the time during which these lines move, they will intersect, as for example at the points *F* and *G*. The collection of points of intersection (called the locus of intersection) defines the curve known as the quadratrix. By definition, the curve has the property that $\angle BAD : \angle EAD = \overset{\frown}{BED} : \overset{\frown}{ED} = AB : FH$. Let *K* be the point of intersection of the quadratrix and the radius *AD*. Then it can be proved (by a complex exhaustion proof, whose details I omit) that $\overset{\frown}{BED} : AB = AB : AK$, so that the radius *AB* is a mean proportional between $\overset{\frown}{BED}$ and the line *AK*.[38] The problem of quadrature is then easily solved. Because $AK : AB = AB : \overset{\frown}{BED}$, we can use the lines *AK* and *AB* to construct a straight line *YZ* equal to the arc of the quadrant; that is to say, $AK : AB :: AB : YZ$, where *YZ* is equal in length to $\overset{\frown}{BED}$. The line *YZ* is one-fourth the circumference of the circle, so that taking a line four times the length of *YZ* will give a straight line equal to the circumference; but this new line and the radius can be used to construct a right

38. For the details and a discussion of the quadratrix see Knorr 1986, 82–83 and chap. 6, and Heath [1921] 1981, 1:227–28. The construction of the quadratrix and proofs of its properties can also be found in Pappus [1875] 1965, 1:252–58.

triangle equal in area to the circle, by proposition 1 of Archimedes' *On the Measurement of the Circle*.

The quadratrix is not a properly geometric curve in the strictest classical sense because it cannot be constructed with compass and rule. The definition of the curve employs the consideration of two uniform motions—one circular and the other rectilinear—and it further requires that the ratio of equality between a circular arc and a straight line be unproblematically assigned throughout the generation of the curve. Indeed, the question of whether lengths of circular arcs can be exactly compared with straight lines is the very question at issue in the quadrature of the circle. It is possible to construct any number of points on the curve by finding the intersection of circular radii with lines parallel to *BC*, but such "pointwise" construction falls short of a general construction of the curve.[39]

1.2.2.2 THE TRISECTION OF THE ANGLE   The problem of trisecting an arbitrary angle presumably has its roots in the project of constructing regular polygons and solids.[40] Euclidean means permit the construction of regular polygons of 2^n sides, for n integer > 1. In addition, planar methods permit the construction of the equilateral triangle and regular pentagon. These means are exploited in the *Elements* to construct the five "Platonic solids" (tetrahedron, cube, octahedron, dodecahedron, and icosahedron), and the question naturally arises whether other regular figures and solids may be constructed. The

39. Clavius, interestingly enough, admits the pointwise construction as fully geometrical while rejecting the definition in terms of motions. He considers the case where the side *AB* is divided into eight equal segments and the curve *BK* approximated (I have altered his labeling of the lines in his diagram to conform to those I am using). "But because these two uniform motions (one of which is through the circumference *[BD]* and the other through the right lines *[BC]* and *[CD]*) cannot be effected unless the proportion between the circular arc and the right line is known, this construction is justly reprehended by Pappus, since when the proportion is unknown, so also is that which is to be investigated by this line. Thus we will describe the same quadratrix geometrically in this way. Let the arc *[BD]* be divided in any number of equal parts, and let both sides *[AB], [BC]* be divided in as many equal parts. The easiest division will be if the arc *[BD]*, and both sides *[AB], [BC]* are first divided in half, and then both of these halves are again bisected, and each of these parts is again bisected, and this process is continued as far as one might wish. And the more divisions are effected, the more accurately the line will describe the quadratrix. To avoid confusion, we have divided both the arc *[DB]* and the two sides *[AD], [BC]* in eight equal parts" (1612 1:296). See Mancosu 1996, 74–77, for an account of Clavius's approximation procedure, which is very similar to the νεῦσις constructions taken up in section 1.2.2.2 of this chapter.

40. Regular polygons are those that have equal sides and angles; a regular solid is one whose faces are equal regular polygons.

division of angles is an essential part of the project of constructing regular figures, and if it were possible to divide the circle into any number of equal arcs, then by joining the successive points of division by chords one could produce any desired regular polygon.

The bisection of any angle is an elementary problem (*Elements* 1, prop. 9), and continued application of this result allows the construction of any regular polygon with 2^n sides. But the problem of angular trisection is more difficult. As Proclus notes in commenting on the Euclidean problem of bisecting any given angle:

> To divide in any ratio that might be chosen—as into three, or four, or five equal parts—goes beyond the present means of construction. We can divide a right angle into three parts by using some of the theorems that follow, but we cannot thus divide any acute angle without resorting to other lines that are mixed in kind. This is shown by those who have applied themselves to the problem of trisecting a given rectilinear angle. Nicomedes made use of conchoids—a form of line whose construction, kinds, and properties he has taught us, being himself the discoverer of their peculiarities—and thus succeeded in trisecting the angle generally. Others have done the same thing by means of the quadratrix of Hippias and that of Nicomedes, they too using mixed lines, namely, the quadratrices. Still others have started from the spirals of Archimedes and divided a given rectilinear angle in a given ratio. The thoughts of these men are difficult for a beginner to follow, and so we pass them by here. (Proclus 1970, 211–12)

As Proclus indicates, the quadratrix also permits the trisection of any angle, and indeed it allows an arbitrary angle to be divided into any number of equal parts. For example, suppose that $\angle EAD$ in figure 1.4 is any acute angle and let FH be divided at F' so that $FF':F'H$ is the desired ratio.[41] Draw $B''C''$ parallel to AD and through F' to cut the curve at $L,$ and join AL to meet the circle at N. Then $\angle EAN:\angle NAD$ will be in the ratio $FF':F'H$ because $\angle EAN:\angle NAD = \overset{\frown}{EN}:\overset{\frown}{ND} = FF':F'H$.

Another approach to the problem of angle trisection uses a technique called "reduction to a νεῦσις." The Greek term "νεῦσις" means "verging" or "approach," and a νεῦσις problem is one in which a line

---

41. The restriction of $\angle EAD$ to an acute angle imposes no loss of generality on the theorem: a right angle can be trisected, and any obtuse angle can be divided into a right angle and an acute angle.

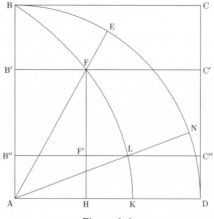

Figure 1.4

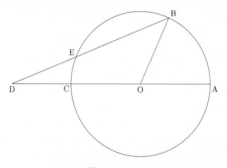

Figure 1.5

is inserted into a construction in such a way that it verges or inclines toward a point that has a desired property. For example, take the angle *BOA* in figure 1.5. We trisect it by first constructing the circle *ABC* with center *O*, then draw a line from *B* to intersect the diameter *OC* extended in the point *D* such that *DE* is equal to the radius of the circle. Then ∠*EDC* will be one-third of ∠*BOA*.⁴² The difficulty in the construction is finding the point *D*, which cannot be constructed with the means available to classical geometry. By trial and error, points on

42. That ∠*EDC* is one-third of ∠*BOA* follows from the properties of isosceles triangles. If we connect *OE* to form the isosceles triangles *OED* and *OEB*, we have ∠*COE* = ∠*EDC*, and ∠*BEO* = ∠*EBO* = 2(∠*EDC*); Similarly, ∠*BOA* = ∠*EBO* + ∠*EDC* = 3(∠*EDC*) and ∠*EDC* trisects ∠*BOA*. The theorem needed for this proof is *Elements* (1, prop. 32): "In any triangle, if one of the sides be produced, the exterior angle is equal to the two interior and opposite angles, and the three interior angles of the triangle are equal to two right angles."

the circle can be investigated to see whether they satisfy the condition of the construction, and with sufficient effort the location of D can be pinned down within any desired degree of precision (hence the term "νεῦσις"), but the actual construction of a line through D cannot be effected by planar means or conic sections.[43]

1.2.2.3 THE DUPLICATION OF THE CUBE   The third and final great problem from antiquity is that of constructing a cube double in volume to a given cube.[44] This is often known as the "Delian problem," because of the legend that the problem arose when the oracle at Delos commanded that the cubiform altar in the temple was to be doubled. Theon of Smyrna, a second-century neo-Platonist, records the legend in his handbook of mathematical material relating to the study of Plato. He reports that:

> when the god pronounced to the Delians in the matter of deliver-
> ance from a plague that they construct an altar double of the
> one that existed, much bewilderment fell upon the builders who
> sought how one was to make a solid double of a solid. Then there
> arrived men to inquire of this from Plato. But he said to them
> that not for want of a double altar did the god prophesy this to
> the Delians, but to accuse and reproach the Greeks for neglecting
> mathematics and making little of geometry. (Knorr 1986, 21)

The problem reduces to the finding two mean proportionals be-
tween a line and its double. Let the side of the given cube be S_1; then the problem is to find a line S_2 such that $(S_2)^3 = 2(S_1)^3$. If we construct two lines X and Y such that $S_1{:}X = X{:}Y = Y{:}2(S_1)$, then by compound-
ing ratios we have $(S_1{:}X)^3 = (S_1{:}X)\,(X{:}Y)\,(Y{:}2(S_1))$, so that $(S_1)^3{:}X^3 = S_1{:}2(S_1)$ and thus $X^3 = 2(S_1)^3$ so that X is the line S_2 that we seek.[45] The unsolvability of the Delian problem (relative to Euclidean means) derives from the same kind of difficulty we encountered earlier in the case of the quadrature of the circle. As before, there are more complex curves that can permit a solution, but the use of "planar" means does not permit the problem to be solved. The class of magnitudes con-
structible from Euclidean operations includes only those that can be

43. For an interesting elementary proof of this result, see Quine 1990.

44. See Knorr 1986, 17–25, and Saito 1995 for more detailed accounts of this problem.

45. In fact, the problem easily generalizes to cover cases of lines in any ratio, rather than just the ratio of 1:2. In the general case, the cubes of the lines S_1 and S_2 may be in any ratio whatever.

obtained by a finite application of the operations of addition, subtraction, multiplication, division, and the extraction of square roots. But the magnitudes required to solve the Delian problem cannot be so obtained.

This concludes our account of the great problems to which Hobbes claimed solution, but we must now consider the more important mathematical developments of his own era. As we will see, the attempt to find solutions to these classical problems led to an important expansion of mathematical methods, and it is essential that we understand the relationship between these new methods and Hobbes's conception of mathematics.

1.2.3 The Seventeenth-Century Mathematical Background

Hobbes's life spanned an enormous transformation in European mathematics. When he was born in 1588 the study of mathematics centered on the classical works of Greek antiquity and most of the mathematical work of the period was devoted to producing editions and commentaries on classical authors. By the time of Hobbes's death in 1679 the mathematical landscape had changed dramatically: Isaac Newton had long since circulated his first papers on the method of fluxions, G. W. Leibniz had applied himself to mathematics, and the great advancement now known as the calculus was well under way. The two most significant mathematical changes during this period were the development of analytic geometry and the rise of infinitesimal techniques, particularly the method of indivisibles. Indeed, it is no exaggeration to say that modern mathematics begins with the advent of these two methods. Hobbes's own conception of mathematics cannot be understood without reference to these methods, and my aim in this section is to outline the fundamental ideas behind them while indicating the controversies they provoked. In general, these controversies were the product of two conflicting motivations shared by many in the mathematical community: the desire to pursue new results (and particularly the solution of the outstanding classical problems) and the desire to maintain the standards of rigorous demonstration upheld by Greek geometry of the classical period.

1.2.3.1 THE ANALYTIC ART AND THE STATUS OF GEOMETRY   The publication of Descartes's *Géométrie* in 1637 is usually regarded as the advent of analytic geometry, and is generally hailed as a significant mathematical advance. Hobbes, however, disparaged analytic geometry and saw it as something akin to a mathematical perversion: in his estimation, the "modern analytics" were a corruption that impeded the progress of geometry. Hobbes's grounds for this judgment will be examined in chapter 5, but in order to clarify matters we must first investigate some of the history of analytic geometry. The fundamentals of the analytic approach are familiar and need no detailed account here.[46] The essential point is that algebraic operations (addition, subtraction, multiplication, division, and the extraction of roots) are interpreted as geometric constructions and curves are then identified with indeterminate equations in two unknowns. Then, algebraic techniques for solving equations are applied to the investigation of geometric curves, and the nature of a curve can be systematically explored by examining the structure of the equation with which the curve is identified. François Viéte's *Isagoge in Artem Analyticem* (Viéte 1646) is a key text in the development of analytic geometry, most notable for its expansion of algebraic techniques and their application to geometric problems.[47] In Britain, William Oughtred's 1631 treatise *Clavis Mathematicae* introduced a generation of mathematicians to algebraic methods and their use in geometry, while the unpublished letters and manuscripts of Thomas Harriot furthered the progress of the subject. It is difficult to discern the extent to which Descartes is indebted to Viéte, Oughtred, or anyone else, but there is no question that all three authors promoted the incorporation of algebraic techniques into geometry.[48] This fusion of algebraic and geometric methods allows

46. See Boyer 1956 for the standard account of the development of the subject, as well as Mahoney 1994, chap. 3. Historians of mathematics also credit Fermat with the invention of analytic geometry, but his work is not prominent in Hobbes's polemics against analytic methods. I will therefore ignore Fermat's contributions in the course of this study. The most accessible recent study of Descartes's geometry is Mancosu 1992, which can be supplemented by Mancosu 1996, chap. 3.

47. Indeed, Viéte's work is so fundamental to the development of modern algebra that Jacob Klein called him "the true founder of modern mathematics" (Klein 1968, 5). For more on seventeenth-century algebra and its philosophical interpretation, see Mahoney 1980 and Krämer 1991, 124–51.

48. Descartes is famously reticent about sources for his mathematical work and never acknowledged a significant debt to other mathematicians, as improbable as this may seem. For purposes of our investigation, the best statement of Descartes's attitude comes from a letter of Pell to Cavendish dated 12/22 March 1646 and reporting a conversation with Descartes on mathematical matters. Pell writes: "I perceive he demonstrates not willingly. He says he hath penned very few demonstrations in his life (understand after ye style of ye old Grecians which he affects not) THAT he never had an Euclide of his owne but in 4 dayes, 30 yeares agoe. . . . Of all ye Ancients he magnifies none but Archimedes, who he sayes, in his bookes de Sphaera & Cylindro and a piece or two more, shows himselfe fuisse bonum Algebraicum & habuisse vere-magnum ingenium. I

systematic and relatively simple solutions to problems that require elaborate constructions with compass and rule in classical geometry. Moreover, analytic geometry can classify curves by their characteristic equations and study curves that are more complex than those accessible to classical investigation. The role of algebraic analysis in the new geometry is summed up in Descartes's remark in book 2 of the *Géométrie:*

> I could give here many other means for tracing and conceiving curves, which means would become more and more complex by degrees to infinity. But to bring together all those that are in nature and to distinguish them in order into certain genres, I know of no better way than to say that all those points of curves we can call geometric, that is to say that have a precise and exact measure, must necessarily have a relation to all the points of a right line, which can be expressed by some single equation. (*AT* 6:392)

The Cartesian program for geometry classifies as properly geometric (as opposed to "mechanical") any curves that have a "precise and exact" measure. This rather vague criterion is elucidated slightly by Descartes's declaration that geometric curves are those that can be described by a regular motion or series of motions (*AT* 6:390). Descartes implicitly assumes that any curve representable algebraically could be traced by such continuous motions and therefore admitted as geometric, and conversely that any geometric curve could be represented by an equation.[49] In the solution of problems, Descartes constrains the choice of curves by the requirement that the simplest curve (i.e., the curve whose equation is of lowest degree) be employed. This restriction has a clear antecedent in Pappus's distinction between planar, solid,

---

will not trouble you of what he said of Vieta, Fermat and Roberval and Golius: Of Mr. Hobbes I durst make no mention to him." British Library MS. Add. 4280, f. 117ʳ. The letter is partially reprinted in *AT* 4:729–32 and fully reprinted in Hervey 1952, 77–79.

49. See Bos 1981 for a discussion of Descartes's program for geometry and the difficulties surrounding his classification and representation of curves. As Bos notes, Descartes "could not simply take as 'geometric' all curves that admit an algebraic equation; if he were to adopt this criterion, Descartes could no longer claim that he was doing geometry" (Bos 1981, 305). The result is that the use of a purely algebraic criterion for geometric curves is merely implicit in Descartes's work. Mancosu 1996, chap. 3, contains an insightful study of this topic with respect to Descartes's mathematical epistemology.

and linear problems, although it is now formulated in terms of the relative complexity of equations.

Another important difference between analytic and classical methods concerns the manner in which algebraic operations are interpreted in geometry. Classically, the geometric multiplication of two lines yields a rectangle, or the product of three lines a solid. But Descartes interprets multiplication as an operation that leaves the dimension of the product homogeneous with that of the multiplicands. Just as the product of two numbers is a number, Cartesian analytic geometry treats the product of two lines as a line. And in general, all operations in analytic geometry are operations on line segments that result in new line segments. This is the import of Descartes's declaration at the beginning of the first book of the *Géométrie:*

> All problems of geometry can easily be reduced to such terms that there is need only to know the length of certain right lines in order to construct them. Just as all of arithmetic is composed of only four or five operations, which are addition, subtraction, multiplication, division, and the extraction of roots (which we can take for a sort of division), so also in geometry we need do nothing more to find the sought lines than to add or subtract other lines; or else, having one line that I shall call unity (in order to relate it as closely as possible to numbers), and which one can ordinarily be taken arbitrarily, and having two other lines given, to find a fourth line that is to one of the given lines as the other stands to unity, which is the same as multiplication; or again to find a fourth line which is to one of the given lines as unity is to the other, which is the same as division; or finally to find one, two, or more mean proportionals between unity and some other line, which is the same as extracting the square root, or cube root, etc. And I will not hesitate to introduce these terms from arithmetic into geometry, in order to make matters more intelligible. (*AT* 6:370)

The approach at work here exploits an algebraic treatment of line segments to allow the product of two magnitudes to be compared with either of its multiplicands. Given an arbitrarily assigned unit segment, the proportional relation 1:a :: b:ab makes all four quantities in the proportion mutually comparable as in figure 1.6. Moreover, the quantity ab can be interpreted as a line segment constructed from two similar triangles with corresponding sides 1, b and a, ab respectively.

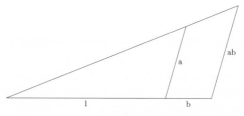

Figure 1.6

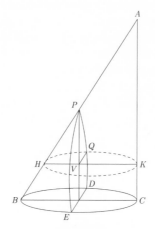

Figure 1.7

This conception of geometry is underwritten by a strong thesis on the unity of arithmetical and geometric magnitudes. Descartes sees nothing peculiarly arithmetical about the operation of addition, or anything uniquely geometrical about the extraction of roots. The resulting application of algebra to geometry therefore treats algebra as a science of magnitude in general, and the specifically geometric content of a problem is removed (and, at least in Descartes's view, the problem is rendered more intelligible) when it is represented as a relation among various abstract magnitudes.

The project of studying geometric objects by inquiring into the relationship between right lines obviously did not originate with Descartes. The classic theory of conic sections, for example, begins with the oblique cone *ABC*, with vertex *A* and circular base *BC* (figure 1.7). Through any point *P* on the element *AB*, we pass a plane cutting the base in the chord *ED*. The section *DPE* of the cone is a parabola, which has the property that products of the lines *QV*, *PV*, *PA*, *BC*, *AC*, and *BA* all stand in a specific ratio. In this case, taking any point *Q* on the parabola, we have the circle HKQ with diameter HK and perpendiculars $VQ, PV,$ such that $QV^2 : PV = (PA \times BC^2):(AC \times BA)$. Where Descartes differs from such a traditional understanding is first in presenting all curves in a simple two-dimensional plane rather than as constructions from solids, which thereby allows the curve to be represented as the solution to an algebraic equation in two unknowns forming the axes of the coordinate system. He also introduces the assumption of the homogeneity of the magnitudes under study, so that lines can be multiplied together any number of times to form curves of fourth and higher dimension.

The advent of analytic methods provoked a philosophical debate on the question whether arithmetic or geometry was the genuinely foundational discipline in mathematics.[50] Classical mathematicians distinguished discrete quantity ("number") from continuous quantity ("magnitude"), declaring the former to be the object of arithmetic and the latter to be the proper object of geometry. Classically, then, geometry and arithmetic are distinct sciences with no common object, so there is no need to ask which is the more fundamental science. This situation changed with the development of analytic geometry. Many interpreted algebra as a kind of generalization of arithmetic, and it was often characterized as the "arithmetic of species," in which variables such as x or ξ were taken as general representatives of kinds or species of quantities.[51] In this scheme, the basic principles of algebra were seen as deriving from arithmetic, and the prominence of algebraic methods in analytic geometry led some to conclude that geometry must, in some important sense, be based on arithmetic.

Wallis argued for the primacy of arithmetic over geometry in his 1657 *Mathesis Universalis*. This work (which began as Savilian lectures) marshals philosophical, historical, and philological arguments to show that all of mathematics is ultimately founded upon arithmetic. Indeed, Wallis's point of view is evident in the full title of the *Mathesis Universalis,* which promises (among other things) "a complete arithmetical work, presented both philologically and mathematically, encompassing both the numerical and the specious or symbolic arithme-

50. Pycior 1987 and Sasaki 1985 are valuable studies of this debate, which can be supplemented by the detailed study of Barrow and Wallis in Maierù 1994 and the account of the battle between ancients and moderns in Hill 1997.

51. Viéte, for example, declares that his analytic method "no longer limits its reasoning to numbers, a shortcoming of the old analysts, but works with a newly discovered logic of species, which is far more fruitful and powerful than numerical logistic for comparing magnitudes with one another" (Viéte 1646, 1).

tic, or geometric calculus."[52] Part of this program to elevate arithmetic to the status of "universal mathematics" involves arguing that geometric results can be achieved more perspicuously and naturally by the use of arithmetical arguments. Thus, for example, Wallis devotes the twenty-third chapter to a series of "arithmetical" demonstrations of results from the second book of Euclid's *Elements,* which he takes to illustrate his contention that the important results in geometry are ultimately founded upon arithmetical principles. In his philosophical case for the primacy of arithmetic, Wallis admits that such geometric terms as *root, square,* and *cube* appear in algebra, but denies that this should lead to the conclusion that algebra is based on geometry. Although some authors have drawn this conclusion, Wallis claims that geometry ultimately takes its principles from arithmetic because the principles of arithmetic are presupposed in any geometrical demonstration.[53] Part of his reasoning on this point is the argument that universal algebra is fundamentally arithmetical and not geometrical. He insists:

Indeed many geometric things can be discovered or elucidated by algebraic principles, and yet it does not follow that algebra is geometrical, or even that it is based on geometric principles (as some would seem to think). This close affinity of arithmetic and geometry comes about, rather, because geometry is as it were subordinate to arithmetic, and applies universal principles of arithmetic to its special objects. For, if someone asserts that a line of three feet added to a line of two feet makes a line five feet long, he asserts this because the numbers two and three added together make five; yet this calculation is not therefore geometrical, but clearly arithmetical, although it is used in geometric measurement. For the assertion of the equality of the number five with the numbers two and three taken together is a general assertion, applicable to other kinds of things whatever, no less than to geo-

52. The full title of the work in Wallis's *Opera Mathematica* of 1693 is *Mathesis Universalis: Sive, Arithmeticum Opus Integrum, Tum Philologice, tum Mathematice tràditum, Arithmeticam tum Numerosam, tum Speciosam sive Symbolicam complectens, sive Calculum Geometricum; tum etiam Rationum Proportionumve traditionem; Logarithmorum item Doctrinam; aliaque, quae Capitum Syllabus indicabit* (OM 1:11).

53. "Because some take the geometric elements for the basis of all of mathematics, they even think that all of arithmetic is to be reduced to geometry, and that there is no better way to show the truth of arithmetical theorems than by proving them from geometry. But in fact arithmetical truths are of a higher and more abstract nature than those of geometry. For example, it is not because a *two foot line added to a two foot line makes a four foot line* that *two and two are four,* but rather because the latter is true, the former follows" (*MU* 11; *OM* 1:53).

metrical objects. For also two angels and three angels make five angels. And the very same reasoning holds of all arithmetical and especially algebraic operations, which proceed from principles more general than those in geometry, which are restricted to measure. (*MU* 11; *OM* 1:56)

Isaac Barrow, the first Lucasian Professor of Mathematics at Cambridge, rejected this reasoning. In fact, he responded directly to Wallis's argument and attempted to show that geometric principles must be presupposed as the foundation of all mathematics. In the third of his *Lectiones Mathematicae* Barrow considers Wallis's argument for the priority of arithmetic and issues the following rebuttal:

> To this I respond by asking, How does it happen that a line of two feet added to a line of two palms does not make a line of four feet, four palms, or four of any denomination, if it is abstractly, i.e., universally and absolutely true that two plus two makes four? You will say, This is because the numbers are not applied to the same matter or measure. And I would say the same thing, from which I conclude that it is not from the abstract ratio of numbers that two and two make four, but from the condition of the matter to which they are applied. This is because any magnitude denominated by the name *two* added to a magnitude denominated *two* of the same kind will make a magnitude whose denomination will be *four*. Nor indeed can anything more absurd be imagined than to affirm that the proportions of magnitudes to one another depend upon the relations of the numbers by which they may be expressed. (*LM* 3, 53)

Barrow's case for the primacy of geometry hinges on the claim that numbers, far from being self-subsistent objects, are mere symbols whose content derives from their application to continuous geometric magnitude. To put it another way, there are no "numbers in the abstract" to serve as the object of arithmetic, except those that arise from the consideration of homogeneous magnitude and its division.[54] Bar-

54. "I say that mathematical number is not something having existence proper to itself, and really distinct from the magnitude it denominates, but is only a kind of note or sign of magnitude considered in a certain manner; so far as the magnitude is considered as simply incomposite, or as composed out of certain homogeneous equal parts, every one of which is taken simply and denominated a unit. . . . For in order to expound and declare our conception of a magnitude, we designate it by the name or character of a certain number, which consequently is nothing other than the note or symbol of such magnitude so taken. This is the general nature, meaning, and account of a mathematical

row took this case to the extreme of denying that algebra is a mathematical science at all, classifying it as simply a part of logic or a set of rules for manipulating symbols. He distinguishes two branches of algebra, analytics and logistics. The former, he says "seems to be no more proper to mathematics than to physics, ethics, or any other science. For it is only a part or species of logic." The latter is no part of mathematics "because it has no object distinct and proper to itself, but only presents a kind of artifice, founded on geometry (or arithmetic), in which magnitudes and numbers are designated by certain notes or symbols, and in which their sums and differences are collected and compared" (*LM* 2, 46). The difference of opinion between Barrow and Wallis is significant for understanding Hobbes's relationship to seventeenth-century mathematics. Hobbes developed views on this particular question that are close to Barrow's, and his account of the nature of mathematics becomes more intelligible if we see him as responding to the concerns that produced this dispute over the relative priority of arithmetic and geometry. This will emerge more clearly in chapter 3, when we investigate Hobbes's philosophy of mathematics.

1.2.3.2 THE METHOD OF INDIVISIBLES   The development of the method of indivisibles was another pivotal advance in seventeenth-century mathematics, notable both for providing a wealth of new results and its share of controversy. The first exposition of the new method was in Bonaventura Cavalieri's *Geometria indivisibilibus continuorum nova quadam ratione promota* (Cavalieri 1635). The method plays upon the intuition that we can reason about the area of a figure by considering the lines it contains, which Cavalieri calls the indivisibles of the figure.[55] Cavalieri was cautious about claiming that these indivisibles actually compose the figure, although he did seek analogies between the composition of cloth out of threads and the relationship of a figure to its indivisibles.[56] Instead of simply taking indivis-

---

number" (*LM* 3, 56). See Mahoney 1990, 186–89, for more on Barrow's account of number.

55. In a perfectly analogous manner the motion of a plane through a solid could produce all the planes of the solid, or the indivisibles of the solid. For more on Cavalieri and his method see Andersen 1985, De Gandt 1991, Festa 1992, and Giusti 1980.

56. In his *Exercitationes Geometricae Sex,* he declares that "it is manifest that we can conceive of plane figures in the form of cloth woven out of parallel threads, and solids in the form of books, which are built up out of parallel pages." He nevertheless quickly adds: "But the threads in a cloth and the pages in a book are always finite and have some thickness, while in this method an indefinite number of lines in plane figures (or planes in solids) are to be supposed, without any thickness" (Cavalieri 1647, 3–4).

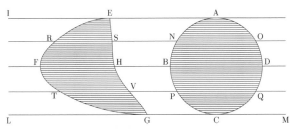

Figure 1.8

ibles as infinitely small components of finite magnitudes, Cavalieri sought to introduce indivisibles as a new species of magnitude that could be brought within the purview of the classical theory of ratios.

In Cavalieri's terminology, "all the lines" of the plane figures *ABCD* and *EFGH* in figure 1.8 are produced by the transit of the line *LM* (called the *regula*) through the figures. Significantly, Cavalieri avoids the question of whether there are an infinite number of indivisibles produced by the transit of the regula *LM* or whether these indivisibles are infinitely small when compared with the figures, apparently hoping that his method would be acceptable on any resolution of the problems surrounding the infinite. He speaks vaguely of an "indefinite" number of lines contained within a figure, and stresses that the ratios can be compared either "collectively" (as one collection of indivisibles to another), or "distributively" by comparing corresponding lines singly.[57] His appeal to continuous motion arises from similar concerns: he seems to have regarded this as a relatively unproblematic concept that can sidestep thorny questions concerning infinity. After all, anyone will admit that a line can pass through the figure, and the intersection of the line and figure will produce, in some fairly innocuous sense, "all the lines" contained in the figure.

Given this starting point Cavalieri treated "all the lines" of the figure as a new species of geometric magnitudes that could be dealt with according to the theory of magnitudes in book 5 of Euclid's *Elements*. His strategy is to establish a ratio between the indivisibles of two fig-

---

57. These two different presentations of the method appear more clearly in his 1647 *Exercitationes Geometricae Sex,* although the second is also contained in the sixth and last book of the *Geometria*. He explains the distinction between the two procedures thus: "The first method proceeds by the first kind of reasoning, and compares aggregates of all the lines of a figure or all the planes of a solid to one another, however many they may be. But the second method uses the second kind of reasoning, and compares single lines to single lines and single planes to single planes, lying in the same direction" (Cavalieri 1647, 4).

ures (either distributively or collectively), and then to conclude that the same ratio holds between the areas of the figures. This is the import of the "very general rule" he announces in the first of his *Exercitationes Geometricae Sex:*

> From these two [ways of comparing indivisibles] a single and most general rule can be fashioned, which will be a summary of all this new geometry, namely this: *Figures, both plane and solid, are in the same ratio as that of their indivisibles compared with one another collectively or ... distributively.* (Cavalieri 1647, 6–7)

Cavalieri's evident caution on foundational matters was not shared by other mathematicians of the seventeenth century, most notably Wallis.[58] In Wallis's treatment, geometric problems are represented analytically and solved by determining the relationship between the infinite sums of infinitely small indivisibles that compose the figures. As an example consider his approach to the quadrature of the cubic parabola in his *Arithmetica Infinitorum.* He begins with arithmetical results in proposition 39, observing that:

$$\frac{0+1}{1+1} = \frac{1}{2} = \frac{1}{4} + \frac{1}{4}$$

$$\frac{0+1+8}{8+8+8} = \frac{9}{24} = \frac{3}{8} = \frac{1}{4} + \frac{1}{8}$$

$$\frac{0+1+8+27}{27+27+27+27} = \frac{36}{108} = \frac{4}{12} = \frac{1}{4} + \frac{1}{12}$$

$$\frac{0+1+8+27+64}{64+64+64+64+64} = \frac{100}{320} = \frac{5}{16} = \frac{1}{4} + \frac{1}{16}$$

From these initial cases, Wallis concludes "by induction" that as the number of terms in the sums increases, the ratio approaches arbitrarily near to the ratio 1:4. Proposition 41, which he takes to follow obviously from proposition 39, asserts that:

58. Andersen 1985, sec. 10; De Gandt 1992, Giusti 1980, 40–65; Jesseph 1989; Malet 1997a; Mancosu 1996, chap. 2; and Wallner 1903 all discuss various reactions to Cavalieri's method and the various changes in the fundamental concepts. For our purposes, the most significant feature of Wallis's reaction is his use of infinite sums of infinitely small parallelograms where Cavalieri had relied upon finite ratios of "all the lines" of one figure compared with another.

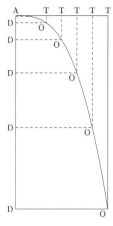

Figure 1.9

If an infinite series is taken of quantities in triplicate ratio to a continually increasing arithmetical progression, beginning with 0 (or, equivalently, if a series of cube numbers is taken) this will be to the series of numbers equal to the greatest and equal in number as one to four. (*AI* 41; *OM* 1:382–83)

Given this result, Wallis turns to the quadrature of the cubic parabola in proposition 42, treating it as an infinite sum of lines forming a series of cubic quantities as in figure 1.9:

And indeed let *AOT* (with diameter *AT*, and corresponding ordinates *TO, TO*, etc.) be the complement of the cubic semiparabola *AOD* (with diameter *AD* and corresponding ordinates *DO, DO*, etc.). Therefore (by proposition 45 of the *Treatise of Conic Sections*), the right lines *DO, DO*, etc. or their equals *AT, AT*, etc. are in subtriplicate ratio of the right lines *AD, AD*, etc. or their equals *TO, TO*, etc. And conversely these *TO, TO*, etc. are in triplicate ratio of the right lines *AT, AT*, etc. Therefore the whole figure *AOT* (consisting of the infinity of right lines *TO, TO*, etc. in triplicate ratio of the arithmetically proportional right lines *AT, AT*, etc.) will be to the parallelogram *TD* (consisting of just as many lines all equal to *TO*) as one to four. Which was to be shown. And consequently the semiparabola *AOD* (the residuum of the parallelogram) is to the parallelogram itself as one to four. (*AI* 42; *OM* 1:383)

Here, Wallis takes the figure as literally composed of indivisibles and unhesitatingly applies arithmetical principles to the solution of geometric problems. The "induction" that leads to his main arithmetical result is clearly not consonant with the classical standard of rigor, nor is his procedure of taking a ratio between two infinite series. But note also that his use of the method of indivisibles departs from the classical approach to geometry because it fails to observe the distinction between discrete and continuous magnitudes. In treating a continuous geometric figure as composed of sums of discrete points or lines, the method of indivisibles simply ignores the classical distinction. Wallis held that the method of indivisibles was essentially equivalent to the classical technique of exhaustion, although he admitted that it must be "applied with due caution" in order to avoid paradox.[59]

Where Wallis thought that the rigor of the method of indivisibles could be guaranteed by using it cautiously, others were convinced that the method was essentially ungeometrical and subjected it to searching criticisms. In particular, Paul Guldin attacked the method as ill-founded and unreliable, and his *Centrobaryca* (Guldin 1635–41) contained a long polemic against Cavalieri in which he argued that the method of indivisibles offends against the principle that there can be no ratio between infinities. As he observes, the attempt to find a ratio between "all the lines" of two figures can be understood only as an attempt to compare one infinite totality with another—but this is explicitly barred by the most basic canons of mathematical intelligibility.[60]

Guldin was not alone in his reservations about the method: others, including Galileo and the Jesuit mathematician André Tacquet, found fault with it and argued that it could lead to false results.[61] It is worth

59. In his *Mechanica,* Wallis declares "this doctrine of indivisibles (now everywhere accepted, and after Cavalieri, approved by the most celebrated mathematicians) replaces the continued adscription of figures of the ancients; for it is shorter, nor is it less demonstrative, if it is applied with due caution" (*Mechanica* 4; *OM* 1:646).

60. Guldin objects, "All the lines and all the planes of one and another figure are infinite and infinite; but there is no proportion or ratio of an infinite to an infinite. Therefore, etc. Both the major and minor premises are clear to all geometers, and so do not need many words" (1635–41, 4:341). Mancosu 1996, 50–64, is a detailed and enlightening study of Guldin's objections, which connects them to his broadly philosophical differences with Cavalieri over the epistemology of mathematics. Other studies of Guldin's attack on Cavalieri are Andersen 1985, sec. 10, and Giusti 1980, sec. 3.

61. Galileo's objections appear largely in letters to Cavalieri, which are studied in Andersen 1985, Giusti 1980, and De Gandt 1991. Tacquet's objection is contained in his 1651 work *Cylindrica et Annularia* (Tacquet 1669, 3:38–39). Mancosu 1996, 119–29, places these and similar objections in their mathematical and philosophical context.

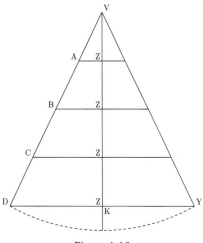

Figure 1.10

considering Tacquet's objection briefly, as it (and attempts to overcome it) brings out some of the significant tensions in the theory of indivisibles. Tacquet argues that, according to the method of indivisibles, the cone DVY (in figure 1.10) will have a surface area equal to that of a circle with radius KD, since the area can be thought of as consisting of all the peripheries of the circles with radii ZA, ZB, ZD, etc. This yields πKD^2 for the surface area, contrary to the established result of $\pi KD \cdot \sqrt{KD^2 + KV^2}$. Barrow took up this objection in the second of his *Geometrical Lectures*, and his reaction to it is intriguing:

I respond that this is the wrong way to undertake the calculation, for in computing the peripheries of which the [curved] surface consists we must use a different way of reasoning from that by which the lines making a plane surface are computed, or the planes from which a body is formed. That is, the number of peripheries constituting the curved surface produced by the revolution of the line VD must be judged from the multitude of points which are in this generative line itself. This is because only a single periphery passes through each point on the line, nor could any more pass through it, whatever the axis may be, or whether the line is further from the axis or nearer to it, for it is the axis alone, according to its various remoteness or nearness and vary-

My treatment of these sorts of difficulties with the theory of indivisibles is correspondingly brief.

ing position which determines the magnitude of the said peripheries. But the multitude of lines of which the plane DVK is supposed to consist, and the multitude of planes of which the solid DVY consists, is to be judged by the number of points in the axis VK; nor indeed can there be more parallel right lines perpendicular to VK and contained within the limits VK, or can more such parallel planes be drawn, than are equal in number to the multitude of those points [in the axis VK]. In observing this difference (which is to be carefully minded) we shall avoid all error *and I judge we shall search out the surfaces generated by the rotation of such curves by the simplest method that the nature of these things admits.* (Barrow 1860, 2:183–84)

The oddity of this response is evident, since it commits Barrow to the absurd thesis that the line VK must contain fewer points than the line VD because VK is shorter than VD, i.e., he contends that the number of points on a line is a function of its length.

Concern about the foundations of the method led Barrow to some further torturous reasoning intended to show that although "there is no greater fault in geometry than to seek or assert the ratio of heterogeneous quantities to one another" (*LM* 16, 260) the method of indivisibles does not actually offend against this principle. He explains

It is true that among those who in the solutions of problems or demonstrations of theorems apply the method of indivisibles, such expressions often occur. All of these parallel lines are equal to such a plane, the sum of these parallel planes constitutes a solid; but they explain their meaning and say that by lines the understand nothing other than certain parallelograms with a very small and (pardon the expression) inconsiderable altitude; by planes they understand likewise nothing but prisms or cylinders of an altitude not to be computed. Or at least by the sum of lines and planes they do not indicate a certain finite and determinate sum, but an infinite or indefinite sum equal to the number of points on some right line. (*LM* 16, 260–61)

Barrow then asserts that anyone who actually claimed that such a sum of "inferior homogeneals" could compose an object of heterogeneous nature "could easily be convinced of his error." Notwithstanding Barrow's assertions of the fidelity of the method to the classical criteria of rigor, one of the key projects in seventeenth-century mathematics was to sort out the conditions under which the method of indivisibles could be used, and ultimately to render it as secure as classical methods.

Hobbes was familiar both with the method and the controversies surrounding it, and as we will see in chapter 3, his own approach to the problem of quadrature is strongly influenced by the method of indivisibles. Hobbes also criticized Wallis's presentation of the method of indivisibles as methodologically suspect, while arguing that only his own metaphysics offered the hope of securely grounding the method and of introducing powerful new mathematical techniques that would enable the solution of any mathematical problem. Before turning to an exposition of Hobbes's philosophy of mathematics, we must first examine the broader social and philosophical background to the dispute, which is the subject of the next chapter.

CHAPTER TWO

The Reform of Mathematics and of the Universities

Ideological Origins of the Dispute

> For seeing the Universities are the Fountains of Civill, and Moral Doctrine, from which the Preachers, and the Gentry, drawing such water as they find, use to sprinkle the same (both from the Pulpit, and in their Conversation) upon the People, there ought certainly to be great care taken, to have it pure, both from the Venime of Heathen Politicians, and from the Incantation of Deceiving Spirits.
> —Hobbes, *Leviathan*

As we have seen, Hobbes prided himself upon having reformed some of the principles of mathematics. But his reforming spirit was hardly confined to this subject, for his philosophy was intended to transform social institutions as well. Indeed, in an age when the rhetoric of reformation appears throughout scientific, theological, and philosophical literature, Hobbes stands out as one of the most committed reformers. The aim of the present chapter is to explain how Hobbes's account of the role of the university in a well-ordered commonwealth brought him into conflict with such notable university men as Ward and Wallis. This undertaking will help to uncover the origins of the dispute between Hobbes and Wallis, and it can also help explain at least some of the vehemence and doggedness with which the dispute was conducted. I begin with an overview of anti-Hobbes opinion provoked by the publication of *Leviathan* and then proceed to a brief summary of Hobbes's doctrines as they relate to the role of the universities. In section three I explore how Hobbes's views on this topic are related to other seventeenth-century critiques of the university system. With this background in hand, I then argue that the original impetus to the quarrel, and much of the fervor with which it was conducted, can be accounted for by seeing Wallis as a defender of the university in a contemporary debate over the status of universities in the commonwealth.

2.1 HOBBES AND HIS ENEMIES

The publication of *Leviathan* in 1651 made Hobbes's name anathema to much of the learned public in England. Where he had previously enjoyed a considerable reputation as a man of letters, the doctrines set forth in his masterwork aroused a storm of opposition, and critiques of Hobbes and his principles remained a fixture in English letters until well after his death. Even some who had previously held Hobbes in high regard were so repulsed by what they found in *Leviathan* that they took pains to dissociate themselves from the name of the notorious Monster of Malmesbury. Seth Ward is a notable case in point.

Ward was an important figure in mid-seventeenth-century British science, having been appointed to Oxford's Savilian Chair of Astronomy in 1649 at the age of thirty-two.[1] Indeed, the same Parliamentary board of visitors that intruded Wallis into the Savilian Professorship of Geometry was prepared to place Ward into the other Savilian chair, notwithstanding his known Royalist and Anglican sympathies.[2] The appointment naturally required Ward to profess allegiance to the new Commonwealth and abjure his allegiance to the Church of England—the same church in which he would acquire the rank of bishop after the restoration of Charles II. There was nothing unique in Ward's manifestly cynical attitude, and few seem to have held it against him. He was prominent in the intellectual life of Oxford in the 1650s, particularly in his promotion of scientific and mathematical learning. Together with Wallis and John Wilkins (who was named warden of Wadham College in 1649), Ward was instrumental in bringing the new "mechanical philosophy" to Oxford, and it is largely through the efforts of their circle that Oxford could claim a place of importance in the scientific world of the day (Shapiro 1969). Sir Kenelm Digby could refer without irony to Ward, Wallis, and Wilkins as exercising a "worthy Triumvirate . . . in literature and all that is worthy,"[3] and these three were without question among the most important scientific figures at Oxford in the 1650s.

1. For biographical details on Ward, see Pope 1697 and Fletcher 1940; the entries in the *Dictionary of Scientific Biography* and *Dictionary of National Biography* are also useful.

2. In fact, Ward had been deprived of his fellowship at Cambridge in 1644 for collaborating in the publication of an anti-Puritan tract, a project in which he assisted Isaac Barrow.

3. The reference is in a letter from Digby to Wallis, dated 1/11 August 1657 and published as part of Wallis's *Commercium Epistolicum de Quaestionibus quibusdam Mathematicis nuper habitum* (Wallis 1658, 10–12).

It is in his capacity as a promoter of scientific learning that Ward first came into contact with Hobbes, and it seems that he actively supported the publication of Hobbes's treatise *Humane Nature* in 1650. This work contains the first thirteen chapters of the treatise later published as *The Elements of Law* and deals with the general principles of scientific investigation, particularly as applied to human psychology.[4] Ward apparently contributed the "publisher's preface" to the book, in which he commends Hobbes for having "written a body of philosophy, upon such principles and in such order as are used by men conversant in demonstration" (*EL* preface; *EW* 4:xi).[5] In 1651, upon Hobbes's return to England, Ward also made a special trip to London to make his acquaintance, and there is no doubt that he regarded him as an important contributor to the advancement of science and philosophy.

Ward was concerned with more than the advancement of science during his years at Oxford, however. He pursued his considerable theological interests as well, taking the degree of doctor of divinity from Oxford in 1654. It is on theological issues that he initially broke with Hobbes, as is evident from the preface to his 1652 *Philosophicall Essay Towards an Eviction of the Being and Attributes of God, the Immortality of the Souls of Men, The Truth and Authority of Scripture.*[6] Speaking of himself in the third person, Ward declares

He must needs acknowledge, that before the edition of this he hath seen M. Hobs his Leviathan, and other Bookes of his, wherein that which is in this Treatise intended as the main Foundation whereon the second Discourse (Of the Souls Immortality) insists, is said to imply a contradiction, viz. That there are any

4. The full title, as published in 1650 reads *Humane Nature, Or: The fundamental Elements of Policie. Being a Discoverie of the Faculties, Acts, and Passions of the Soul of Man, from their original causes; According to such Philosophical Principles as are not commonly known or asserted.*

5. The grounds for attributing the preface to Ward are strong, but not completely certain. Anthony à Wood reports that "Seth Ward writ the epistle to the reader in the name of Francis Bowman bookseller, before this book" (Wood [1813] 1967, 3:column 1209). Hobbes himself shared this opinion, and in addressing Ward in the *Six Lessons* remarked that "Which [part of my book that concerneth policy merely civil], if you, the Astronomer, that now think the Doctrine unworthy to be taught, were pleased once to honour with praises printed before it, you are not very constant nor ingenuous. But whether you did so or not, I am not certain, though it was told me for certain" (*SL* 6; *EW* 7:336).

6. The role of Ward's *Philosophicall Essay* in Hobbes's controversy with Ward and Wallis has been amply investigated by Probst (1993); my account of the matter differs minimally from his.

such things as Immateriall or Incorporeall substances. Upon which occasion he thought good onely to say, That he hath a very great respect and a very high esteem for that worthy Gentleman, but he must ingenuously acknowledge that a great proportion of it is founded upon a belief & expectation concerning him, a belief of much knowledge in him, and an expectation of those Philosophicall and Mathematicall works, which he hath undertaken; and not so much upon what he hath yet published to the world, and that he doth not see reason from thence to recede from any thing upon his Authority, although he shall avouch his discourse to proceed Mathematically. That he is sure he hath much injured the Mathematicks, and the very name of Demonstration, by bestowing upon it some of his discourses, which are exceedingly short of that evidence and truth which is required to make a discourse able to bear that reputation. (Ward 1652, preface, A3)

It is unclear which "other Bookes" besides *Leviathan* Ward may have had in mind in writing this passage, but the general message is clear enough: the radical materialism of *Leviathan* is anathema to him, and this principally because it does not allow for the immortality of the soul.[7]

It is worth pointing out that *Humane Nature* (which Ward had seen and praised) differs from *Leviathan* in not explicitly denying the existence of immaterial substances. Hobbes's earlier work gives a mechanistic account of human sensation and argues for the subjectivity of all sensory qualities, concluding that "*whatsoever accidents* or qualities our senses make us think there be in the *world,* they be *not* there, but are *seemings* and *apparitions* only: the things that really *are* in the world without us, are those *motions* by which these seemings are caused" (*EL* 1.2.10; *EW* 4:8). But *Humane Nature* does not assert that the human mind is itself an assemblage of bodies in motion, and Hobbes even seems to leave room for an immaterialistic conception of the mind in the first chapter when he distinguishes faculties of the body from faculties of the mind. As a matter of fact, other authors of the period (such as Descartes) could combine a mechanistic treatment of sensation with a thoroughly immaterialistic understanding of the hu-

---

7. Ward summarizes his principal argument for immortality as follows: "Now the substance of all that I shall speak towards the demonstration of the souls Immortality shall be summarily comprised in this one syllogism. Whatsoever is incorporeall is immortal. But the souls of men are incorporeall substances, *Ergò*" (Ward 1652, 35).

man soul or intellect, and before the publication of *Leviathan* Ward was presumably unaware of the extent of Hobbes's commitment to materialism.[8]

The materialism of *Leviathan* was attacked by others besides Ward, and the polemics against Hobbesian materialism were mounted principally for theological reasons.[9] By denying the coherence of the concept of an immaterial substance, Hobbes challenged the very foundation of traditional theology in which souls, angels, and God himself were conceived of as essentially immaterial beings. Aubrey put the matter succinctly with the report that "the divines say, 'Deny spirits, and you are an atheist'" (Aubrey 1898, 2:318). In rejecting the standard opposition between the material and the spiritual, Hobbes therefore set himself the task of providing an alternative reading of scriptural passages dealing with such (allegedly immaterial) beings as spirits or angels. This led to one of the most controversial parts of *Leviathan*— chapter 34, which treats "*Of the Signification of* SPIRIT, ANGEL, *and* INSPIRATION *in the Books of Holy Scripture*." The chapter attempts to show that all biblical references to spirits or angels can be understood materialistically, and this is one of the parts of *Leviathan* that Hobbes admitted "may most offend" because it contains "certain Texts of Holy Scripture, alleged by me to other purpose than ordinarily they use to be by others" (*L* epistle; *EW* 3:ii).

I will investigate Hobbes's theological views, and particularly his

---

8. This suspicion is strengthened by Ward's remark in the preface to the *Philosophicall Essay* that Hobbes's denial of immaterial spirits

> can rationally import no more but this, That he himself hath not an apprehension of any such beings, and that his cogitation (as to the simple object of it) hath never risen beyond imagination, or the first apprehensions of bodies performed in the brain; but to imagine that no man hath an apprehension of the God-head, because he may not perhaps think of him so much as to strip off the corporeal circumstances wherewith he doth use to fancy him; or to conclude every man under the sentence of being non-sensicall, whosoever have spoken or written of Incorporeal substances, he doth conceive to be things not to be made good by the Authority of M. Hobbs. (Ward 1652, preface, sig. A4)

The opposition here between imagination on the one hand (which depends upon sensation, which itself is explained materialistically in terms of motions in the brain) and intellectual apprehension on the other (which "strips off" corporeal circumstances and enables the conception of immaterial beings) suggests that Ward accepted a broadly Cartesian picture of the mind, where the "lower" faculties depend upon the body, while the "higher" intellectual faculties are immaterial. I will have more to say about these matters as they relate to Hobbes's account of reasoning in chapter 5.

9. See Mintz 1962, chaps. 4–5, for a summary of the antimaterialist and theological attacks on Hobbes. I will return to this issue in chapter 7.

alleged atheism, in greater detail in chapter 7, but it is worthwhile stressing the point that his contemporaries saw a seamless connection between materialistic metaphysics, the denial of an immortal soul, a moral philosophy of pure egoism, and atheism. Hobbesian philosophy was therefore easily caricatured as the creed of the godless libertine who denies the possibility of an afterlife and takes the satisfaction of selfish desires as the summum bonum. Moreover, Hobbes's frequent claims to have grounded his philosophy in demonstration and his characterization of alternative systems as nonsensical Scholastic jargon gave him a reputation for dogmatism.

This picture of Hobbes as a dogmatic atheist and libertine emerges quite clearly in a letter from Wallis to Thomas Tenison (later archbishop of Canterbury) written approximately a year after Hobbes's death (the letter is dated 30 November/10 December 1680). Tenison had attacked Hobbes's philosophy with his 1670 book *The Creed of Mr. Hobbes Examined,* and he was evidently planning to write a life of Hobbes that would show the erroneous nature and harmful consequences of Hobbes's philosophy and thus serve as an admonition to all who might be tempted to embrace Hobbesian principles and thereby fall into impiety. In the letter, Wallis approves Tenison's project and offers his own opinion of Hobbes and his philosophy. He claims to have had no personal acquaintance with Hobbes, but that as a result of their controversy Wallis formed the firm opinion that the Monster of Malmesbury "was not a man of strong Reason; but only of a bold daring phansy, which, with his magisteriall way of speaking, did (not convince, but) sway those that loved to be Atheists, and were glad to have any body dare boldly to say, what they wish'd to be true" (Bodleian Library MS. Add. D.105, f. 70^r). Wallis concludes that Hobbes's principles must be those of a man who secretly dreads the punishments of the afterlife even as he denies the very existence of an afterlife:

> In summe; I can hardly believe y^t Mr Hobbes himself (nor perhaps any other pretenders to it) was so much an Atheist, as he would fain have been: but did really dread a future state: otherwise, he would not have been so dreadfully afraid of death, as ye concurrent testimony of those who knew him do represent him. (Bodleian Library MS. Add. D.105, f. 70^v)

This morbid fear of death was, according to Wallis, confirmed by an incident related to him "divers years agoe," in which "a great Lady" reported that Hobbes had declared "if he were master of all y^t was to dispose of, he would give it to live one day." To the reproach that "a person of his knowledge, & who had so many friends to oblige or gratify would not deny himself one days content of things if thereby to be able to gratify them with all ye World," Hobbes is supposed to have evinced both his utter selfishness and his terror of death with the reply "Madame, what should I be the better for that, when I am dead? I say again If I had all ye World to dispose of, I would give it to have one day" (Bodleian Library, MS. Add. D.105, 70^v).

Wallis found further evidence for this view of Hobbes and his philosophy "in y^t sermon at ye Funeral of ye late Earl of Rochester: Who could talk Atheisticall things with as much briskness & as much wit as Mr Hobbes, and trip most at sense & reason: yet could not thoroughly beleeve it; but was galled (we are to understand) with a recoiling conscience" (Bodleian Library, MS. Add. D.105, 70^v). In addition to his acknowledged literary gifts Rochester was one of the most notorious libertines associated with the court of Charles II (a venue not known for its modest living). However, shortly before his death he embraced Christianity and repented his former life of dissolution.[10] The sermon preached by Robert Parsons at the funeral of Rochester singled out Hobbes as the man whose principles had done the most to encourage the life of atheistic debauchery. Speaking of the departed Rochester, Parsons declared:

> How remarkable was his Faith, in a hearty embracing and devout confession of all the Articles of our Christian Religion, and all the Divine mysteries of the Gospel? Saying, that *that absurd and foolish Philosophy, which the world so much admired, propagated by the late Mr.* Hobbs, *and others, had undone him, and many more of the best parts in the Nation;* who, without Gods great mercy to them, may never, I believe, attain such a Repentance. (Parsons 1680, 26)

We need not presume that Hobbes's enemies had the correct interpretation of his philosophy, its consequences, or its author's opinions on matters of morals or the afterlife. Nevertheless, it is clear that he had acquired the reputation as an atheist, a libertine, and the proponent of views that were inimical to religion and morals.

Quite aside from the thoroughgoing materialism that led many to regard it with horror, *Leviathan* was also controversial for its political

---

10. On Rochester, his conversion, and his relationship to Hobbes, see Mintz 1962, 140–42.

doctrines. The central theme in Hobbes's political theory is the necessity of absolute, undivided sovereignty. Sovereign power is not to be shared among competing persons or institutions, for (on Hobbes's analysis) such divided power must inevitably produce conflicting claims upon the loyalties of subjects, and these conflicts must eventually lead to civil war. Similarly, sovereignty cannot be limited to a particular area of competence, because the imposition of limits on a sovereign's power supposes the existence of a supreme authority over and above that of the sovereign. One important consequence of this theory is that the civil and ecclesiastical sovereign are "consolidated" in one supreme authority, and such sovereigns "have all manner of Power over their subjects, that can be given to a man, for the government of mens externall actions, both in Policy and Religion; and may make such Laws, as themselves shall judge fittest, for the government of their own Subjects, both as they are the Common-wealth, and as they are the Church: for both State, and Church are the same men" (*L* 3.42, 299; *EW* 3:546) It is the sovereign, and only the sovereign, who may lawfully resolve questions of church governance, scriptural interpretation, liturgy, and other ecclesiastical matters.

Although such a doctrine could find favor among committed Royalists who desired to see the crown's power absolute and unchallenged, it was bitterly opposed by those who insisted that spiritual matters must be entrusted to ecclesiastical authorities distinct from the civil sovereign. These included Anglicans who favored an episcopal system granting broad powers to bishops, Presbyterians who would abolish episcopacy and invest ecclesiastical authority in a body of clergymen and lay elders, and other varieties of Protestant thought that denied the authority of the civil sovereign in religious matters. Hobbes provoked the wrath of these churchmen with his version of ecclesiastical history in part 4 of *Leviathan*. He there describes the clergy as part of the "Kingdome of Darknesse" that through various deceitful practices has conspired to usurp the civil power. The Catholic Church is the most notable agent of darkness in this Hobbesian demonology, but Anglican bishops and Presbyterian doctors of divinity are also condemned for their attempt to establish a power contrary to that of the rightful sovereign. In a memorable passage in chapter 47 of *Leviathan* Hobbes describes the sovereign as having been bound by knots tied by ambitious clergymen, which (at least in England) have finally been untied by a continuing process of reformation that leads to "the dissolution of the praeterpoliticall Church Government in England":

First, the Power of the Popes was dissolved totally by Queen Elizabeth; and the Bishops, who before exercised their Functions in Right of the Pope, did afterwards exercise the same in Right of the Queen and her Successours; though by retaining the phrase of *Jure Divino,* they were thought to demand it by immediate Right from God: And so was untyed the first knot. After this, the Presbyterians lately in England obtained the putting down of Episcopacy: and so was the second knot dissolved: And almost at the same time, the Power was taken also from the Presbyterians: And so we are reduced to the Independency of the Primitive Christians to follow Paul, or Cephas, or Apollos, every man as he liketh best: Which, if it be without contention, and without measuring the Doctrine of Christ by our affection to the Person of his Minister . . . is perhaps the best. (*L* 4.47, 385; *EW* 3:696)

This endorsement of "Independency" may seem surprising coming from a man who holds that the civil sovereign has the absolute power to compel conformity in all matters of worship. The Independents and the Presbyterians were the two principal factions among supporters of the Parliamentary cause in the English Civil War. They were united in opposition to the Anglican system of episcopacy, but disagreed about what system should replace it. As the name suggests, the Independents favored a relatively loose form of church governance in which local congregations could resolve liturgical or doctrinal issues independently, while the Presbyterians insisted upon the imposition of a strict system of church organization that would enforce conformity throughout the nation. Hobbes's suggestion that Independency is "perhaps the best" reflects, in part, his concern that any centralized system of church government could pose a threat to the sovereign's power. Moreover, the events of the Civil War clearly suggested that, although the sovereign has the right to impose conformity in matters of worship, it is not always in his interest to enforce a policy of strict conformity. A wise sovereign could best avoid potential challenges to his rule by permitting a fairly wide latitude in belief and worship, reserving his intervention in religious affairs to discipline those who claim a power in opposition to his own.[11]

---

11. Richard Tuck, in commenting upon this aspect of Hobbes's theory notes that "in Hobbes's eyes, the sovereign would not have the same kind of reasons for enforcing particular dogmas upon his citizens as churches historically had acted on in controlling their members. His right to enforce doctrine was essentially negative, and intended above all to stop *non-sovereigns* from claiming such a right" (Tuck 1989, 88). For more on

It should by now be evident that the publication of *Leviathan* had the remarkable effect of offending nearly every side in the struggles taking place in England in 1651. This is not to say that Hobbes was entirely without a readership or admirers, but he clearly had done enough to guarantee that he would have a plentiful supply of enemies. One of the most important sources for opposition to Hobbes was the universities, and particularly the University of Oxford. It is therefore advisable that we turn our attention more closely to Hobbes's doctrines as they relate to the universities in order to understand the background to his war with Wallis.

2.2 HOBBES AND THE UNIVERSITIES

In holding that all institutions in a commonwealth must be under the absolute control of the sovereign, Hobbes thereby requires that the university be subordinated to the state. No doctrines may be legitimately taught, except those explicitly permitted by the authority of the sovereign, and the sovereign is the sole judge of which teachings are permissible (*L* 2.18, 91; *EW* 3:164–65). This principle extends, ironically enough, to the teaching of geometry, for Hobbes holds that the sovereign may ban the teaching of geometry and order "the burning of all books of Geometry" if he should judge any geometric result contrary to his right of dominion.[12]

Hobbes directed some of his most vitriolic polemics at the universities, largely because he saw them as endangering civil peace by challenging the authority of the sovereign. Although some parts of his critique of the universities were intended to apply to Continental as well as English universities, it is clear that Oxford and Cambridge were his principal targets. He condemns the universities as agents of the "Kingdome of Darknesse" and traces their history from the schools of the ancient Jews and Greeks. In his portrayal, the "unprofitable" schools of ancient times have been succeeded by equally useless and benighted universities, which are so far from being centers of learning that they

---

Hobbes's account of church, state, and religious order see Sommerville 1992, chaps. 5 and 6.

12. Hobbes declares, "For I doubt not, but that if it had been a thing contrary to any mans right of dominion, or to the interest of men that have dominion, *That the three Angles of a Triangle should be equall to two Angles of a Square;* that doctrine should have been, if not disputed, yet by the burning of all books of Geometry, suppressed, as farre as he whom it concerned was able" (*L* 1.11, 50; *EW* 3:91).

teach neither genuine philosophy nor (until very lately) even the rudiments of geometry. A university, Hobbes declares, is

> a Joyning together, and an Incorporation under one Government of many Publique Schools, in one and the same Town or City. In which, the principall Schools were ordained for the three Professions, that is to say, of the Romane Religion, of the Romane Law, and the Art of Medicine. And for the study of Philosophy it hath no otherwise place, then as a handmaid to the Romane Religion: And since the Authority of Aristotle is onely current there, that study is not properly Philosophy, (the nature whereof dependeth not on Authors,) but Aristotelity. And for Geometry, till of very late times it had no place at all; as being subservient to nothing but rigide Truth. And if any man by the ingenuity of his owne nature, had attained to any degree of perfection therein, hee was commonly thought a Magician, and his Art Diabolicall. (*L* 4.46, 370; *EW* 3:670)

The charges Hobbes brings against the universities are grave indeed, but three are of particular concern in this context. These include: obscurantism in teaching nonsensical scholastic metaphysics, the fomenting of civil war by training preachers in theological doctrines that undermine the sovereign power, and the promotion of civil discord by teaching men that the lawful government is an illegal tyranny.

To understand these charges properly we must first observe that Hobbes regarded the university as an institution of central importance in maintaining civil order. He was acutely aware that a subject's loyalty to his sovereign is not an inborn trait, but rather something that must be carefully inculcated by an aggressive system of education. By nature people desire power for themselves, and their pride or self-conceit leads them to resent any authority placed over them. For any system of social organization to work, people must therefore be brought to see that their interests lie in submission to an all-powerful central authority, and their natural rebelliousness must be curbed by the threat of sanction from the sovereign. It is therefore only through the combination of education and discipline that the fundamental problem of the state of nature can be solved on a continuing basis.

The alternative to civil peace is civil war, and Hobbes traces the root of civil war to the prevalence of private judgment, and especially private judgment about matters of political or religious obligation. In the state of nature, conflicting private judgments abound and are the source of the war of all against all. Hobbes holds that "so long as every man holdeth this Right of doing any thing he liketh; so long are all men in the condition of Warre" (L 1.14, 65; EW 3:118) and "if their actions be directed according to their particular judgements, and particular appetites, [men] can expect thereby no defence, nor protection, neither against a Common enemy, nor against the injuries of one another" (L 2.17, 86; EW 3:155). But even in the state of society there is a danger of lapsing back into chaos if citizens are allowed to determine for themselves what is to count as lawful, or which obligations they have. Hobbes lists several "Opinions, contrary to the peace of Man-kind," that tend toward civil war, including:

> That men shall Judge of what is lawfull and unlawfull, not by the Law it selfe, but by their own Consciences; that is to say, by their own private Judgements: That Subjects sinne in obeying the Commands of the Common-wealth, unlesse they themselves have first judged them to be lawful: That their Propriety in their riches is such, as to exclude the Dominion, which the Common-wealth hath over the same; That it is lawfull for Subjects to kill such, as they Call Tyrants; That the Soveraign Power may be divided, and the like. (L 2.30, 179; EW 3:330–31)

The task of the sovereign therefore includes that of keeping private judgment in check. But this can be accomplished only by getting citizens to relinquish their dangerous opinions and to accept the authority of the sovereign, and Hobbes sees the universities as the principal tool by which this is to be effected.

As Lloyd has noted, Hobbes "holds a decidedly trickle-down theory of education and influence," in which principles taught at the universities will be accepted by clergymen trained there and thus disseminated throughout the society (Lloyd 1992, 193–94). Hobbes himself declares it "manifest, that the Instruction of the people, dependeth wholly, upon the right teaching of Youth in the Universities" (L 2.30, 180; EW 3:331). His approach to political theory consequently requires that one primary use of the sovereign's power is to coerce the universities into reforming their principles so that the doctrines taught there are conducive to social stability. Not surprisingly, Hobbes held out the hope that "at one time or other, this writing of mine, may fall into the hands of a Soveraign, who will consider it himselfe . . . without the help of any interested or envious Interpreter; and by the exercise of entire Soveraignty, in protecting the Publique teaching of it, convert this Truth of Speculation, into the Utility of Practise" (L 2.31, 194; EW 3:358). In particular, Hobbes hoped that such an enlightened sovereign might see in *Leviathan* the ideal text to be assigned to students in the universities. Thus, in the "Review and Conclusion" to *Leviathan* he observed, "I think [this book] may be profitably printed, and more profitably taught in the Universities, in case they also think so, to whom the judgement of the same belongeth. For seeing the Universities are the Fountains of Civill, and Morall Doctrine, . . . there ought certainly to be great care taken, to have [these fountains] pure, both from the Venime of Heathen Politicians, and from the Incantation of Deceiving Spirits" (*L* review and conclusion, 395; *EW* 3:713). This declaration became the basis for Wallis's repeated accusations that, as an enemy of the universities and a vain dogmatist in love with his own theories, Hobbes intended to have the doctrines of *Leviathan* set up as official teaching at Oxford and Cambridge.

The specific errors with which Hobbes charges the universities are closely related to his conception of the university as an institution crucial for the maintenance of civil peace. Subjects can be misled by prideful and ambitious teachers into accepting doctrines that work against social order, even if the connection between the false doctrine and the resulting social instability is not immediately apparent. Aristotelian metaphysics is a case in point. According to Hobbes, the Scholastic presentation of Aristotelian hylomorphism and the doctrine of separated essences are mistaken doctrines with politically dangerous consequences. Those who are taught these doctrines are led to think that a person's soul can exist apart from the body, that bread and wine can be transubstantiated into flesh and blood, and that virtues can exist separated or apart from humans and be deposited or "blown in" to the soul. These and other doctrines derived from Aristotelian teachings,

> serve to lessen the dependance of subjects on the sovraign power of their countrey. For who will endeavour to obey the laws if he expect obedience to be powered or blown into him? Or who will not obey a priest, that can make God, rather than his sovraign; nay than God himselfe? Or who, that is in fear of ghosts, will not bear great respect to those that can make the holy water that drives them from him? (*L* 4.46, 373; *EW* 3:675)

There is nothing novel in Hobbes's claim that the universities are intellectual backwaters corrupted by the undue authority of Aristotle. Condemnations of Aristotle and the alleged prevalence of Aristotelianism at the universities are a commonplace among seventeenth-century "moderns" such as Descartes, Bacon, and Galileo. But Hobbes is certainly unique in seeing Scholastic obscurantism as productive of social disorder, rather than simply confusion, error, or bad philosophy.

Hobbes's attacks on the seditious doctrines promulgated by the clergy are related to this condemnation of Aristotelianism in the universities. The Catholic doctrine of the pope's supreme ecclesiastical power was naturally anathema to Hobbes, but he equally opposed the Presbyterians' insistence that temporal authority must be subservient to spiritual authority. According to Hobbes, Catholics and Presbyterians share a common error of believing that the kingdom of God is present in the form of the church, and consequently that the civil sovereign is subject to the dictates of the church.[13] As he puts it, "The greatest, and main abuse of Scripture, and to which almost all the rest are either consequent, or subservient, is the wresting of it, to prove that the Kingdome of God, mentioned so often in the Scripture, is the present Church" (L 4.44, 334; EW 3:605). The principal effect of this error is to undermine the sovereign's power by dividing it: if the kingdom of God is the present church, the temporal authority of the sovereign is placed in opposition to the spiritual authority of the church, and power rightfully belonging to the civil sovereign is usurped by divines. The universities, as centers of religious authority and teaching, stand naturally to benefit from this usurpation of civil power, and they endeavor to protect their position by employing "the frivolous Distinctions, barbarous Terms, and obscure Language of the Schoolmen," in order to "keep these Errors from being detected, and to make men mistake the *Ignis fatuus* of Vain Philosophy, for the Light of the Gospel" (L 4.47, 383; EW 3:693).

Hobbes's third principal indictment of the universities is that they have taught men "to call all manner of Common-wealths but the Popular ... *Tyranny*" (L 4.46, 377; EW 3:682). Here again, he detects the corrupting influence of Aristotle, combined with other "heathen politicians" such as Cicero. The teaching of their false doctrines of right and obligation has resulted in a number of dangerous errors, not

---

13. Lloyd summarizes this aspect of Hobbes's thought nicely: "Indeed, for Hobbes, Presbyterians and papists are reverse sides of the same coin: Both contend that the kingdom of God presently exists; both counterpose temporal and spiritual government, asserting the latter to have primacy; both engage in what Hobbes sees as esoteric and dangerous doctrinal disputes; and each takes itself to be the only true church. In Hobbes's view, excepting their commendable attack on popish superstition, Presbyterians are just as bad as Catholics—and much more dangerous given English commitments, because much more influential" (Lloyd 1992, 196).

least of which is the opinion "that in a wel ordered Common-wealth, not Men should govern, but the Laws." From these false tenets, the ambitious doctors in the universities "induce men, as oft as they like not their Governours, to adhaere to those that call them Tyrants, and to think it lawfull to raise warre against them" (*L* 4.46, 377–78; *EW* 3:683–84). Hobbes thus takes the English Civil War to be a university-instigated conflict, and one principally caused by an ambitious Presbyterian clergy, whose pursuit of spiritual and temporal power led to disaster. But once the causes of the rebellion have been uncovered, civil peace can be ensured by reforming the universities and subordinating them to the power of the lawful sovereign.

This is all well and good, but one might ask what bearing it has on Wallis and the quadrature of the circle. For present purposes it is sufficient to note that, in addition to holding a position of considerable power within the University of Oxford, Wallis was a prominent Presbyterian who served as a secretary to the Westminster Assembly of Divines, and that his appointment as Savilian professor of geometry in 1649 was effected by the victorious Parliamentary forces who had recently executed Charles I. Indeed, he was exactly the kind of man Hobbes detested: an ambitious doctor of divinity with close ties to the Puritan establishment. In the *Six Lessons,* Hobbes could complain that his political doctrine "is generally received by all but those of the Clergy, who think their interest concerned in being made subordinate to the Civil Power; whose testimonies therefore are invalid," and insist that because "so much as could be contributed, to the Peace of our Country, and the settlement of Soveraign Power without any Army, must proceed from Teaching; I had reason to wish, that Civill Doctrine were truly taught in the Universities" (*SL* 6; *EW* 7:333, 335). There were few, if any, in England who had more reason to fear and detest Hobbes and his teachings, and Wallis was not a man to work quietly behind the scenes to defend himself and his position against a perceived threat.

2.3 THE UNIVERSITY REFORM DEBATE

In calling for a fundamental reform of the universities, Hobbes took part in an important seventeenth-century debate over the status of universities in England. Many Puritan critics of Oxford and Cambridge had also campaigned for an overhaul of the universities, with the intent of making their practices consonant with the principles of true government and religion. It was widely agreed both by the critics and defenders of the universities that their central function was to provide ministers for the state church, and Puritan calls for reform were generally intended to ensure that these parsons would be staunch defenders of the newly reformed state church. With the Parliamentary party's victory in the Civil War and the abolition of monarchy, the universities were purged of the most committed Anglicans and Royalists and left largely in the hands of the two main Puritan groups: the Presbyterians and the Independents. In this process, Wallis was appointed Savilian professor of geometry at Oxford in 1649 and John Owen (the "Atlas of Independency" and a close associate of Cromwell) was named vice-chancellor of Oxford in 1652.

The more radical elements in the English revolution (including such sects as the Levellers, Ranters, Diggers, and the like) were not content with such measures, and questioned the very desirability of a state church and universities dedicated to it.[14] The radicals opposed the system of tithes that supported the state church and insisted that autonomous, freely assembled local congregations could and should support their chosen ministers. These "mechanick preachers" need not be university-educated men who had mastered the obscure subtleties of "school divinity," but instead plain men called to preach by the direct inspiration of the Holy Spirit. In the view of the radicals, the reform of the universities involved much more than simply ensuring that the properly reformed theology was taught or ejecting Royalists and Anglicans from their university livings. They campaigned rather for reorienting the university away from its historic mission of training preachers and bringing it to the service of a broader population by promoting "useful learning" in the arts and sciences.

2.3.1 *John Webster's* Academiarum Examen

John Webster (1610–82) was a radical with close ties to Cromwell's army, where he served as both surgeon and chaplain. His 1654 treatise *Academiarum Examen* set forth the radical program for the universities, calling for nothing less than expunging theology from the university curriculum and replacing the doctrines of the schools with a new learning that combined Baconian principles with occult and alchemical researches into "natural magic." According to Webster, the universities are burdened by their heritage of Scholasticism, so much so that

14. See Hill 1975 for a study of the radical program for the universities. Debus 1970 deals with the critique of universities and its relationship to the scientific developments of the period. For an overview of the different radical Puritan sects see McGregor and Reay 1984 and Mullett 1994.

the opinion of Aristotle holds sway in all matters touching on natural philosophy. In language that could have come straight out of Hobbes's *Leviathan*, Webster laments that "the *Philosophy* which the *Schools* use and teach, being meerly *Aristotelical*, [has been] imbraced and cryed up more than all others [so that] he should be accounted the Prince of *Philosophers*, the Master-piece of Nature, the Secretary of the Universe, and such an one beyond whose knowledge there is no progression" (1654, 52). This alleged monopoly of Aristotelianism has prevented the universities from applying themselves to the study of nature and has reduced natural philosophy to a literary exercise in the interpretation of Aristotle:

> This School Philosophy is altogether void of true, and infallible demonstration, observation, and experiment, the only certain means, and instruments to discover, and anatomize natures occult and central operations; which are found out by laborious tryals, manual operations, assiduous observations, and the like, and not by poring continually upon a few paper Idols, and unexperienced Authors. (1654, 68)

The remedy for this ill is to banish Aristotelian natural philosophy from the universities and institute a program of study in "Pyrotechny or Chymistry," grounded particularly in the hermetic-magical tradition of Paracelsus, Jan Baptist van Helmont, and Robert Fludd. The anticipated result of this program is "that the Schools therefore would leave their idle, and fruitless speculations, and not be too proud to put their hands to the coals and furnace, where they might find ocular experiments to confute their fopperies, and produce effects that wou'd be beneficial to all posterities" (1654, 71).

Webster finds fault with the theology of the schools as well. The influence of Scholastic teachings in theology has corrupted the simple, pure message of the Gospel and produced "a confused *Chaos*, of needless frivolous, fruitless, triviall, vain, curious, impertinent, knotty, ungodly, irreligious, thorny, and hel-hatch't disputes, altercations, doubts, questions and endless janglings, multiplied and spawned forth even to monstrosity and nauseousness" (1654, 15). In fact, Webster finds the enterprise of school divinity opposed to true Christian principles precisely because it keeps the Gospel and its interpretation in the hands of vain and ambitious doctors of divinity who have monopolized the spiritual life of England. In some of the seventeenth century's more purple prose, Webster complains that

[f]rom this putrid and muddy fountain [of school divinity] doth arise all those hellish and dark foggs and vapours that like locusts crawling from this bottomlesse pit have overspread the face of the whole earth, filling men with pride, insolency, and self-confidence, to aver and maintain that none are fit to speak, and preach the spiritual & deep things of God, but such as are indued with this *Scholastick,* & mans *idol-made learning,* and so become fighters against God, and his truth, and persecutors of all those that speak from the principle of that wisedome, *that is from above, and is pure and peaceable:* not confessing the nothingness of creaturely wisedom, but magnifying, and boasting in that which is *earthly, sensual,* and *devillish.* (1654, 12)

The new university envisaged by Webster would not train preachers, because "the teaching of spiritual and Gospel knowledge is onely and peculiarly appropriated and attributed unto the Spirit of God" (1654, 4). Instead, Webster's properly reformed university would eschew theology altogether and teach "onely humane science," which is now regarded as too low and common for the attention of educated men. The result will be a university that takes as its main function the expansion of knowledge of nature and the implementation of this knowledge for practical improvements in the lives of the generality of mankind.

2.3.2 William Dell

William Dell, whom the Parliamentarians had appointed master of Gonville and Caius College in Cambridge, was every bit as radical as Webster in his proposed reforms of the university and the abolition of tithes. Indeed, his writings against university learning have a strong "levelling" strain that is ironic in a man who held a position of considerable power within the university.[15] Like Webster, Dell thought that Oxford and Cambridge paid insufficient attention to the practical consequences of learning, and he proposed that new universities be set up in every major city to instruct the common folk in the arts and sciences. Once the monopoly of Oxford and Cambridge was broken, the benefits of education could be extended to the more deserving lower orders, while righteous men called to preach the Gospel and freely chosen by their congregations could replace the caste of state-supported

15. See Hill 1975, 136–42, and Webster 1973 for accounts of Dell's radical program for the universities. Walker 1970 is a biography that downplays Dell's radicalism and treats his writings on the universities as an unfortunate aberration.

ministers. As he explains in his sermon on "The Right Reformation of Learning, Schools and Universities,"

> [I]f human learning be so necessary to the knowledge and teaching of the scriptures, as the universities pretend, they surely are without love to their bretheren, who would have these studies thus confined to [Cambridge and Oxford], and do swear men to read and teach them no where else: certainly it is most manifest, that these men love their own private gain, more than the common good of the people. (Dell 1816, 588–89)

Dell's educational remedy proceeds from the principle that "it would be more suitable, and more advantageous to the good of all the people, to have universities or colleges, one at least, in every great town or city in the nation" (Dell 1816, 589).

Dell also agreed with Webster in holding that true ministers of the gospel are not produced by the Scholastically tainted theological training of the universities, and he was as harsh in his condemnation of the errors produced by such theological training. In his "Testimony from the Word against Divinity Degrees in the University," Dell complains that

> the university, through power received from antichrist, [gives] men, chiefly for money, divinity degrees; and through those degrees, it gives authority and privilege to batchelors in divinity to expound part of the scriptures, and to doctors to expound and profess all the scriptures; and they that gain these degrees to themselves, are (as there is good reason) the great men in account with the university, and also with the carnal people of antichrist, how destitute soever they be of the faith and Spirit of the Gospel. (Dell 1816, 564)

The universities "are those antichristian soldiers, who put a reed into Christ's right hand instead of a sceptre: and their reed is philosophy, that vain deceit" (Dell 1816, 500). Thus, by expunging the writings of heathen philosophers (i.e., Aristotle) from the university, and eradicating the institution of divinity degrees, Dell hopes to reform the university in essentially the same manner as Webster.

The program of the radicals was anathema to Hobbes in many respects. In particular, the radicals' insistence upon immediate inspiration (rather than school divinity) as the only true path to the interpretation of Scripture is precisely the kind of emphasis upon "private judgement" that he denounced. Nevertheless, Hobbes and the radicals shared a diagnosis of what was wrong with the universities and a general prescription of how it should be fixed. Specifically, they agreed that the universities were so heavily under the influence of Scholastic teachings that they could not be trusted to reform themselves, and they distrusted the capacity of cynical and ambitious doctors of divinity to cooperate with any move toward reform. Rather than rely upon a relatively autonomous university to cast off the pernicious influence of Aristotle, Hobbes and the radicals insisted that state power be brought to bear in forcing the universities to change their ways.

2.3.3 Oxford's Response

These criticisms of the universities did not pass without notice. Those in control of the universities were threatened by the calls for radical reform, and they had evident reason to fear the consequences their institutions might face. Vice-chancellor Owen, addressing Cromwell in December of 1654, reports that "we live among men ignorant of our ways, making our position slippery and uncertain. Then it came about that by listening carefully we heard a calumny. Once the pride and delight of the nation, we were soon almost a laughing stock" (Owen 1970, 79). Concern for the future of the universities was heightened in 1654 when the Barebones Parliament entertained a proposal for the outright abolition of the universities (Malcolm 1988, 54). Ward took the offensive against the universities' critics, and in 1654 published a detailed rebuttal to Webster under the title *Vindiciae Academiarum,* which included appendices replying to Hobbes and Dell.[16] There is a quite remarkable irony in the fact that Hobbes was explicitly linked to Webster and Dell. On matters of theology, church government, and politics, Hobbes was diametrically opposed to the views of these reformers; Ward was sensitive to the irony and remarked "how scornfully he will take it to be ranked with a Friar and an Enthusiast" ([Ward] 1654, 51).

The appendix directed at Hobbes was the first shot in the war that would last the next quarter century. In fact, it is virtually certain that Hobbes was led into his long-running mathematical controversy with

16. Although it was an anonymous work of Ward's (with an anonymous preface by Wilkins), we can identify the authors of the different parts of the *Vindiciae Academiarum.* As the *Dictionary of National Biography* entry for Webster notes, "The book *[Academiarum Examen]* was answered by Seth Ward, bishop of Salisbury, under the signature of H. D., the final letters of both his names, with a prefatory epistle by John Wilkins, bishop of Chester, also signed with final letters, N. S., and which has in consequence been assigned to Nathaniel Stephens."

Wallis by the goading of Ward and Wilkins. In the prefatory epistle to *Vindiciae Academiarum* Wilkins portrays Hobbes as "a person of good ability and solid parts, but otherwise highly magisteriall, and one that will be very angry with all that do not presently submit to his dictates, And for advancing the reputation of his own skill, cares not what unworthy reflexions he casts on others" (1654, 6–7). These charges are accompanied by insinuations of plagiarism from the manuscripts of Walter Warner[17] and a challenge to Hobbes to publish his solutions to the great geometrical problems such as the quadrature of the circle.[18] Ward betrays a familiarity with Hobbes's geometrical pretensions, presumably gathered through contact with friends and associates of Hobbes, and he strongly hints that there are those in Oxford who are prepared to debunk his claims to mathematical glory:

Geometry hath now so much place in the Universities, that when Mr *Hobbs* shall have published his Philosophicall and Geometricall Pieces, I assure my selfe, I am able to find a great number in the University, who will understand as much or more of them then he desires they should, indeed too much to keep up in them that Admiration of him which only will content him. (1654, 58)

Ward's familiarity with Hobbes's scientific work makes it highly likely that he had knowledge of the state of his mathematical undertakings, and the unavoidable suspicion is that he hoped to provoke him into publishing an inept quadrature of the circle that would diminish Hobbes's mathematical reputation. In the event, the Monster of Malmesbury took the bait and hastily added chapter 20 to *De Corpore,* in

17. "It were not amisse, if he were made acquainted, that for all his slighting of the Universityes, there are here many men, who have been very well versed in those notions and Principles which he would be counted the inventer of, and that before his workes were published. And though he for his part may think it below him to acknowledge himselfe beholding to Mr *Warners* Manuscripts, yet those amongst us who have seen and perused them must for many things give him the honour of precedency before Mr *Hobbs*" (1654, 7). Prins (1993) has argued that this charge is without merit.

18. "I have heard that M. *Hobbs* hath given out, that he hath found the solution of some Problemes, amounting to no lesse then the Quadrature of the Circle, when we shall be made happy with the sight of those his labours, I shall fall in with those that speake loudest in his praise, in the meane time I cannot dissemble my feare, that his Geometricall designe (as to those high pieces) may prove answerable to a late Opticall designe of his, of casting Conicall glasses in a mould, then which there could not be any thing attempted lesse becoming such a man, as he doth apprehend himselfe to be" ([Ward] 1654, 57).

which he set forth a circle quadrature whose ineptitude probably surpassed Ward's fondest wishes.[19]

Meanwhile, Wallis was lying in wait.[20] He obtained the sheets of *De Corpore* from the printer as it was being printed, and prepared diligently to attack the mathematics in Hobbes's treatise as soon as it appeared in public. *De Corpore* was published in April of 1655, and Wallis's *Elenchus* appeared in August of the same year. Even by Wallis's relatively prolific standards, this is rather remarkable: he seems to have completed over 130 pages of Latin mathematical diatribe in a matter of months, all of it aimed at refuting every mathematical claim in *De Corpore* while interspersing other comments on the Hobbesian enterprise. Ward was busy as well, although he did not match the pace set by Wallis. By evident prior arrangement with Wallis, Ward attacked Hobbes's metaphysics, epistemology, theology, and political theory in his *In Thomae Hobbii Philosophiam Exercitatio Epistolica* (1656)—over 350 pages of anti-Hobbesian polemic.

The evidence assembled to this point should make it appear reasonable that Wallis's attack on Hobbes's geometric work has a significant connection with the contemporary debates over the universities. And, in fact, when we consider Wallis's own pronouncements on the matter, it becomes quite clear that his intention was not simply to expose some flaws in Hobbes's putative circle quadratures, but rather to con-

19. It is difficult to be certain whether Hobbes inserted the quadrature in chapter 20 as a direct result of the challenges issued by Ward. The chapter seems out of place, and reads rather like a late insertion. Hobbes hastily went through at least three versions of the failed quadrature, abandoning two previous efforts after friends had pointed out errors. Wallis (whose detective skills were honed by cryptographic efforts on behalf of the Parliamentarians in the Civil War) noted that chapter 20 had been printed, cut, and revised twice, basing his conclusion on an examination of the gatherings and the figures in the printed version of *De Corpore* (*Elenchus* 2–3, 95–96). I will investigate the background to the circle quadrature in *De Corpore* more closely in chapter 3.

20. Hobbes had the extraordinarily bad timing to publish *De Corpore* at the time when Wallis was preparing his *Arithmetica Infinitorum* for publication and was deeply involved with questions of circle quadrature. One of the principal results of *Arithmetica Infinitorum* is Wallis's method of representing π as an infinite product, and his researches in this area were ideal preparation for a frontal assault on Hobbes's mathematics. Wallis himself acknowledged as much: "having set aside other things, I gathered together these geometric pieces of yours, eager to see those things which mathematicians have sought in vain through every past century, and which you hope at last to have found. And I did this in principal part because this office [of Savilian Professor] did seem to demand it; and in part because I had been engaged in the same arena (nor, plainly, did I work in vain); and indeed I had brought the matter so far that four years ago I discovered the proposition on circle quadrature which I have recently made public" (*Elenchus* 2).

vince the reading public that the entire Hobbesian philosophy was ill founded. Recall that Hobbes had presented *De Corpore* as the first and foundational part of his philosophical enterprise, and he held that his doctrines were sufficiently well grounded that they would enable the solution of all problems. In Hobbes's scheme, the recognition that only body is real leads naturally to the conclusion that mathematics must be a generalized science of body, and a reformation of mathematics can then be effected by recasting the traditional principles of geometry in a more clearly materialistic form. Because the fundamental nature of body is clearly apprehended, this program (so Hobbes thinks) will make short work of the most difficult geometric problems, and the truth and power of the Hobbesian philosophy will be made manifest by its success in the solution of outstanding problems. Wallis intends to turn the tables on Hobbes, taking the demolition of the geometric work as a means to show that the grand program of its master is mistaken.

In a letter to Huygens in January of 1659 Wallis made his motivations for attacking Hobbes explicit, particularly as they relate to the universities. There he wrote:

> But regarding the very harsh diatribe against Hobbes, the necessity of the case and not my manners led to it. For you see, as I believe, from other of my writings how peacefully I can differ with others and bear those with whom I differ. But this was provoked by our Leviathan (as can be easily gathered from his other writings, principally those in English), when he attacks with all his might and destroys our universities (and not only ours, but all, both old and new), and especially the clergy and all institutions and all religion. As if the Christian world knew nothing sound or nothing that was not ridiculous in philosophy or religion; and as if it has not understood religion because it does not understand philosophy, nor philosophy because it does not understand mathematics. And so it seemed necessary that now some mathematician, proceeding in the opposite direction, should show how little he understands this mathematics (from which he takes his courage). Nor should we be deterred from this by his arrogance, which we know will vomit poison and filth against us. (Wallis to Huygens, 1/11 January 1659; *HOC* 2:296–97)

Wallis repeated a similar sentiment at the beginning of the edition of his *Opera Mathematica,* first published in 1693: "Certain works written long ago against Thomas Hobbes (pseudo-geometer) will not be found here, for I would not want to seem to triumph over one now dead. Nevertheless, as things then stood, it was something that had to be done, when he had set himself up in the guise of a great geometer and dared to offer false suggestions to our unsuspecting youth in matters of religion" (*OM* 1:sig. a4v). The structure of Wallis's *Elenchus* reflects his concern with a thorough "root and branch" eradication of Hobbes's philosophy. Rather than confining himself to showing the shortcomings in Hobbes's attempted quadrature of the circle, Wallis argues that Hobbes's entire conception of mathematics is mistaken, and concludes that his opinions on other, weightier matters must be equally unreliable.[21]

In the dedicatory epistle to John Owen in the *Elenchus*, Wallis explicitly raises the question "why, leaving aside theological and other philosophical matters, I should undertake to refute his geometry, when he has made other far more dangerous errors?" His response is revealing, for he expresses a strong concern to defend Oxford against the "sheer calumny which he is ever trying to persuade people, that there are no mathematicians, or at least not in the university." But he is equally concerned

> that these things should be speedily opposed, not, to be sure, because I think that geometry should be protected lest it suffer harm (indeed this of all disciplines has least to fear from logical fallacies), but so that when this balloon has been burst, that man, so full of airy talk, might be quite deflated and that others, less skilled in geometry, may know that there is nothing more to be feared from this Leviathan on this account since its armor (in which he had the greatest confidence) is easily pierced: and also so that outsiders (if they saw him maintain such things unchecked) might not think all men who practice geometry here are like him. (*Elenchus* epistle, sig. A3v–A3r)

It is important to notice that this campaign against Hobbes makes sense only if Wallis has reason to think that Hobbes's views were taken seriously. The project of determining Hobbes's influence is a complex

---

21. Thus, in justifying his attention to Hobbes's mathematics, he declares, "But I do not [confine my attention to the geometric aspects of *De Corpore*] because I think that the remainder is better, for whoever stumbles so horribly in geometry, where demonstrative proofs have a place, can hardly be thought to walk more securely in other matters, where conjectures will often have to be made" (*Elenchus* 4).

and difficult undertaking, and one best left for another day, but it is clear that he must have been a force to be reckoned with.[22]

In the course of their dispute, Hobbes and Wallis debated issues that went far beyond questions of circle quadrature or the nature of mathematics; fine points of Latin grammar (such as the proper use of the ablative case), problems of Greek etymology, and questions of ecclesiastical government were all brought into play. Indeed, their battle assumed a life of its own and continued long after the Restoration had put to rest the debate over university reform from the 1650s. Disputes over the nature and function of the universities form the background for the initial conflict between Hobbes and Wallis, but they were irrelevant to its continuation through the 1660s and beyond. As might also be expected, strategic alliances were formed and disrupted in the course of the battle. The most salient example of this phenomenon concerns Wallis and Owen. They began by making common cause in the defense of the universities, with Wallis dedicating his *Elenchus* to Owen; but they later quarreled when Wallis attacked Independency by publishing a defense of Presbyterian principles.[23]

Having concluded our study of the social and ideological background to the dispute, we must now look at Hobbes's philosophy of mathematics, since an important element in the controversy is the clash between Hobbes's thoroughly materialistic interpretation of mathematics and Wallis's more traditional account of the subject.

22. Malcolm expresses a similar view of Ward and Wilkins' joint attack on Hobbes, when he observes, "It was precisely because Hobbes still appeared in 1654 as an authoritative speaker on behalf of the new science that Wilkins and Ward took such trouble to attack him" (Malcolm 1988, 54). See Rogers 1988 for an interesting and important study of Hobbes's influence and reactions to the Hobbesian philosophy.

23. The particular thesis that bothered Owen was Wallis's negative response to the question "Whether the power of the evangelical minister extends wholly to the members of one particular church?" in a public disputation in 1654 and later published as part of a collection of theological works (Wallis 1657a). These matters will be investigated in chapter 7.

CHAPTER THREE

De Corpore and the Mathematics of Materialism

> I am the first that hath made the grounds of Geometry firm
> and coherent. Whether I have added anything to the Edifice
> or not, I leave to be judged by the Readers.
> —Hobbes, *Six Lessons*

The publication in 1655 of Hobbes's treatise *De Corpore* was the advent of the first part of his complete three-part system of the elements of philosophy. Although first in Hobbes's envisioned logical order of exposition, *De Corpore* was not the first part of the system to be published, appearing as it did four years after *De Cive*—the third part of the system. Taken together, Hobbes's three works *De Corpore, De Homine,* and *De Cive* were intended as an exposition of all the philosophy worth knowing. The structure of the three works reflects Hobbes's conception of the structure of knowledge: beginning with the nature of body, they proceed to the nature of man (i.e., an animated, rational body), and thence to the nature of the artificial body of the commonwealth, formed by the covenants that bind men together. The result is that *De Corpore*—a dissertation on the nature of body—is the foundational work in the Hobbesian system, and its status as a foundation derives from the fact that Hobbes held to a strict materialism in which only body is truly real and all else is to be explained as the consequence of the motion and impact of bodies.

Even a casual perusal of this work reveals that Hobbes's exposition of the elements of philosophy led him into territory far removed from anything we today regard as strictly philosophical. In particular, chapters 12–20 deal with mathematical material that seems out of place in a treatise on the nature of body. These chapters contain an extended discussion of the nature and methods of mathematics and include (among other claimed geometric results) a putative quadrature of the circle. Despite appearances, the presence of this mathematical material in *De Corpore* is by no means an oversight or an anomaly. Rather, Hobbes's excursion into mathematics reflects his conception of the unity of all knowledge, as well as his confidence that a thoroughly materialistic treatment of mathematics will provide the solution to all mathematical problems. Philosophy, as Hobbes declares in the opening chapter of *De Corpore* "*is knowledge of the effects or phenomena acquired by right reasoning from their known causes or generations, and conversely of such generations as may be from known effects*" (DCo 1.1.2; OL 1:3). This broad conception of the philosophical enterprise counts as philosophical any body of knowledge grounded in the consideration of causes, and more specifically, mechanical causes. A key part of Hobbes's program is to set forth the true (mechanical) causes of mathematical objects and thereby make mathematics a branch of philosophy.

The guiding principle in this philosophico-mathematical program is that mathematics is a science whose object is the principal affections of body. We have seen that mathematics was traditionally characterized as the science of quantity, where quantities (whether discrete or continuous) are anything capable of being measured or counted. Hobbes can accept this definition of mathematics, but adds the condition that the three dimensions of body (length, breadth, and depth) are the only genuine subjects of quantity since "there is no Subject of Quantity, or of Equality, or of any other accident but Body" (SL 2; EW 7:226–27). This condition follows quite naturally from his thoroughgoing materialism: because he holds that only body is real, Hobbes must also hold that mathematical quantities (numerical as well as geometrical) are ultimately grounded in the nature of body. The philosophy of mathematics set forth in *De Corpore* (particularly in chapters 12–14) is thus Hobbes's attempt to show how all of mathematics can be interpreted as a science of body.

In adopting this approach Hobbes departs quite radically from accepted doctrines in the philosophy of mathematics, and particularly the traditional understanding of mathematical objects. The standard Scholastic classification of the mathematical sciences relied upon a distinction between pure and "mixed" mathematics in which the pure science studies quantities abstracted from physical or material circumstances. Mixed mathematics deals with actual physical objects that embody or instantiate (perhaps imperfectly) the characteristics of the objects of pure mathematics. This view has Aristotelian origins but was widely accepted in Hobbes's day, not least because it allows mathematics to be integrated into the traditional conception of science in which mathematics is a mean between physics (which must remain at the level of material things) and metaphysics (whose objects are such immaterial things as God, the soul, or being in general). Clavius phrases the issue this way in his commentary on Euclid's *Elements:*

> Because the mathematical sciences treat of things that are considered apart from all sensible matter, although they are themselves immersed in matter, this is the principal reason that they occupy a middle position between the metaphysical and natural sciences. If we consider [the sciences each according to its] subject, as Proclus correctly pronounced, the subject of metaphysics is indeed separated from all matter of any kind. But the subject of physics is always conjoined with some kind of matter. Whence, as the subject of the mathematical disciplines is considered apart from all matter, it is clear that it constitutes a mean between the other two. (Clavius 1612, 1:5)

On this view, the Euclidean line (defined as "length without breadth") is an abstraction in which length is mentally separated from breadth. Although there are no breadthless lengths to be found in nature, pure mathematics is not constrained to deal with things actually existent, and its theorems are true independent of the structure or contents of the physical world. A mixed science such as astronomy then treats physical objects as approximations to the abstractions of pure mathematics: a light ray can be considered as an abstract line, for example, or a planet can be treated as a point with respect to its orbit. According to Barrow's version of the distinction: "The pure or abstract parts of mathematics contemplate the nature and proper affections of magnitude and number; but the mixed or concrete parts consider the same as applied to certain bodies and particular subjects, combined with motive force and other physical accidents. Whence it is that Aristotle calls them φυσικωτέρας (*Physics*, Book 2) and αἰσθήτικας (*Posterior Analytics* 1.33) and others are accustomed to call them physico-mathematical" (*LM* 1, 1:31).[1]

---

1. The Aristotelian passages Barrow has in mind are the following. "Similar evidence is supplied by the more natural of the branches of mathematics, such as optics, harmonics, and astronomy. These are in a way the converse of geometry. While geometry investigates natural lines but not qua natural, optics investigates mathematical lines, but qua natural, not qua mathematical" (*Physics* 2.2; 194a7–11). "The reason why differs from the fact in another fashion, when each is considered by means of a different science. And such are those that are related to each other in such a way that the one is under the other, e.g., optics to geometry, and mechanics to solid geometry, and harmonics to arithmetic, and star-gazing to astronomy" (*Posterior Analytics* 1.13; 78b34–79a2).

In taking body as the fundamental object of mathematics Hobbes rejects this widely held "abstractionist" philosophy of mathematics and sets himself the task of developing an alternative to the received view. Chapters 12–14 of *De Corpore* are his attempt to construct an account of quantities, ratios, and figures consistent with his materialism, while chapters 15–19 are intended to exploit this account in the investigation of motion and the determination of ratios between magnitudes. Chapter 20 contains several ill-fated attempts to apply this conception of mathematics to the problem of squaring the circle.[2] In the course of these investigations, Hobbes covers ground familiar from the history of the philosophy of mathematics, but his aim is always to impose a new set of definitions on traditional subject matter. Although my exposition here concentrates on Hobbes's views as set forth in *De Corpore,* the complexity of the issues will occasionally make it necessary to consider some of his later writings, as well as those of other authors.

3.1 HOBBES ON THE NATURE OF QUANTITY

The foundation of the Hobbesian program for mathematics is set forth in chapter 8 of *De Corpore,* which bears the title "Of Body and Accident." In his discussion of body and its principal accidents, Hobbes introduces several key definitions that he later exploits in his more detailed discussion of the nature of mathematics in chapters 12–14. Article 4 of chapter 8 casually defines the key term *magnitude* with the remark that "the extension of a body is the same as its magnitude, or that which some call 'real space'" (*DCo* 2.8.4; *OL* 1:93).[3] This identification of magnitude with space is pursued further in article 12 of the same chapter, when Hobbes introduces his definitions of three fundamental geometric objects—point, line, and surface. As he declares:

> If the magnitude of a body which is moved (although it must always have some) is considered to be none [*nulla*], the path by which it travels is called a *line* or one simple dimension, and the

2. There are significant differences between the scheme of printed chapters and those that survive in manuscript, which indicates that Hobbes's intentions underwent some significant changes in the course of writing *De Corpore.* I will have more to say about this when we consider the background to the printing of the circle quadratures in the twentieth chapter of *De Corpore.*

3. On Hobbes's conception of space, and particularly the doctrine of "real space," see Leijenhorst 1998, chap. 3, pt. 1.

space it travels along a *length,* and the body itself is called a *point.* This is the sense in which the earth is usually called a point and the path of its annual revolution the ecliptic line. But if a body that is moved is considered now as long and it is supposed to move so that each of its parts is understood to make a line, the path of each and every part of the body is a *breadth* and the space produced is called a *surface,* consisting of two dimensions, *breadth* and *length,* of which the whole of one is applied to the single parts of the other. (DCo 2.8.12; OL 1:98–99)

These definitions extend naturally to cover the generation of solids, and Hobbes connects the nature of solids to the three-dimensionality of space in the remainder of article 12:

Again, if a body is now considered as having a surface, and is understood to be so moved that each of its single parts makes lines, then the path of all these parts is called thickness or depth, and the space made is called a solid, consisting of three dimensions any two of which are entirely applied to all the parts of the third.

But if a body is further considered as a solid, it cannot happen that each of its parts describes single lines. For however it may be moved, the path of the following part will coincide with the preceding part, and the same solid will be made that would have been made by the first surface itself. And so there can be no other dimension of body, as it is a body, beyond the three mentioned. Although, as will be said later, velocity, which is motion through a length, applied to all parts of a solid, makes a magnitude of motion consisting of four dimensions, just as the worth *[bonitas]* of gold computed in all of its parts makes its price *[pretium].* (DCo 2.8.12; OL 1:99)

The significance of these definitions for Hobbes's philosophy of mathematics can scarcely be overstated, and we will see their consequences in nearly all of his mathematical writings. For present purposes it suffices to stress the connection between these definitions and Hobbes's strict materialism. Because his ontology recognizes only body as real, and because he holds that it is only through motion of bodies that anything can be brought about, Hobbes's program in the philosophy of mathematics must found all of geometry upon the principles of matter and motion. That is to say, mathematical objects must be interpreted as bodies or as things produced by the motion of bodies.

As I have indicated, Hobbes's program is a clear departure from the traditional conception of geometric magnitudes. Euclid, for example, defines a point as "that which has no parts," while a line is "breadthless length," and a surface is "that which has length and breadth only" (*Elements* 1, defs. 1, 2, 5). Philosophers and philosophically minded mathematicians after Euclid accepted these definitions and typically interpreted them as expressing the fundamentally immaterial nature of geometric objects, since no physical bodies can satisfy the definitions. Proclus held that the Euclidean definitions articulated the fundamental properties of forms separable from matter, so that geometric definitions cannot accurately be interpreted as true of material things. He explains that

> in the forms separable from matter the ideas of the boundaries exist in themselves and not in the things bounded, and it is because they remain precisely what they are that they become agents for bringing to existence the entities dependent upon them. . . . Matter muddies their precision; the idea of the plane gives the plane depth, that of the line blurs its one-dimensional nature and becomes generally divisible, and the idea of the point ends by becoming bodily in character and extensible together with the thing that it bounds. For all ideas when they flow into matter . . . are filled with their substrates: they forsake their native simplicity for alien combinations and extensions. (Proclus 1970, 87)

Similarly, Clavius's comment on the Euclidean definition of a point contains the remark that "[n]o example of this can be found in material things, unless you mean that the extremity of the sharpest needle expresses some similitude to a point; which nevertheless is wholly untrue, since this extremity can be divided and cut to infinity, but a point must be supposed altogether indivisible *[individuum prorsus]*" (Clavius 1612, 1:13).

Such remarks as these show the prevalence of the idea that (pure) mathematics is not at all concerned with the nature of physical body and underscore just how radically Hobbes's program departs from the traditional understanding of geometry and its object. The extent of his departure from the received view can be gauged by Wallis's apoplectic comments on the definition of *point* in *De Corpore*:

> Who ever, before you, defined a point to be a body? Who ever seriously asserted that mathematical points have any magnitude?

If it is a body, if it has size, then one point added to another will make a greater. For you will not, I think, deny this of bodies, therefore neither will you deny it of points, if indeed a point is a body. But if the addition of one point adds nothing, then neither will the addition of two, three, a hundred, a thousand, or even an infinity of points. Indeed, you elsewhere argue in this way about impetus, so why does the same not hold of magnitude? Perhaps you will say that the magnitude itself, so far as there is any, is nevertheless not considered. Fine. But I now ask why is it still considered to be a body? If it is not, then to what purpose is this mention of body in the definition, or why cannot this corporeity also be excluded, and also this magnitude? Why do you not then define it thus, *A point is a body which is not considered to be a body, and a magnitude [magnum] which is not considered to be a magnitude? (Elenchus* 6–7)

Hobbes retorts that the Euclidean definition of a point is ambiguous and can lead to absurdity. As he notes, we can interpret the talk of a point having no parts to mean either that a point is indivisible or undivided. He maintains that in the first sense a point is simply nothing:

That which is indivisible is no Quantity; and if a point be not Quantity, seeing it is neither substance nor Quality, it is nothing. And if *Euclide* had meant it so in his definition, (as you pretend he did) he might have defined it more briefly, (but ridiculously) thus, *a Point is nothing. (SL* 1; *EW* 7:201)

Defining a point as something undivided but capable of further division will indeed make it a quantity, but such a definition fails to express the essence of a point. For Hobbes, the essence of a point is that its magnitude is not considered in a demonstration, and his alternative definition captures this essential feature of the point and places all of geometry on a firm foundation.

In defense of his definition of *point,* Hobbes even tries to argue that it is consistent with the opinions of Euclid and Proclus, since both classical authors admit that a point cannot be considered in the course of a geometric argument.[4] However, in defining points as bodies, Hobbes

4. Proclus, in commenting upon Euclid's definition of a point remarks that "[i]n geometrical matter, then, the point alone is without parts, and in arithmetic the unit; and the definition of the point, though it may be imperfect from another point of view is perfect as far as the science before us is concerned. . . . It all but clearly says that 'what is without parts is a point for my purposes and a principle for me; and the simplest object is none other than this'" (Proclus 1970, 76–77). Hobbes takes this to mean that

is also committed to the thesis that points have extension (an unextended body being an utterly absurd concept, on Hobbes's or anyone else's principles). The notion of an extended point conflicts mightily with the traditional conception, and it also seems to demand that there be points of different sizes, since there could certainly be different sizes of divisible objects whose parts are not considered in a demonstration.[5]

Hobbes naturally finds the Euclidean definition of a line ("breadthless length") as flawed as the definition of a point, since he holds that a length without breadth is simply nothing. However, there is an alternative tradition according to which the line is produced by the motion of a point. Aristotle, for example, mentions such an understanding of points in *De Anima,* when he remarks that "since they say a moving line generates a surface and a moving point a line, the movements of the units must be lines" (*De Anima* 1.4; 409a4–5). Clavius reports a similar definition when he reports that "mathematicians also, in order to teach us the true understanding of the line imagine a point, as described in the above definition, to be moved from one place to another. Now since the point is absolutely indivisible, there will be left behind by this imaginary motion a certain path lacking in all breadth" (Clavius 1612, 1:13). Proclus rejects such an account as precisely the wrong way to think about geometric lines because it "appears to explain [the line] in terms of its generative cause and sets before us not line in general, but the material line" (Proclus 1970, 79).

Hobbes disagrees with Proclus and finds such a definition of the line congenial to his project. Clearly, the attempt to reduce geometry to motion and extension would be advanced by including the concept of motion in the definition of a line, and Hobbes is quite happy to include an explanation of the generative cause of the line. In this scheme, a line will indeed be the path of a moving point, but the point must be understood in Hobbes's terms:

"no Argument in any Geometricall demonstration should be taken from the Division, Quantity, or any part of a Point; which is as much as to say, a Point is that whose quantity is not drawn into the demonstration of any Geometricall conclusion; or (which is all one) whose Quantity is not considered" (*SL* 1; *EW* 7:201).

5. Hobbes embraces this consequence in *De Corpore* 3.15.2 when he says, "Although just as a point can be compared with a point, so a *conatus* can be compared with a *conatus,* and one can be found to be greater or less than another. For if the vertical points of two angles are compared with one another, they will be equal or unequal in the ratio of the angles, or if a right line should cut many circumferences of concentric circles, the points of section will be unequal in the ratio of these perimeters" (*DCo* 3.15.2; *OL* 1:178). I will have more to say about Hobbes's concept of *conatus* in section 4 of this chapter.

But everyone knows that nothing except body can be moved, nor can motion be conceived of anything except body. And every body in motion traces a path with not only length, but also breadth. Therefore, the definition of a line should be as follows: *a line is the path traced by a moving body, whose quantity is not considered in a demonstration.* (PRG 2; OL 4:393)

Wallis objected to the Hobbesian definition of line because, among other things, the definition includes the concepts of motion and body, which Wallis regards as inessential to geometry.[6] In reply, Hobbes urges that not only do such classical authors as Euclid define geometric objects in terms of the motions by which they are generated, but that geometry can be made part of genuine philosophy only if its definitions contain the generation of the things it studies. Thus, geometric definitions must appeal to the motions by which geometric objects are created; or, as Hobbes puts it: "I say, to me, howsoever it may be to others, it was fit to define a Line by Motion. For the generation of a Line is the Motion that describes it. And having defined Philosophy in the beginning, to be the knowledge of the properties from the generation, it was fit to define it by its generation" (*SL* 2; *EW* 7:215). He further explains that by employing such definitions he has removed the grounds for skeptical doubts about geometry that had been raised by Sextus Empiricus in his treatise *Against the Mathematicians:*

[W]here there is place for Demonstration, if the first Principles, that is to say, the Definitions contain not the Generation of the Subject; there can be nothing demonstrated as it ought to be. And this in the three first Definitions of *Euclide* sufficiently appeareth. For seeing he maketh not, nor could make any use of them in his Demonstrations, they ought not to be numbered among the Principles of Geometry. And *Sextus Empiricus* maketh use of them (misunderstood, yet so understood as the said Professors understand them) to the overthrow of that so much renouned Evidence of Geometry. In that part therefore of my Book where

6. Wallis asks, "Moreover, what mathematician ever expected such a definition of line or length? The things you affirm are indeed true, but they are not definitions. Neither are they reciprocal propositions. What need is there for the idea of motion, or of a body moved, in order that it be understood what a line is? For are there not lines in quiescent bodies, no less than moved ones? Why then this mention of motion? And will not the distance even between two quiescent bodies be a length, no less than the measure by transit? To what purpose then this mention of a body moved?" (*Elenchus* 6). These complaints will be investigated more fully in chapter 4.

I treat of Geometry, I thought it necessary in my Definitions to express those Motions by which Lines, Superficies, Solids, and Figures were drawn and described; little expecting that any Professor of Geometry should finde fault therewith; but on the contrary supposing I might thereby not only avoid the Cavils of the Scepticks, but also demonstrate divers Propositions which on other Principles are indemonstrable. (*SL* epistle; *EW* 7:184–85)

Hobbes returns to the issue of skepticism in mathematics when he charges Wallis with employing definitions that place geometry in danger of succumbing to skeptical challenges, whereas his definitions can "redeem" geometry.[7] It is a disputed question among Hobbes scholars whether the philosopher from Malmesbury was deeply concerned with skepticism and the appropriate response to it, but from such comments as these it is evident that he thought his philosophy of mathematics had the notable advantage of forestalling skeptical attacks on geometry.[8]

The materialistic understanding of points, lines, and surfaces that Wallis finds so objectionable is just the beginning of the Hobbesian program for mathematics. Hobbes more fully expounds his philosophy of mathematics in chapter 12 of *De Corpore,* which bears the title "Of Quantity" and sets forth his leading ideas on the nature of quantities. A quantity, he declares, can be defined only as *"a determined dimension, or a dimension whose termini are known either by their place or by some other comparison"* (*DCo* 2.12.1; *OL* 1:124). The reason for defining quantities in terms of dimensions was touched on before and should be obvious enough: the three dimensions of length, breadth, and depth are fundamental affections common to all bodies, and thus make all body the subject of quantity. Defining quantities as *determined* dimensions requires that quantities be marked off or limited by definite boundaries, since quantities are always appropriate responses to a question about how great some body is. As Hobbes puts it:

7. Hobbes writes: "From this one and first Definition of *Euclide, a Point is that whereof there is no part,* understood by *Sextus Empiricus,* as you understand it, that is to say mis-understood, *Sextus Empiricus* hath utterly destroyed most of the rest, and Demonstrated, that in Geometry there is no Science, and by that means you have betrayed the most evident of the Sciences to the *Sceptiques.* But as I understand it for *that whereof no part is reckoned,* his Arguments have no force at all, and Geometry is redeemed" (*SL* 5; *EW* 7:317–18).

8. See Tuck 1988a and 1988b for a portrayal of Hobbes's system as a response to skepticism, and particularly to Descartes's "hyperbolic doubt." Sorell (1995) takes issue with this sort of interpretation and downplays its significance for Hobbes's philosophy.

When, therefore, it is asked how much something is, for example how long is the road, it is not answered indefinitely "length," nor to one who asks how large is the field, indefinitely "surface," nor if someone asks how great is the bulk *[moles]*, indefinitely "solid." Rather, it is responded determinately, that the road is one hundred miles, the field is one hundred acres, or the bulk is one hundred cubic feet, or at least in some way that the magnitude of the thing asked after can be comprehended by certain limits in the mind. (*DCo* 2.12.1; *OL* 1:123)

The process of determining magnitudes depends in the first instance upon the motions by which dimensions are described, but it can also be performed by the "apposition" of previously determined quantities or the "section" of such determined quantities.[9] There are also two different ways to determine a quantity: either by immediate "exposition to sense" or by expressing it as a multiple of some common measure that itself is "recoverable to sense." The height of a wall, for instance, can be immediately sensed or grasped, or it can be expressed as a number of meters, where the meter itself is a common measure whose magnitude can be immediately sensed.[10]

This account of quantity gives pride of place to the continuous magnitudes of geometry, and it may at first sight seem to provide no means of capturing the classical distinction between discrete and continuous quantities. As I mentioned in chapter 1, classical authors divided mathematics into two sciences (geometry and arithmetic) on the basis of a

---

9. Hobbes writes:

Lines, surfaces, and solids are first exposed by motion, in the way we said they are generated in chapter 8, in such a way that their traces remain; as when they are marked in matter, like a line on paper or carved in durable matter. The second way is by apposition, as when a line is added to a line, that is, a length is added to a length, or a breadth to a breadth, or a thickness to a thickness, which is to describe a line by points, a surface by lines, or a solid by surfaces, except that by points here is to be understood very short lines, and by surfaces very thin solids. The third way lines and surfaces can be expressed is by sections, namely a line may be made by cutting an exposed surface, and a surface may be made by cutting a solid. (*DCo* 2.12.3; *OL* 1:124–25)

10. Hobbes writes, "But quantity is determined in two ways; the one to sense, which may be done by a sensible object, as when a line, surface, or solid of a foot or a cubit is set before the eyes marked in some matter, which way of determining is called that of exposition, and the quantity thus known is called exposed quantity. In the other way quantity is determined by memory, which is done by comparison with an exposed quantity" (*DCo* 2.12.2; *OL* 1:124).

fundamental difference in their objects. Geometry takes as its object the continuous quantities such as lines, angles, and areas, while arithmetic deals with the discrete quantities or "number" that are collections of units. Hobbes admits such a distinction in *De Corpore,* but draws it in terms of the means by which the different kinds of quantities are determined or declared. Continuous quantities are determined by setting out their limits, and such determination ordinarily depends upon motion; numbers or discrete quantities, however, are exposed "by the exposition of points, or also by the numeral names *one, two, three,* etc. And these points should not be so contiguous to one another that they cannot be distinguished by marks *[nullis notis],* but so positioned that they can be discerned. For it is by this that number is called *discrete* quantity, while all quantity which is determined by motion is called *continuous*" (*DCo* 2.12.5; *OL* 1:125). Elsewhere he argues that the discrete quantities of arithmetic can be generated from the uniform division of geometric magnitudes, and he thus shares Barrow's opinion that arithmetic is subordinate to geometry.[11]

The doctrine thus far presented fails to include many things that are ordinarily taken to be quantities—times, masses, velocities, densities, temperatures, etc. are all capable of measure and must therefore be integrated within the general framework of quantities in *De Corpore.* Hobbes's approach to this problem is to allow such further quantities to be "exposed" by the fundamental quantities, and particularly by points and lines in motion. Time, for example, is exposed "when not only some line is exposed, but also a movable thing *[mobile]* that is uniformly moved upon it, or is supposed to be so moved. For since time is the image of motion so far as in it earlier and later, that is succession, are considered, it is not sufficient for the exposition of time that a line is described, but also that there is in the mind an image of some movable thing passing over that line, and this with uniform motion, so that time can be divided and compounded as often as there is need" (*DCo* 2.12.4; *OL* 1:125).[12] Similar expositions of other quanti-

11. As Hobbes declares in the *Examinatio,* "[B]ecause any given continuous magnitude can be divided into any number of equal parts, with its ratio to any other magnitude remaining unchanged, it is manifest that arithmetic is contained in geometry" (*Examinatio* 3; *OL* 4:28).

12. It is worth noting that unless Hobbes takes the concept of uniform motion as undefined and unproblematic, this definition involves a troublesome circularity. To say that time is exposed by the uniform motion of a point across a line requires that the concept of uniform motion be understood; but if uniform motion is defined as motion covering equal distances in equal times, it is taken for granted that we have some way of exposing equal times, which was the very thing to be explained.

ties can then be made in the obvious way, by representing a quantity through lines and motions.[13] This approach thus provides a unified conception of quantity in which such apparently disparate quantities as time and temperature can all be reduced to or represented as aspects of body. There are other kinds of geometric quantities such as figures and angles whose Hobbesian treatment I will consider in section 3.3, but first it is necessary to examine Hobbes's doctrine of ratio and proportion.

3.2 RATIO AND PROPORTION IN *DE CORPORE*

The doctrine of ratios is an essential part of any geometric theory, and Hobbes was concerned to formulate a treatment of the subject that is of a piece with his general account of quantity. There were significant disputes in the seventeenth century over the proper interpretation of the theory of ratios, and it is necessary to touch on some of these contested issues to clarify both Hobbes's position and the doctrines it was intended to supersede. As we will see, Hobbes was familiar with the disputes regarding the nature of ratios and he offered his own theory as a solution to the difficulties encountered by other writers.

The starting point for an understanding of the doctrine of ratios and proportions is the classical theory developed in book 5 of Euclid's *Elements* as a general doctrine applicable to all geometric magnitudes. The principal definitions read as follows:

> 3. A **ratio** is a sort of relation in respect of size between two magnitudes of the same kind.
> 4. Magnitudes are said to **have a ratio** to one another which are capable, when multiplied, of exceeding one another.
> 5. Magnitudes are said to **be in the same ratio,** the first to the second and the third to the fourth, when, if any equimultiples whatever be taken of the first and third, and any equimultiples whatever of the second and fourth, the former equimultiples alike exceed, are alike equal to, or alike fall short of the latter equimultiples respectively taken in corresponding order.
> 6. Let magnitudes which have the same ratio be called **proportional.** (*Elements* 5, defs. 3–6)

13. The exposition of mass is a somewhat different case, since the mass of a body is exposed with the body itself. As Hobbes puts it, "But weight is exposed by any heavy body of any sort of matter, provided that it is always of the same heaviness *[gravitas]*" (*DCo* 2.12.7; *OL* 1:126).

It is important to observe that this understanding of ratios does not treat a ratio as a quotient of two numbers, but rather as a special kind of relation between two magnitudes. These magnitudes are conceived of as falling into species or kinds, and it is only within species that ratios can be constructed or quantities compared. For example, lines, angles, surfaces, and solids are different species of magnitudes, but the restrictions in definitions 3 and 4 prohibit the formation of ratios between lines and surfaces, since there is no number of lines that could ever exceed a surface, and thus there can be no "relation in respect of size" between heterogeneous magnitudes. The sixth definition allows comparison of ratios across species, so it makes sense to say that the ratio between two lines L_1 and L_2 is the same as the ratio between two spheres S_1 and S_2. The proportion $L_1:L_2 :: S_1:S_2$ would therefore be legitimate. Note, however, that the definition of equality of ratios (definition 5 above) does not assert that $A:B :: C:D$ whenever $A \times D = C \times B$, because the relevant magnitudes may be heterogeneous and incapable of being multiplied together. Instead, sameness of ratio is defined by the property of preserving order relations under arbitrary equimultiples, as expressed in the complicated fifth definition above.

The Euclidean theory of ratios explicitly prohibits the introduction of infinitesimal ratios (by definitions 4 and 5), but it does allow for incommensurable magnitudes to appear as terms in ratios. The exclusion of infinitesimal magnitudes is guaranteed by the conditions listed in definition 4, which restricts ratios to the consideration of finite differences between finite magnitudes. Incommensurable magnitudes may still appear as terms in a Euclidean ratio because the first definition allows ratios to be formed between any two magnitudes of the same kind. In particular, this allows any two line segments to be compared, so that the ratio between, say, the side and diagonal of a square can be formed in the classical theory, even though this ratio cannot be expressed as a ratio between two integers.

The classical doctrine (which we can call the "relational" theory of ratios) was not universally accepted in the seventeenth century, for some authors preferred an understanding of ratios that we can call the "numerical" theory.[14] The differences between these can be illustrated by asking whether ratios are quantities. According to the relational theory, the answer is no: ratios are not themselves magnitudes, but

14. I take this terminology from Sylla 1984. The numerical theory did not originate in the seventeenth century, but can be traced back through the Middle Ages to late antiquity. Nevertheless, the clash between the relational and numerical theories was most evident in the seventeenth century.

rather relations that hold between pairs of magnitudes. Just as it is absurd to hold that the numerical relation "less than" is a number, the relational theory of ratios denies that there is any sense to the assertion that ratios are quantities.

In contrast, the numerical theory of ratios assimilates ratios into the general domain of magnitudes. In the numerical presentation, every ratio has a size or "denomination," and two ratios are the same if and only if they have the same denomination. By assigning denominations to ratios, the numerical theory avoids the elaborate Euclidean definition of equality of ratios in terms of arbitrary equimultiples, and this is one of its main attractions. Nevertheless, the theory is not without its difficulties. From the standpoint of the numerical theory it is natural to assume that the ratio $A:B$ is the same as the ratio $C:D$ just in case $A \times D = B \times C$. As we have seen, such a criterion makes no sense under the relational theory because the magnitudes A and D may be of different species and therefore incapable of being compared directly with one another. Thus, a fully developed numerical theory of ratios must require that all ratios be homogeneous and capable of mutual comparison. Further, it is natural to see the denominations as answering to quotients, since a ratio is taken to be a magnitude that arises from the division of the antecedent by the consequent. Yet in identifying ratios with quotients the numerical theory requires that the quotient of two incommensurable magnitudes be defined. The quotient is understood classically as arising from the division of integers; deriving from the Latin *quoties* or "how many," the quotient is taken to indicate how many common units from the denominator are contained in the numerator, and in effect quotients can simply be identified with rational numbers. A full vindication of the numerical theory therefore requires expanding the classical understanding of quotients to include quotients of incommensurable magnitudes. As might be expected, these debates between the proponents of the numerical and relational theories of ratios were a source of considerable interest to Hobbes.

The conflict between these competing theories of ratios led to some significant debates in the seventeenth century, although not all authors explicitly raised the issue. Descartes, for example, employed the numerical conception of ratios in his *Géométrie* without comment,[15] but

15. Grosholz observes that "Descartes's treatment of ratios and proportions clearly belongs to this [numerical] tradition. In his exposition of multiplication on the second page of the *Geometry,* for example, he infers from the proportion BA:BD :: BC:BE that BE × BA = BD × BC without further comment; there is no need to worry about the association of terms in a non-continuous proportion, since their terms and their ratios

other adherents of the theory felt the need to argue for its acceptance. Wallis upheld the numerical theory and took it as part of his philosophical program to make arithmetic the foundation of all mathematics. Indeed, the numerical theory of ratios is tantamount to an assertion of the priority of arithmetic over geometry; at the very least, the project of interpreting all of mathematics as essentially arithmetical will be helped along by reducing the entire theory of ratios (which traditionally had been deemed quintessentially geometrical) to arithmetic. This is obviously part of Wallis's strategy when he asserts that the comparison of magnitudes in ratios renders all ratios homogeneous and places them in the genus of number, so that the doctrine of ratios belongs more to arithmetic than to geometry.[16] The homogeneity of terms desired by Wallis is guaranteed by his assumption that both rational and irrational magnitudes are truly numbers, so that a ratio is simply a number expressing a numerical quotient of its two terms. This aspect of Wallis's treatment of number led Jacob Klein to describe him as the man whose theories are "the final act in the introduction of the new 'number' concept" (Klein 1968, 211). This new concept of number treats all magnitudes as numbers, whereas the Greek doctrine of number (ἀριθμός) is restricted to positive integers or "collections of units."

Barrow argued for the relational theory as the best alternative to the "corruptions" introduced by the numerical theory. In fact, he devotes the last seven of his twenty-three *Lectiones Mathematicae* to an exhaustive analysis and defense of the Euclidean theory against all objections urged on behalf of contemporary rivals. He acknowledges that his veneration for the classical account of ratios was something of a minority view, but he was convinced that Euclid's presentation of the relational theory was the only philosophically defensible account of ratios. As he explains:

> Nevertheless I must dare to oppose so many and such great men and contradict [ἀντιβλέπειν] such an illustrious authority. Truth needs some sort of protection against such powerful enemies (at

---

(quotients) and products are all homogeneous in a strong sense. Of course, Descartes chooses line segments, not numbers, to be his homogeneous set of terms" (1991, 53).

16. Wallis says, "But where a comparison of quantities according to ratio is made, it frequently happens that the ratio of the compared quantities leaves the genus of magnitude of the compared quantities and is transferred into the genus of number *[transit in genus numerosum]*, whatever that genus of the compared quantities may be. . . . And this is the principal reason I affirm that the doctrine of ratios belongs rather to the speculations of arithmetic than geometry" (*MU* 25; *OM* 1:136).

least in my judgment). This opinion surely seems to be to be not only false, but wholly harmful, because it engenders and fosters a good many useless controversies and introduces false reasonings [ἀκυρολογίας] and errors into the doctrine of proportions. A few quarrels will be checked, I judge, some difficulties removed, errors avoided, and clouds dispersed, by maintaining that a ratio is not a genus of quantity, nor something subject to quantity, and that it is by no means ever properly attributed to quantity directly and in itself, except by catachresis [κατάχρησιν] or metonymy [μετωνυμίαν]. (LM 20, 318)

Barrow's case is simplicity itself: because a ratio is a "pure, perfect relation" it cannot "pass into another category and become a genus of quantity" or be the subject of quantity. He admits that the relational theory of ratios permits such expressions as "the ratio A:B exceeds the ratio C:D," but insists that this can be understood without making ratios magnitudes. On Barrow's analysis, one ratio is greater than another when the antecedent of the one is greater than the antecedent of the other, *provided that the ratios have common consequents.* Similarly, the addition and subtraction of ratios proceeds on the assumption that the ratios have been expressed with common consequents.[17] Barrow's attempted defense of the orthodox Euclidean point of view ultimately failed to win the day, as the numerical conception of ratios displaced the older tradition. Nevertheless, it is enlightening to read Hobbes's comments on the nature of ratios and proportions against the background of this dispute.

The Hobbesian theory of ratios is expressed most completely in chapter 13 of *De Corpore,* which bears the title "Of Analogism, or The Sameness of Ratio." This treatment depends upon a brief definition of the term *ratio* in chapter 11, where Hobbes declares:

the similarity, dissimilarity, equality, or inequality of any body to any other body is called its relation; and these bodies are therefore called relata or correlata. Aristotle calls them τὰ πρὸς τι, the first of which is usually called the antecedent and the second the consequent. But the relation of the antecedent to the consequent

17. Barrow writes, "Whatever is commonly attributed to ratios, only truly and properly agrees with the denominators of ratios, that is, to their antecedents reduced to a common consequent. The quantity that others assign to ratios is nothing other than the quantity and ratio of the denominators, and when they think they add or subtract ratios themselves, they only add or subtract these denominators, and it is the same thing when they multiply or compound, divide or resolve them" (LM 20, 325).

according to magnitude, that is to say its equality, excess, or defect, is called the ratio and proportion of the antecedent to the
consequent; and so ratio is nothing other than the equality or
inequality of the antecedent compared to the consequent according to magnitude. (*DCo* 2.11.3; *OL* 1:118–19)

The most salient feature of this definition is that it takes bodies to be
the terms of any relation or comparison, and hence it assumes that all
ratios are, in essence, ratios of bodies. This should come as no surprise,
given what we have already seen of Hobbes's materialistic treatment of
mathematics, but it is worth observing the thoroughness with which
Hobbes pursued his program for a philosophy of mathematics based
on the concepts of body and motion.

Having defined ratios as relations of magnitudes (whether relations
of excess, defect, or equality), Hobbes draws a further distinction between two fundamentally different kinds of ratios, grounding the distinction in two different ways of comparing the excess or defect of one
quantity to another. The *arithmetical* ratio of two quantities is formed
by considering the simple difference between them, while the *geometrical* ratio of two quantities compares them as part to whole.[18] For example, 5:7 and 11:13 are the same arithmetic ratio because the consequents differ from the antecedents by two; in contrast, 1:3 and 4:12
are the same geometrical ratio because in each case the antecedent is
one-third of the consequent. Because geometrical ratios are of much
greater interest in the tradition, the term *ratio* is usually taken to be
synonymous with *geometrical ratio* and Hobbes generally follows
this usage.

Hobbes concludes from this treatment that ratios of excess or defect
(whether arithmetical or geometrical) are quantities, while the ratio of
equality is not a quantity. The reason for this somewhat unusual view
is quite simple: "a ratio of excess, as well as of defect, is a quantity,
since it is capable of greater and less; but a ratio of equality, because it

---

18. Hobbes writes:

Further, every one of these ratios is twofold. For if it is asked how great any given
magnitude is, it can be replied by two ways of comparing it. First, if it is said that
it is greater or less than another magnitude by so much, as seven is less than ten
by three units. And this ratio is called arithmetical. Second, if it is said that it is
greater or less than another magnitude by such a part or parts of it, as seven is
less than ten by three tenths parts of the same ten. And although this ratio cannot
always be expressed in numbers, it is nevertheless a determinate ratio and of a
different kind from the former, and it is called a geometric ratio, or more commonly simply a ratio. (*DCo* 2.13.1; *OL* 1:129)

is capable of neither greater nor less, is not a quantity. And so ratios of inequality can be added together, or one subtracted from another, or multiplied together and by numbers, and divided; but not so ratios of equality" (*DCo* 2.13.3; *OL* 1:130). This doctrine did not originate with Hobbes, but is instead something that he adopted from Mersenne. In his *Cogitata Physico-Mathematica,* Mersenne argued (against the opinion of Clavius) that the proportion of equality is of no quantity, while the proportion of a greater to a lesser magnitude is a genuine quantity, and the proportion of a lesser to a greater is an *antiens* (literally, anti-being) or a quantity less than nothing.[19] Hobbes makes no claim for originality on this point and admits that he first heard of the doctrine from Mersenne during a conversation in Paris while the Minim friar was preparing his *Cogitata Physico-Mathematica* for the press.[20] Wallis found such a doctrine absurd, and in addition to dismissing Hobbes's views as nonsense he was moved to append a brief polemic against Mersenne to his 1657 critique of Marcus Meibom's dialogue on proportions.[21]

The centerpiece of Hobbes's conception of ratios is his alternative definition of sameness of ratio or proportionality.[22] Where Euclid had

19. The doctrine is expressed in the general preface to the *Cogitata,* where Mersenne argues that "[t]he proportion of equality bears a similitude to nothing, a proportion greater than equality is raised above nothing and is similar to being, while the proportion less than equality is pushed down below nothing and can be compared to anti-being *[antienti comparari potest]*" (Mersenne 1644, general preface, sect. 14).

20. As Hobbes says in the *Six Lessons,* "The first time I heard it argued, was in *Mersennus* his Chamber at *Paris,* at such time, as the first volume of his *Cogitata Physico-Mathematica* was almost printed: In which, because he had not said all he would say of Proportion, he was forced to put the rest into a Generall Preface; which as was his custom, he did read to his Friends, before he sent it to the Press" (*SL* 3; *EW* 7:235).

21. Thus, in *Due Correction* Wallis remarks that

[y]ou tell us, Chap. 13 art. *3 That the proportion of Inequality is Quantity, but that of Equality is not.* Which *I* said was very absurd; and that the one did no more belong to the Praedicament of Quantity than the other; and that it is to bee, of both equally, either denied or affirmed: And that your argument for it, (That *One equality is not greater or lesse then another; but of proportions of inequality, one may be more or less unequall:*) might as well conclude that *Oblique angles,* be quantities, but not *Right angles,* for these be all equal, and equally Right, but not those. . . . What you alledge out of Mersennus, was but his mistake. (*Due Correction* 68)

The appendix to Wallis's critique of Meibom bears the title "A Criticism of a Passage in Mersenne" (*OM* 1:289–90) and makes essentially the same argument.

22. Substantial linguistic confusions are inherent in any discussion of this topic, since Hobbes uses the Latin *ratio* and *proportio* indifferently for what are usually called

defined proportions in terms of arbitrary equimultiples of paired terms in two ratios, Hobbes seeks a definition that will establish the theory of proportions on the basis of the concepts of body and motion. These considerations lead him to propound the following definition: "One geometric ratio is the same as another geometric ratio when some cause can be assigned that, producing equal effects in equal times, determines both ratios" (DCo 2.13.6; OL 1:132).

In taking this idiosyncratic approach to the doctrine of ratios Hobbes attempts to steer something of a middle course between the numerical and relational theories. He rejects Euclid's account of ratios as vague, demanding that the Euclidean "sort of relation between magnitudes" be replaced with a treatment of ratios founded on principles of body and motion. Furthermore, in accepting the principle that ratios (or at least ratios of excess and defect) are quantities, Hobbes rejects a key element in Barrow's defense of the relational theory of ratios. But he also opposes the fundamental thesis of the numerical theory, namely that ratios are quotients. Where Wallis had subsumed the doctrine of ratios within the province of arithmetic by identifying ratios with quotients, Hobbes insists that quotients can be formed only by commensurable magnitudes. As he explains the matter in his *Six Lessons:* "In Lines incommensurable there may be the same Proportion, when nevertheless there is no Quotient; for setting their Symboles one above the other doth not make a Quotient; for Quotient there is none, but in *aliquot parts*. It is therefore impossible to define Proportion universally, by comparing Quotients" (SL 3; EW 7:241).[23] Here Hobbes plainly allies himself with the classical understanding of quotients. He insists that it is impossible to define ratios as quotients, since this will permit only rational numbers as ratios and leave the "universal" treatment of ratios out of mathematics.[24] He further explains that his own

---

ratios, and uses *analogismo* or *eadem ratio* for what is normally called proportion. These difficulties persist in his English discussions of this topic. Context, however, usually makes Hobbes's declarations unambiguous. I use the terms "ratio" and "proportion" in their usual sense in discussing Hobbes's doctrines and translating his Latin, even where this does not match his English usage.

23. A number K is said to be an aliquot part of a number M, when for some integer n, $M = Kn$. As Hobbes uses the term, he is not claiming that only integers can be quotients, but rather that a quotient can be formed only when the numerator and denominator have "aliquot parts," i.e., when both can be expressed as multiples of a common unit.

24. Thus, Hobbes can retort to Wallis:

This incommensurability of Magnitudes was it that confounded *Euclide* in the framing of his Definition of Proportion at the fifth Element. For when he came to numbers, he defined the *same Proportion* irreprehensibly thus, *Numbers are*

account of ratios and proportions was developed with the intent of accommodating incommensurable magnitudes within a theory that avoids the obscurity of Euclid while retaining sufficient generality to permit ratios of incommensurables. Thus, Hobbes declares:

> I thought it necessary to seek out some way, whereby the Proportion of two Lines, Commensurable, or Incommensurable, might be continued perpetually the same. And this I found might be done by the Proportion of two Lines described by some uniform motion, as by an Efficient cause both of the said Lines, and also of their Proportions. Which motions continuing, the Proportions must needs be all the way the same. And therefore I defined those Magnitudes to have the same Geometricall Proportion, *when some cause producing in equall times, Equall Effects, did determine both the Proportions.* (SL 3; EW 7:241–42)

This treatment of ratios was one of the principal targets of Wallis's attack, and I will be concerned in chapter 4 with a more extensive analysis of that part of the controversy that concerns the nature of ratios. For present purposes it is sufficient to observe that one key charge that Wallis levels against Hobbes's theory is that it defines ratios strictly in terms of the difference between two quantities, and thus restricts all ratios to the case of arithmetical ratios. When Hobbes declares in article 5 of chapter 11 of *De Corpore* that "the ratio of an antecedent to a consequent consists in the difference, that is in that part of the greater by which it exceeds the less, or in the remainder of the greater, having taken away the lesser; but not in itself simply, but as it is compared with either of the things related" (*DCo* 2.11.5; *OL* 1:119), Wallis uncharitably reads this as defining ratios by the subtraction of a greater magnitude from a lesser. He then concludes that (according to Hobbes) "wherever there is the same excess or defect, there is the same ratio: and vice versa, wherever there is not the same excess or defect, there is not the same ratio" (*Elenchus* 15). From this Wallis concludes that Hobbes's doctrines can only account for arithmetical ratios. In fact, however, Hobbes does make the traditional distinction

---

then Proportional, when the first of the second, and the third of the fourth are equimultiple, or the same part, or the same parts; and yet there is in this Definition no mention at all of a Quotient. For though it be true that if in dividing two Numbers you make the same Quotient, the Dividends and the Divisors are Proportionall, yet that is not the Definition of the same Proportion, but a Theoreme Demonstrable from it. But this Definition *Euclide* could not accommodate to Proportion in Generall, because of incommensurability. (SL 3; EW 7:241)

between arithmetical and geometrical ratios, and Wallis's complaint on this point derives from a deliberate misrepresentation intended to make his opponent's doctrine look more outlandish than it actually is. The crucial point is the manner in which the excess or defect is compared to the quantities related, and since (as Hobbes observes) there are two ways to effect the comparison, there are two kinds of ratio.

It is ironic that even Barrow, who shared Hobbes's distaste for the numerical theory of ratios, found the Hobbesian conception of ratios thoroughly objectionable and subjected it to an extensive critique in his *Lectiones Mathematicae* (*LM* 23, 370–73). Barrow was particularly bothered by the reliance upon physical concepts in Hobbes's account, and declared "if anyone should read through the writings of all mathematicians (both ancient and recent), I think he would never come across anything an author undertakes to illustrate that is made more obscure, and nothing laboring under more or graver errors" (*LM* 23, 371). Interestingly, Barrow attributes the shortcomings in Hobbes's doctrines to an excessive infatuation with the doctrines of Galileo. According to Barrow, Hobbes "seems more intent upon those things Galileo has written concerning uniform motion, or is fixed upon the contemplation of physical motions, when he refers everything concerning magnitude and quantity to certain preconceived ideas on motion" (*LM* 23, 373). The connection between Hobbes and Galileo will be of greater concern to us when we discuss the third part of *De Corpore,* and particularly the sixteenth chapter. For now, however, we can leave the issue of ratios aside and move on to a consideration of Hobbes's account of figures and angles.

3.3 FIGURE AND ANGLE ACCORDING TO HOBBES

We have seen that Hobbes's materialism leads him to view geometry as a science whose objects are produced by motion and extension, and we have considered some of his proposals to replace the traditional definitions with ones that treat geometry as a science of body. This campaign extends to cover the definition of such important geometric objects as lines, figures, and angles; in this section I investigate Hobbes's attempts to define such objects.

The place to start is with the definition of a line. We have seen that Euclid's definition declares a line to be "breadthless length" (*Elements* 1, def. 2). This definition was taken to apply both to straight and curved lines, so a special definition of straight (or right) line is needed. Euclid defines a right line as "a line which lies evenly with the points on itself" (*Elements* 1, def. 4)—a definition notorious for its obscurity.[25] Hobbes dismisses it as "inexcusable" and wonders "How bitterly, and with what insipide jests, would [Wallis] have reviled *Euclide* for this, if living now he had written a Leviathan" (*SL* 1; *EW* 7:203). He therefore offers his own definition in terms of motion, but not simply the motion of a point. He defines a right line as a special case of a line, namely one whose termini cannot be understood to be drawn apart without altering its magnitude. The magnitude of a line is "judged by the greatest distance that can be between its termini" (*DCo* 2.14.1; *OL* 1:154), so the image here is that a straight line cannot have its termini drawn farther apart, but the endpoints of a curved line can be separated while the line retains the same length. The result is that we must consider two kinds of motions in defining a right line: first the motion of a point (by which a line *simpliciter* is traced), and then a motion drawing the termini of the line apart from one another. If the second kind of motion cannot be conceived without altering the magnitude of the line, it is a right line.

We have already seen that Hobbes defines surfaces and solids in terms of the motions of bodies. The natural extension of this doctrine is the definition of plane figures as quantities determined within a plane, or a portion of a surface determined by extreme points or boundaries. As Hobbes defines it: "a figure is the quantity determined by the location or position of all its extreme points" (*DCo* 2.14.22; *OL* 1:174). The determination of these extreme points can be carried out in any number of ways, but the motion of lines and points within a surface is obviously the principal means of such determination of figures. It is important to observe that Hobbes does not define a surface or plane figure as the product of two lines; in his view, the "drawing of lines into lines" is a quintessentially geometric operation that must not be conflated with the arithmetical operation of multiplication. Although the area of a rectangle can be computed by multiplying the lengths of its sides, this is not the process by which the figure is actually created and therefore cannot be part of its definition.

A figure of special interest is the circle, which Euclid defines as "a plane figure contained by one line such that all the straight lines falling upon it from one point among those lying within the figure are equal to one another" (*Elements* 1, def. 15). Hobbes agrees that this defini-

---

25. Heath's commentary (Euclid [1925] 1956, 1:165–69) gives a good overview of the difficulties with the definition and suggested alternatives, which can be supplemented by Federspiel 1991.

tion tells us something true about the circle, but argues that it is not adequate as a definition because it fails to give a causal account of its generation. As he puts it in the first of his *Six Lessons:*

> But if a man had never seen the generation of a Circle by the motion of a Compass or other aequivalent means, it would have been hard to perswade him, that there was any such Figure possible. It had been therefore not amiss first to have let him see that such a Figure might be described. Therefore so much of Geometry is no part of Philosophy, which seeketh the proper passions of all things in the generation of the things themselves. (*SL* 1; *EW* 7:205)

On Hobbes's account, the proper definition of a circle is in terms of the rotation of a line about one of its termini.[26] This definition not only gives the cause of the circle (which can be shown by the use of a compass) but thereby enables us to investigate its properties. Hobbes's methodology holds that once the cause of something has been identified, it is possible to establish demonstrative truths about that thing, and there is nothing that need remain hidden or unexplained. In fact, it is Hobbes's faith in such definitions by causes that led him to think that such notorious problems as the quadrature of the circle could be easily solved once one had grasped the nature of the circle. Hobbes specifically mentions knowledge of the properties of the circle as an example of how "the knowledge of any effect may be gotten from its known generation" (*DCo* 1.1.5; *OL* 1:6). According to Hobbes, it would be impossible to tell by sense whether a figure that closely resembles a circle is, in actuality, a circle or (say) an ellipse that differs very minutely from a true circle. But if we are told that the figure was

26. In *De Corpore,* the definition reads, "If a right line in a plane is so moved that, while one endpoint remains in place, the whole line is carried about until it returns to the place from which it began to move, a plane surface terminated in every direction will be described by the curved line that is described by that endpoint of the line that was carried around. This surface is called a circle, and the unmoved point is called the center of the circle, and the curved line that terminates it is called the perimeter, and any part of it is called the circumference or arc" (*DCo* 2.14.4; *OL* 1:157). It might be objected that this definition presupposes a good deal of knowledge of the circle, since it implicitly assumes that the line, in being "rotated" or "carried about" will return to the same position in which it started. Indeed, to define the circle in terms of its generation *by circular motion* looks plainly circular, if such a bad pun can be excused. Curiously, although Wallis raised any objection he could think of to Hobbes's definitions, he could only fault the definition of the circle for requiring the point at the center of the circle remain unmoved: "But if it is unmoved, it is not a point, since you defined a point (in chapter 8, article 12) as 'a body so moved,' etc." (*Elenchus* 26).

generated by the circumduction of a line about one of its endpoints, there will be no difficulty in identifying it as a genuine circle. This identification of the cause will therefore enable other effects or properties of the circle's generation (such as its area in comparison with a square) to be known.

Angles are the final class of geometric objects to consider. Hobbes was of the opinion that the understanding of angles requires a prior understanding of the circle, and he was convinced that his principles allow for an account of angles that is vastly superior to that in Euclid. Euclid defines an angle as "the inclination to one another of two lines in a plane which meet one another and do not lie in a straight line" (*Elements* 1, def. 8). Hobbes dismisses this definition as both vague and poorly understood.[27] To overcome the difficulties he finds with the Euclidean definition, he proposes an alternative. The first step here is to recognize that the classical account of angles allows for angles to be formed by the concourse of right lines as well as curves. Thus, all of the examples in figure 3.1 could be classified as angles in the Euclidean sense. These include the familiar case of rectilinear angles, curvilinear angles formed by two curves, and the "mixtilinear" case formed by the concurrence of a right line and a curved line. Hobbes was aware of problems posed by the case of some angles involving curves, and it is clear that he fashioned his account of angles to avoid these difficulties. I will have more to say about these matters in chapter 4 when I investigate the problem of the angle of contact, but for now it is sufficient to note that in *De Corpore* Hobbes begins his discussion of angles by drawing a distinction between lines (both curved and right) that touch one another and those that cut one another. As he phrases the distinction:

> Two lines are said to *touch* one another if, when they are drawn to the same point, and produced any amount (produced, I say, in the same manner by which they are generated), still will not cut one another. And so two right lines, if they touch one another at any point, will be contiguous throughout their whole length. (*DCo* 2.14.6; *OL* 1:159)

One line cuts another if it falls on both sides of it, and this distinction thus introduces two different ways in which lines that form angles can meet in a point.

27. As Hobbes writes in *De Principiis et Ratiocinatione Geometrarum*: "Few have impressed in their minds the idea or image of the angle, of which so much is said by

Figure 3.1

Hobbes argues that the most general definition of the term *angle* depends upon the fact that angles are formed where lines share a common point. As he phrases it, the "most general" definition of the term is *"the quantity of that divergence, when two or more lines or surfaces come together in a single point and diverge everywhere else"* (DCo 2.14.7; OL 1:159). We can leave aside the case of angles formed by surfaces and confine our attention to the case of angles in a plane. The general definition of *angle* allows for two species of angles, which are distinguished by "the two ways by which lines may diverge from one another":

> For two right lines applied to one another are contiguous throughout their whole length, but one can be pulled apart from the other, leaving their concurrence in one point, either by circular motion (the center of which is that point of concurrence) so that each line retains its rectitude, and the quantity of the separation or divergence is an *angle* simply so called; or one line may be pulled apart by continual flexion or curving in every imaginable point; and the quantity of this separation is called the *angle of contingence*. (DCo 2.14.7; OL 1:159–60)

Circular motion is essential to the generation of both kinds of angles, but simple angles arise when the lines forming the angle remain rigid through the separation, while "angles of contingence" originate from the uniform (i.e., circular) flexion of one line as it is pulled away from the other. Observe also that in the first case the lines that form the angle cut one another, while in the second they do not.

The quantity of a simple angle can be measured from the amount of rotation of a circle's radius, as in the following example. Let *AB* in

geometers. For the most part, whatever in a surface is open wide at one end, and ends in a narrow end, is commonly called an angle" (PRG 8; OL 4:399).

figure 3.2 be a radius of circle *BCD*, and let *E* be any point on *AB*. Consider the circles *BCD* and *EFG*. We note that if an arbitrary point *H* is taken on *BCD*, $\overparen{BH}$ and $\overparen{EI}$ will be formed. The angle ∠*BAH* is clearly of the same magnitude as ∠*EAI*, but the measure of the angle cannot be defined by the areas of the sectors *AEI* and *ABH* or by the size of $\overparen{EI}$ and $\overparen{BH}$. But the circles are generated by the motion of the radius *AB*, and in generating these circles, ∠*BAH* and ∠*EAI* are formed. Hobbes's idea is that the angle can be defined in terms of the construction of the circle by declaring the angle to be the amount of "circumlation" or rotation required by radius *AB* to generate the sectors *BAH* and *EAI*. Using this definition, we can take as the measure of the angle ∠*BAH* the ratio of $\overparen{BH}$ to the circumference of the circle *BCD*, since $\overparen{BH}$:*BCD* :: $\overparen{EI}$:*EFG*.

This account underscores the fact that "neither the length nor the equality or inequality of the lines that comprehend it contributes anything to the quantity of an angle," and leads to the definition that *"the quantity of an angle is the quantity determined by the ratio of the arc or circumference of the circle to the whole perimeter"* (DCo 2.14.9; OL 1:161). In the example used thus far, we have only considered rectilinear angles, but Hobbes's account of simple angles also includes angles formed by curved lines that cut one another. To measure such angles, it is necessary that the measuring circle be constructed from the tangents to the curved lines at the point where they cut. Thus, if the curve *AB* cuts the curve *CD* at *E* (as in figure 3.3), Hobbes defines the angle at *E* from the tangents *Eα* and *Eβ*. As Hobbes puts it: "the angle that two curved lines make is the same as that which is made by

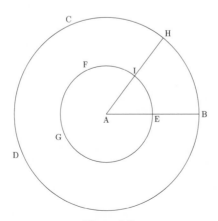

Figure 3.2

two right lines that touch them" (*DCo* 2.14.9; *OL* 1:162). Angles of contingence fail to satisfy this definition, but Hobbes does not conclude that they are not measurable. Instead, he argues that they are heterogeneous to ordinary angles and properly measured in terms of degree of curvature. His definition is: "The angle of contact is the quantity of curvature that is in the arc of the circle made by continual and uniform bending of a right line" (*PRG* 8; *OL* 4:406). To understand this definition, recall that the angle of contact is formed by the intersection of the circle and a right line. If *BCD* is a circle and *BK* a line tangent to the circle (figure 3.4), the angle of contact is the angle *CBK*. Hobbes insists that such angles must be quantities, since the circle can have a greater or lesser divergence from the straight line, and anything that can be ordered by a relation of greater and less must be a quantity. Angles of contact can therefore be compared with one another by observing the amount of curvature in $\overset{\frown}{BC}$, since the greater the curvature, the greater the angle. As a measure of this degree of curvature, we can take the inverse of the radius of the circle, since the degree of curvature varies inversely as the radius of the circle *BCD*.[28]

This conception of angles will occupy our attention in chapter 4 when we consider Hobbes's and Wallis's polemics over the problem of the angle of contact between a curve and tangent. For now, however, we can leave this part of the Hobbesian enterprise and move to a study of those chapters of *De Corpore* that are directed at the solution of problems of quadrature and the rectification of curvilinear arcs.

3.4 MAGNITUDE, MOTION, AND THE GENERAL METHOD OF QUADRATURE

Hobbes intended his program for mathematics to do more than simply substitute one set of definitions for another. Definitions are certainly

---

28. As Hobbes puts it:

Since the angle in general is defined as the opening or divergence of two lines that come together in a point, and since one opening is greater than another also in the generation of the angle of contingence, it cannot be denied that this angle is a quantity, for wherever there is greater and less, there is also quantity. But this quantity consists in greater and lesser flexion. For as the circumference of the circle is greater, its circumference approaches the nature of a right line; inasmuch as the whole curvature (which is made when the periphery of a circle is traced by a right line) is greater when applied to a lesser right line. And therefore, when many circles are tangent to one right line, the angle of contingence made with a lesser circle is greater than that made with a greater circle. (*DCo* 2.14.16; *OL* 1:170)

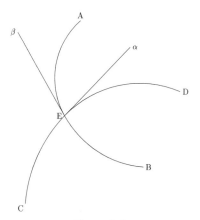

Figure 3.3

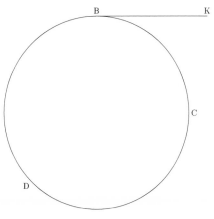

Figure 3.4

important, but the purpose of Hobbes's rewriting of Euclid was to enable the solution of outstanding problems and the resolution of controversies.[29] The fruits of Hobbes's definitional efforts were supposed to
be put on display most prominently in the third part of *De Corpore*
(comprising chapters 15 through 24), which bears the title "On Ratios

29. Du Verdus was enthusiastic in his praise for Hobbes's definitions. It must have
pleased the philosopher greatly to read:

Before your definitions appeared, one of two things happened: either people did
not understand the matter at all, or they just talked gibberish about it. For it was
quite possible, for example, to know what a straight line was; but as for having
a clear and distinct idea corresponding to the words "which lies evenly with the

of Motions and Magnitudes" and contains Hobbes's presentation of a general method of investigating ratios between magnitudes by considering the motions through which such magnitudes are generated. In the opening article of the fifteenth chapter, Hobbes announces that "I thought I should advise the reader coming to this place that he should take into his hands the writings of Euclid, Archimedes, Apollonius, and others, both ancient and modern. For why should I do what has already been done? The little I will therefore say of geometric matters in some of the following chapters will be new, and such as is chiefly of service to natural philosophy" (*DCo* 3.15.1; *OL* 1:175–76). Not surprisingly, part 3 also contains the Hobbesian attempt to square the circle, but these chapters also have a close relationship to Galileo's analysis of uniformly accelerated motion and to Cavalieri's presentation of the method of indivisibles. Indeed, this part of *De Corpore* clearly shows Hobbes's debt to the scientific and mathematical work of seventeenth-century Italian sources. In this section I outline the Hobbesian doctrines derived from Galileo and Cavalieri; more particularly, I will concentrate on chapters 15–18 of *De Corpore* as well as their relationship to Galileo's *Two New Sciences* and Cavalieri's *Exercitationes Geometricae Sex*. I will also investigate the connection between Hobbes's methods and those of Roberval, particularly in connection with the problem of determining the arc length of the Archimedean spiral.

The key to Hobbes's approach to the general problem of quadrature in part 3 of *De Corpore* is his concept of *conatus*.[30] As Hobbes defines it, *conatus* is essentially a point motion, or motion through an indefinitely small space: "*conatus* is motion through a space and a time less than any given, that is, less than any determined whether by exposition or assigned by number, that is, through a point" (*DCo* 3.15.2; *OL* 1:177). Naturally, Hobbes employs his own conception of points here, and in explicating the definition he remarks that "it should be recalled

---

points on itself", I for one had no corresponding idea at all. But when you write, it is Nature herself who is speaking. Above all, Sir, you have the Honour, which none can take away from you, of having been the first true founder of geometry. You are its creator. (du Verdus to Hobbes 13/23 December 1655; *CTH* 1:227)

30. The term *conatus* derives from the Latin verb *conor,* meaning to strive or attempt. The English translation of *De Corpore* uses the term "endeavour," which was apparently Hobbes's preferred English equivalent. Nevertheless, I will retain the Latin usage, as it has gained wide currency. For studies of this doctrine see Barnouw 1992; Brandt 1927, chap. 9; and Lasswitz 1890, 2:214–24. Bernstein 1980 studies Hobbes's conception of *conatus* in relation to Leibniz's early doctrines of motion.

that by a point is not understood that which has no quantity, or which can by no means be divided (for nothing of this sort is in the nature of things), but that whose quantity is not considered, i.e., neither its quantity nor any of its parts are computed in demonstration, so that a point is not taken for indivisible, but for undivided. And as also an instant is to be taken as an undivided time, not an indivisible time" (*DCo* 3.15.2; *OL* 1:177–78).

This definition allows for a further concept of *impetus,* or the instantaneous velocity of a moving point; the velocity of the point at an instant can be understood as the ratio of the distance moved to the time elapsed in a *conatus.* In Hobbes's terms *"impetus is this velocity* [of a moving thing] *but considered in any point of time in which the transit is made.* And so impetus is nothing other than the quantity or velocity of this *conatus"* (*DCo* 3.15.2; *OL* 1:178).[31]

The concepts of impetus and *conatus* can be applied to the case of geometric magnitudes as well as to moving bodies. Because magnitudes are generated by the motion of points, lines, or surfaces, it is possible to inquire into the velocities with which they are generated, and this inquiry can be extended to the ratios between magnitudes and their generating motions. For example, we can think of a curve as being traced by the motion of a point, and at any given stage in the generation of the curve, the point will have a (directed) instantaneous velocity. This, in turn, can be regarded as the ratio between the indefinitely small distance covered in an indefinitely small time; this ratio will be a finite magnitude that can be expressed as the inclination of the tangent to the curve at a point. Take the curve $\alpha\beta$ as in figure 3.5. The *conatus* of its generating point at any instant will be the "point motion" with which an indefinitely small part of the curve is generated; the impetus at any stage in the curve's production will be expressed as the ratio of the distance covered to the time elapsed in the *conatus.* If we represent the curve in the familiar coordinate axis system, the instantaneous impetus will be the ratio between the instantaneous increment along to the y-axis to the increment along the x-axis. The tangent to the curve

31. The English version of this passage is significantly different: "I define IMPETUS, or *Quickness of Motion,* to be the *Swiftness* or *Velocity of the Body moved, but considered in the several points of that time in which it is moved; In which sense* Impetus *is nothing else but the quantity or velocity of Endeavour. But considered with the whole time, it is the whole velocity of the Body moved, taken together throughout all the time, and equal to the Product of a Line representing the time, multiplyed into a Line representing the arithmetically mean* Impetus *or Quickness"* (*EW* 1:207). These modifications allow Hobbes to distinguish the instantaneous impetus of a body at a point from the entire impetus acquired over an interval of time.

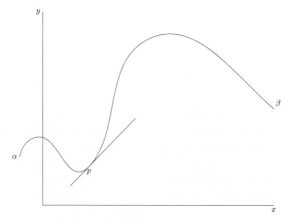

Figure 3.5

at the point p can then be analyzed as the right line that continues or extends the *conatus* at p (equivalently, the tangent is the dilation or expansion of the point motion into a right line).[32]

This approach can also be extended to the general problem of quadrature, with the area of a plane figure being analyzed as the product of a moving line and time. In the very simplest case, the whole impetus imparted to a body throughout a uniform motion will be representable as a rectangle, one side of which is the line representing the instantaneous impetus while the other represents the time during which the body is moved. More complex cases can then be developed by considering nonuniform motions produced by variable impetus. Hobbes applies the concept of impetus to problems of quadrature in chapters 16 and 17 of *De Corpore*. In chapter 16 he considers uniform and accelerated motions of bodies, while in chapter 17 he deals with "deficient figures" produced by the motion of lines. Both of these chapters are of great interest, not for the quality of the mathematical work (which is often sloppy and occasionally quite inept), but because they show the sources of Hobbes's conception of mathematics and its method.

3.4.1 *Galileo, Hobbes, and Chapter 16 of* De Corpore

As I have indicated, Galileo's work on naturally accelerated motion is prominent as part of the background for Hobbes's analysis of motion

32. There is a significant similarity between this approach and Leibniz's formulation of the infinitesimal calculus. I explore the possibility of a connection between Hobbes's mathematics and Leibniz's thoughts on the calculus, via the concept of *conatus*, in Jesseph 1999a.

and magnitude. In fact, it is fair to consider chapter 16 as Hobbes's attempt to incorporate Galileo's work into his own program for a thoroughly materialistic mathematics. Galileo's treatise *Two New Sciences* is devoted to the exposition of a mathematical analysis of the science of mechanics and the science of local motion. In particular, the Galilean treatment of local motion investigates naturally accelerated motion by applying the concept of impetus and taking acceleration to be the accumulation of successive impetuses over time, so that in uniformly accelerated motion the increment in speed is proportional to the increment in time. Galileo defines uniformly accelerated motion in his *Two New Sciences* to be that "which, abandoning rest, adds on to itself equal momenta of swiftness in equal times" (Galileo 1974, 162).[33] The analysis of uniform acceleration leads to two key theorems, the first of which is the "mean speed theorem":

> The time in which a certain space is traversed by a moveable in uniformly accelerated movement from rest is equal to the time in which the same space would be traversed by the same moveable carried in uniform motion whose degree of speed is one-half the maximum and final degree of speed of the previous, uniformly accelerated, motion. (Galileo 1974, 165)

Galileo's proof of this theorem begins by representing the time by the line *AB* and the space traversed by the line *CD* as in figure 3.6. The line *EB* drawn perpendicular to *AB* represents the final velocity of the moving body, and the parallels between the lines *AB* and *AE* represent instantaneous velocities of the body, increasing uniformly along the time line *AB*. The triangle *ABE* thus represents the accumulated impetus of the accelerated motion. Bisecting *BE* at *F* and drawing *FG* parallel to *AB* yields a rectangle *ABFG* whose area is equal to that of the triangle *ABE*. This rectangle can be thought of as representing the motion of a body, which begins with a velocity one-half that of *AE*, and continues unaccelerated throughout the time *AB*. The distance covered will be equal in both cases, since that traveled by the unaccelerated body is the product of velocity and time, while that of the accelerated body is treated as the aggregate of instantaneous impetus over the same time. Galileo does not want to assert that the areas of the rectangle and triangle actually represent the distances, since the distance is

---

33. The expression *momenta of swiftness* is equivalent to Hobbes's term *impetus*. As Stillman Drake notes, *impetus* is "freely interchangeable with *momento*" (Galileo 1974, xxxiv).

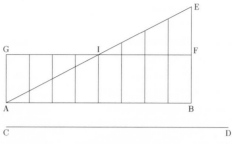

Figure 3.6

supposed to be given by the line *CD* (which cannot be compared to an area). Nevertheless, he can assert that the ratio between the distances is the same as that of the areas, and this suffices for the theorem.

The second key theorem in Galileo's account of accelerated motion relates the distance traversed to the elapsed time, and establishes that the ratio between any two distances covered by accelerated motion is the same as the ratio of the squares of the times. In Galileo's words: "If a moveable descends from rest in uniformly accelerated motion, the spaces run through in any times whatever are to each other as the duplicate ratio of their times; that is, as the squares of those times" (Galileo 1974, 166). The details of the proof need not detain us here,[34] for the important point is the connection between Hobbes's and Galileo's approaches to the mathematics of motion. As we will see, Hobbes is eager to exploit this result as part of his program in the philosophy of mathematics.

A final Galilean result of importance for understanding chapter 16 of *De Corpore* is the theorem that a projectile traces a parabolic path. We can follow Galileo in understanding a projectile as "a moveable [whose] motion is compounded from two movements; that is, when it is moved equably and is also naturally accelerated" (Galileo 1974, 217). In particular, projectile motion arises when a body is moved laterally with a uniform velocity, but naturally accelerated downwards. The principal result of interest to us is the first theorem of the fourth-

34. They can be found in Galileo 1974, 166–67. Essentially, the proof invokes the previous result to show that the ratio of the distances at any two points can be compared by considering equivalent distances covered with uniform velocity. Galileo then invokes the principle that the distances covered by equable motion are in the ratio compounded from the speeds and times, from which he concludes that the distances covered by uniformly accelerated motion must be as the squares of the elapsed times. For a detailed study of Galileo's mathematization of naturally accelerated motion see Clavelin 1974, 298–323.

day dialogue in the *Two New Sciences*. It asserts that "[t]he line described by a heavy moveable, when it descends with a motion compounded from equable horizontal and natural falling [motion], is a semiparabola" (Galileo 1974, 221). The parabola is a line with the property that the distance along the vertical y-axis varies as the square of the distance along the horizontal x-axis (to use the terminology of coordinate axes). Galileo's theorem asserts that if we assume the movable to have uniform motion along the x-axis (line *AB* in figure 3.7) and to be naturally accelerated downward along the y-axis (line *AC*), then the combination of these two motions will trace the path *AD*, in which uniform horizontal motion combines with accelerated downward motion increasing as the square of the time. The result is one-half of the parabola $y = x^2$, and the path *AD* is thus a semiparabola.

Scholars have long recognized that Hobbes was influenced by Galileo, and the link between the two thinkers is well established by Hobbes's remark in the "dedicatory epistle" to *De Corpore* that Galileo "was the first who opened to us the first gate of universal physics, which is the nature of motion" (*DCo* epistle; *OL* 1:sig. h4v). Legend even has it that a conversation with Galileo in 1635 or 1636 inspired Hobbes to pursue the goal of presenting moral and political philosophy in a rigorously geometrical method.[35] It is also worth observing

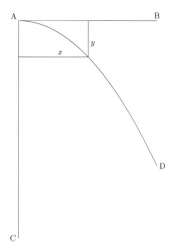

Figure 3.7

35. Tönnies 1975, 85–91, reports the evidence of a personal connection between Hobbes and Galileo. The source for Galileo's alleged direct influence on Hobbes's program in moral and political philosophy is A. G. Kästner's *Geschichte der Mathematik,*

that Mersenne was active in bringing the work of Galileo to the attention of the learned circles in Paris, and Hobbes's exposure to Galileo's thought may well have been mediated by his contact with Mersenne. Whatever its full extent may have been, Hobbes's indebtedness to Galileo is evident in chapter 16 of *De Corpore,* which bears the title "On Accelerated and Uniform Motion, and on Motion by Concourse." The approach pursued here takes uniform motion to be the result of a constant impetus, so that a body moved uniformly traverses equal distances in equal times. The velocity of a body, as Hobbes understands it, "has its quantity determined by the sum of all the several quicknesses or impetus[es], which it hath in the several points of the time of the body's motion" (*DCo* 3.16.1; *EW* 1:218).[36] Representing the motion of a body by a figure, one side of which designates the time and the other the impetus generated in the corresponding instant, we have a rectangle (such as *ABCD* in figure 3.8) in the case of uniform motion. The uniform motion starts with a given impetus *AB,* which remains unaltered as the body moves through time *BC.* On the other hand, motion uniformly accelerated from rest is represented by the triangle *BCD,* in which the impetus increases continually with the time.[37] Finally, the calculation of distances covered by such motions proceeds in essentially the same manner as that used by Galileo—a body uniformly accelerated from rest covers distances that are in the duplicate ratio of the elapsed times, so that the distance dispatched is as the square of the time. (*DCo* 3.16.3; *OL* 1:186–88)

---

which reports that "John Albert de Soria, former teacher at the university in Pisa, assures us it is known through oral tradition that when they walked together at the grand-ducal summer palace *Poggio Imperiale,* Galileo gave Hobbes the first idea of bringing moral philosophy *[Sittenlehre]* to mathematical certainty by treating it according to the geometrical method" (Kästner 1796–1800, 4:195).

36. The Latin version of this passage is less clear than the English. At the beginning of chapter 16 Hobbes has the definition: "Velocitas cujuscunque corporis per aliquod tempus moti tanta est, quantum est quod fit ex impetu (quem habet in puncto temporis) ducto in tempus ipsius motus" [The velocity of any body moved through some time is equal to the product of the impetus that it has at a point of time and the time in which it is moved] (*DCo* 3.16.1; *OL* 1:184). Wallis attacks this definition for the obvious problem, namely that it does not specify at which point of time the product of impetus and time should be computed (*Elenchus* 39).

37. As Hobbes puts it, "If the impetus is everywhere the same and any right line is taken for the measure of time, the impetuses applied ordinately to this right line will designate a parallelogram, which will represent the velocity of the whole motion. If the impetus increases uniformly from rest, that is, always in the same ratio as the elapsed times, the whole velocity of the motion will be represented by a triangle, one side of which is the whole time, and the other the greatest impetus acquired in that time" (*DCo* 3.16.1; *OL* 1:185).

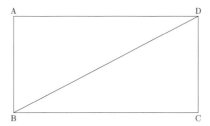

Figure 3.8

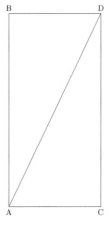

Figure 3.9

The investigation of "motion by concourse" in the remainder of chapter 16 of *De Corpore* is another part of Hobbes's Galilean inheritance. In exact analogy with Galileo's treatment of projectile motion, Hobbes treats motion of a body acted upon by two forces as a complex "concourse" of the two simpler motions. In the simplest case, where the combined motions are both uniform, the resulting motion will be a straight line equal to the diagonal of a parallelogram whose sides are proportional to the two component velocities. Thus, in figure 3.9, the compound motion arising from the concourse of motions *AC* and *AB* will be the diagonal motion *AD*. This result can then be extended to consider cases of nonuniform velocity. In particular, Hobbes asserts that

> [i]f a moving body *[mobile]* is carried by two movements together, meeting at any given angle, of which the first moves uniformly and the other by a motion uniformly accelerated from rest

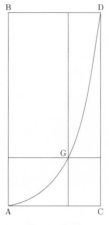

Figure 3.10

(that is, so that the impetuses are in the ratio of the times; that is, so that the ratio of the lengths is the duplicate ratio of the times) until the greatest impetus acquired by acceleration is equal to the impetus of uniform motion, the line in which the moving body is carried will be a semiparabola, whose base is the impetus last acquired. (*DCo* 3.16.9; *OL* 1:196)

This result is, of course, identical to Galileo's theorem that projectile motion describes a semiparabola, although it is phrased somewhat differently. In this case, the semiparabola *AGD* in figure 3.10 arises from the concourse of the uniform motion *AC* and the accelerated motion *AB*. As Hobbes presents the case, he imagines the straight line *AB* to be moved uniformly to *CD,* while in the same time the line *AC* is moved with uniform acceleration to *BD*. The locus of intersection of the two moving lines will then be the semiparabola *AGD*. This differs slightly from the presentation of projectile motion in Galileo, but is clearly closely related to it.

Although there is an obvious similarity between Hobbes's and Galileo's analysis of motion, there are also important differences. Galileo was concerned with the mathematical analysis of motion, and the results presented in his *Two New Sciences* are confined to such phenomena as arise from naturally accelerated and projectile motion. Hobbes, however, pursues a more general and abstract inquiry in *De Corpore,* for he also considers motions that correspond to nothing in nature. He extends, for example, the case of projectile motion to the case of motion accelerated as the triplicate ratio of the times, and then considers a general rule for describing and investigating the quadrature of "parabolasters" of higher degree that correspond to accelerations of any degree whatever.[38] This extension of Hobbes's investigations beyond the case of natural motions led Frithiof Brandt to see a certain incongruity between these chapters of *De Corpore* and Hobbes's general program for a thoroughly mechanistic science of nature, and he regarded it as strange that Hobbes should have bothered to consider motions that do not arise in nature.[39] However, there is really no deep difficulty here. Hobbes was clearly concerned to present an account of mathematics that could extend to the most abstract and general classes of problems, and his program for a thoroughly materialistic conception of mathematics would naturally have to extend beyond those parts of mathematics that have an immediate physical application. Notwithstanding his declaration that his contributions to geometry were to be "such as are chiefly of service to natural philosophy" (*DCo* 3.15.1; *OL* 1:176), Hobbes was also concerned with problems in pure mathematics. If nothing else, he was hardly alone in seeking a general approach to problems of quadrature.

Wallis was unrelenting in his critique of Hobbes's sixteenth chapter, and he devoted a remarkable 28 pages of his 136-page *Elenchus* to a detailed criticism of it. Of the twenty articles that initially made up this chapter, he denounced all but three as unsound, for reasons ranging from sloppy formulation of principles to technical errors in Hobbes's demonstrations. Wallis was happy to supply suggestions for

---

38. As Hobbes puts it, "By the same method it can be shown what line it is that is made by a moving body driven *[actum]* by the concourse of any two motions, of which one is moved uniformly, while the other is accelerated, but according to ratios of spaces and times that are explicable by numbers, such as duplicate, triplicate, and other such ratios, or such ratios as can be designated by any fraction whatever" (*DCo* 3.16.11; *OL* 1:198). Hobbes uses the term *parabolaster* for curves of higher degree than the simple parabola.

39. As Brandt puts it:

That this is far removed from Galileo, seems to us fairly obvious. The latter did not occupy himself at all with the possibilities posited by Hobbes, of course because they have no relation to experience. Result: *Hobbes both directly and indirectly shows a conspicuous lack of interest in mathematics as applied to the motions occurring in experience. He shows this directly by making no mention whatever of the fact that the formula* $s = at^2$ *is the law for the free fall and that the parabola of projection is compounded of the uniform motion and the accelerated motion on the assumption* $s = at^2$. *And he shows it directly by placing the aforementioned possible formulae of motion on an equal footing. This is mathematically warrantable but seems strange in a philosopher who is driving at a mathematico-mechanical conception of Nature.* (Brandt 1928, 320)

the mending of such errors, and he recommended his own *Arithmetica Infinitorum* as a superior treatment of these matters.[40] The general tone of his complaints against Hobbes's procedures can be gathered from his remarks on article 18:

> What need is there for this whole apparatus you have, and for so many interspersed paralogisms and other suppositions which have nothing sound about them, in order to prove a thing so easy, and which no man would ever doubt? Unless perhaps someone had come across it in your demonstration, and seeing so many false and putrid things in it, he should assume that anything needing such things for proof must contain something wrong. (*Elenchus* 63)

Such a harsh tone is not altogether unwarranted, for Hobbes's work contains a considerable number of errors, both of omission and commission. The general intention of the chapter is clear enough and it has a sufficient connection to other mathematical work of the period to be reasonably intelligible, but it is hardly the magnum opus that Hobbes envisaged. In fact, the English version of this chapter differs from the Latin original, and many of the differences involve Hobbes unacknowledged incorporation of Wallis's criticisms.[41]

3.4.2 Cavalieri and Hobbes

Cavalieri's influence upon Hobbes's mathematical work is at least as strong as Galileo's, as we can see from chapter 17 of *De Corpore.*

---

40. Thus, in commenting upon the fourth article in chapter 16, Wallis remarks, "What should be put here in place of all of this you may learn from propositions 23, 69, and 71 of my *Arithmetica Infinitorum*. For what is treated universally in that place can easily be accommodated to this business" (*Elenchus* 46). Later, he suggests that a similar course of study would benefit Hobbes's presentation of parabolic curves: "But if you desire to understand this, you may learn it and things greater than it from propositions 23 and 64 of my *Arithmetica Infinitorum*" (*Elenchus* 51).

41. To take one example, Hobbes's original formulation of article 3 of chapter 16 (on the comparison of ratios of impetus in accelerated motion) was altered in the English translation to meet some of Wallis's objections, as Wallis himself points out in his response to the *Six Lessons:* "Now these two Objections [in the *Elenchus*] were clear and full, (and did destroy your whole demonstration;) and this you discerned well enough, though you did not think fit to make any reply or confession; (but invent some other objections, which I never made, that you might seem to answer to somewhat.) And therefore in the English, without making any words of it, you mend it" (*Due Correction* 97–98). Notwithstanding the changes, Wallis still found the English version of the article unacceptable, and even charged Hobbes with introducing more errors than were contained in the Latin original.

Hobbes was clearly familiar with Cavalieri's work from early on, for among his surviving papers is a notebook in which he copied extracts and summaries of Cavalieri's *Exercitationes Geometricae Sex.*[42] The notebook was probably used during Hobbes's extended stay in France in the 1640s—the period in which he assembled his *De Corpore* and when Cavendish enlisted his aid in Pell's campaign against Longomontanus.[43] Chapter 17 of *De Corpore* on the measure of "deficient figures" comes almost straight out of the *Exercitationes Geometricae Sex,* as we can see by comparing its second article with proposition 23 of part 4 of the *Exercitationes.* In Hobbes's parlance, the deficient figure *ABEFC* in figure 3.11 is produced by the motion of the line *AB* through *AC,* while *AB* diminishes to a point at *C.* The "complete figure" corresponding to the deficient figure is the rectangle *ABDC,* produced by the motion of *AB* through *AC* without diminishing. The complement of the deficient figure is *BDCFE,* the figure that, when added to the deficient figure, makes the complete figure. Hobbes's task is to find the ratio of the area of the deficient figure to its complement, given a specified rate of decrease of the quantity *AB.* He concludes that the ratio of the deficient figure to its complement is the same as the ratio between corresponding lines in the deficient figure and their counterparts in the complement. His statement of the theorem reads:

> A deficient figure, which is made by a quantity continually decreasing until it vanishes, according to ratios everywhere proportional and commensurable, is to its complement as the ratio of the whole altitude to an altitude diminished at any time is to the ratio of the whole quantity, which describes the figure, to the same quantity diminished in the same time. (*DCo* 3.17.2; *OL* 1:209)

---

42. Chatsworth House, Library, Hobbes MS. C.I.5.

43. The dating of the manuscript is not certain, but the hand and subject matter put it in the same period as MS. A.10 in the Hobbes archive at the Library of Chatsworth House. This manuscript is a draft of *De Corpore,* and it seems that Hobbes copied out the extracts from Cavalieri while he was writing *De Corpore.* The *Exercitationes* were published in 1647, and both Hobbes and Sir Charles Cavendish must have sought it eagerly. In a letter to Pell on 2/12 August 1648, Cavendish wrote that "Mr: Hobbes hath nowe leisure to studie and I hope we shall have his within a twelve-month." He then adds "I saw a booke at Paris of the excellent Cavalieros lately printed, concerning Indivisibles; whom you know was the Inventor or Restorer of that kinde of Geometrie; I had no time to reade it before I came awaye, and they are not to be bought; Mr: Careavin comming latelie from Italie brought this with him" (British Library Add. MSS. 4278, f. 273r). It is therefore no great stretch of the imagination to think that Hobbes was reading Cavalieri as he put together the mathematical sections of *De Corpore* in 1648 and 1649.

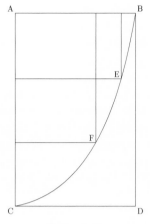

Figure 3.11

Thus, if the rate of diminution of *AB* is uniform, the line *ABEFC* will be a right line (indeed, the diagonal of the rectangle), and the deficient figure will be to its complement as one to one. In more complex cases, as when *AB* decreases as the square of the diminished altitude, the area of the deficient figure will be twice that of its complement. And, in general, if the line *AB* decreases as the power *n*, the ratio of the deficient figure to its complement will be *n*:1. Hobbes's proof procedure for this theorem involves the consideration of ratios between "all the lines" in the deficient figure and its complement, and shows a significant similarity to Cavalieri's famous procedure in the fourth of his six *Exercitationes Geometricae*—an exercise entitled *On the Use of Indivisibles in Cossic Powers*.[44]

There, Cavalieri pursued a result that historians of mathematics generally characterize as the attempt to prove the geometric equivalent of the theorem $\int_0^a x^n dx = a^{n+1}/(n + 1)$. Except for differences in diagrams and terminology, Cavalieri's fourth *exercitatio* delivers the same results as Hobbes's account of deficient figures. In proposition 23 of *exercitatio* 4, Cavalieri asserts his version of the theorem we saw earlier from *De Corpore*:

44. Hobbes's proof in the original Latin version differs from that in the English version. In fact, the English version has two separate proofs, the second of which is a reworked version of the Latin original. Beyond this, the edition of *De Corpore* in Hobbes's 1668 *Opera* contains yet another variation on the same proof. Despite these differences, the proofs all depend upon concepts taken from Cavalieri. These various proofs are examined in section 2 of the appendix.

In any parallelogram such as *BD* [as in figure 3.12], with the base *CD* as *regula*, if any parallel to *CD* such as *EF* is taken, and if the diameter *AC* is drawn, which cuts the line *EF* in *G*, then as *DA* is to *AF*, so *CD* or *EF* will be to *FG*. And let *AC* be called the first diagonal. And again as DA^2 is to AF^2, let *EF* be to *FH*, and let this be understood in all the parallels to *CD* so that all of these homologous lines *HF* terminate in the curve *AHC*. Similarly, as DA^3 is to AF^3, let also *EF* be to *FI*, and likewise in the remaining parallels, to describe the curve *CIA*. And as AD^4 is to AF^4, let *EF* be to *FL*, and likewise in the remaining parallels to describe the curve *CLA*. Which procedure can be supposed continued in the other cases. Then *CHA* is called the second diagonal, *CIA* the third diagonal, *CLA* the fourth diagonal, and so forth. Similarly the triangle *AGCD* is called the first diagonal space of the parallelogram, the trilinear figure *AHCD* is the second space, *AICD* the third, *ALCD* the fourth, and so on. I say therefore that the parallelogram *BD* is twice the first space, triple the second space, quadruple the third space, quintuple the fourth space, and so forth. (Cavalieri 1647, 279)

In its details Cavalieri's proof of this result departs from Hobbes's efforts. Cavalieri introduced the concept of "all the squares" of a figure, as well as "all the cubes," "all the square-squares," etc. He then argued that all the lines of a parallelogram are double all the lines of the triangle that is half the parallelogram, while all the squares of the same parallelogram are triple those of the triangle, and all the cubes of the parallelogram are quadruple all the cubes of the triangle. This, together with some other principles governing the comparison of lines in a fig-

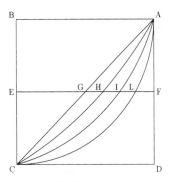

Figure 3.12

ure, yields Cavalieri's result.[45] Hobbes tried to establish the result by a more direct (if ultimately unsuccessful) argument that does not employ such devices as "all the squares" of a figure. These various attempts to prove this result are examined in the appendix and can be left aside here. Nevertheless, despite the differences in the details of the proofs, it is clear that Cavalieri's general approach was the model that Hobbes used in his treatment of deficient figures.

Such similarities between Hobbes and Cavalieri on this point are certainly striking, and it was obvious to Hobbes's contemporaries that this approach to quadratures had its roots in Cavalieri's work. In fact, Wallis eventually accused Hobbes of plagiarizing from Cavalieri, although when he wrote the *Elenchus* he could only voice his suspicion that much of chapter 17 of *De Corpore* must have been taken from someone else (*Elenchus* 83–84). In his *Due Correction* Wallis made the charge of plagiarism more specific by printing an excerpt from a letter alleging that "those propositions which Mr. Hobs had concerning the measure of the parabolasters were not his own, but borrowed from somebody else without acknowledging his author" and that "they were to be found demonstrated in an exercitation of Cavalerius *De Usu Indivisibilium in Potestatibus Cossicis*" (*Due Correction* 7).[46] Charges of plagiarism figure prominently in the dispute between Hobbes and Wallis, as we will see in more detail when we consider the question of Roberval's influence on Hobbes. For present purposes it is

45. The main points are argued in propositions 19–21 (Cavalieri 1647, 272–78). For an outline and analysis of the argumentation in the fourth *Exercitatio,* see Giusti 1980, 76–79.

46. The author of the letter is known to us only as a British gentleman by the name of Vaughan. In a letter to Hobbes of 19/29 December 1655 Stubbe reports, "Yesterday there came to mee one M:^{r} Vaughan of Jesus colledge desireing to know whether yo^u intended to answer Wallis. I assured him of it. whereupon hee told mee y^t ye Gentleman whose letters in English d:^{r} W: had inserted was his brother & ye occasion of ye writeing y^m was this. When D:^{r} W's Elenchus came out, it was by him sent to his brother being a louer of ye Mathematiques: who upon ye perusall of it, sent him in a private letter y^t remarque, y^t D:^{r} W's had guessed aright, y^t such & such propositions were out of Cauallerius. Mr Vaughan being in company with Wallis, did let him know this. whereupon ye D:^{r} upon ye edition of his Arithmetica infinitorum, sent him one, together with a letter. The gentleman haueing receiued it, delayed an answer, ye letter not requireing any; untill his brother did reminde him that ye D:^{r} might expect a reply, and howeuer it would bee civill to returne him ye complement. Upon this account hee wrote y^t letter you see, not intending ye publication of it, or y^t euer hee should bee engaged in ye defence of a booke hee had not then examined, when hee wrote y^t. hee is very much dissatisfyed with ye publishing thereof, & would bee glad to see him corrected according to his demerit, & so would his brother too, he sayth" (*CTH* 1:394–95).

enough to note that there is sufficient evidence to make these charges quite plausible in the case of Hobbes and Cavalieri.

Aside from the obvious similarities in their methods, there are other important points of similarity between Hobbes's and Cavalieri's approaches to indivisibles. Both conceive of indivisibles as described by the motion of a line through a figure, and both rely upon the calculation of ratios between lines that is rooted in the Euclidean theory of proportions rather than the algebraic treatment of magnitudes characteristic of analytic geometry. This means that they do not attempt to represent the area of a figure as an infinite sum of infinitesimal elements. In contrast, Wallis does not appeal to motion, relies heavily upon arithmetical and algebraic methods, and does not use the painstaking comparison of ratios we find in Hobbes and Cavalieri. Most significantly, he is happy to analyze the area of a figure as an infinite sum, the terms of which are infinitely small in comparison with the total area. Some of these issues will reemerge in chapter 4 when we turn our attention to Hobbes's and Wallis's conceptions of the infinite.

3.4.3 Roberval, Hobbes, and the Rectification of the Spiral

The recognition of Hobbes's intellectual debt to Galileo and Cavalieri on these matters by no means rules out the possibility that his mathematical work was influenced by others. In particular, his knowledge of the method of indivisibles may also have come in part from his contact with Roberval in Paris during the 1640s. Roberval is an enigmatic character in the history of mathematics.[47] He held the chair in mathematics at the Collège Royal in Paris from 1634 until his death in 1675 and was a pioneer in the application of the method of indivisibles, but he actually published very little during his lifetime.[48] The mathematical chair at the Collège Royal was awarded on the basis of triennial competitions, and Roberval's reluctance to publish may well have derived from the fact that his continued livelihood could be ensured by keeping his methods and discoveries secret. Although he published very little, Roberval was nevertheless active in Parisian intellectual life; he was a

47. For more on Roberval see Auger 1962, Costabel and Martinet 1986, and Vita 1973. Walker 1932 is a dated but useful study of Roberval's principal work in the method of indivisibles.

48. Pierre Costabel observed that "[t]he discoveries that this mathematician and logician made, while being at once jealous of his priority and preoccupied with keeping his means secret, are the expression of a work much more important than that of the two books published in his lifetime" (Costabel and Martinet 1986, 23).

leading member of the "Mersenne circle," and several of his discoveries were first made public in some of Mersenne's publications.

Roberval's most important single work is the *Traité des Indivisibles,* which remained unpublished during his lifetime but appeared in 1693 as part of the *Divers Ouvrages de Mathématique et de Physique, par Messieurs de L'Academie Royale des Sciences* (Roberval 1693, 190–245). His procedures are very similar to those of Wallis, which is to say that Roberval considers plane surfaces as composed out of infinitely many indivisible parts, and the quadrature of a figure is typically effected by analyzing infinite sums of these indivisible elements. Hobbes's friend du Verdus was a pupil and close associate of Roberval, and he wrote a brief treatise outlining Roberval's methods that was later published as *Observations sur la composition de mouvements et le moyen de trouver les tangentes aux lignes courbes* in the *Divers Ouvrages* of 1693.[49] The *composition of motions* is Roberval's term for the method of analyzing a curve as the product of a point in motion, and particularly for the idea of considering the motion at any instant as a complex composition of two motions, such as Galileo's construction of the parabola from uniform rectilinear motion along one axis and uniformly accelerated motion along another.

It is the mention of Roberval in Mersenne's *Cogitata Physico-Mathematica* that led Wallis to include a specific allegation of plagiarism against Hobbes at the very end of the *Elenchus.*[50] Wallis reports that as he was finishing the *Elenchus,* and while much of it was already printed, he "unexpectedly came across some passages in Mersenne, which abundantly confirmed that which I have occasionally indicated earlier, namely that when something true is included among these things of yours, it is not really your own, but taken from somewhere else" (*Elenchus* 132). Earlier in the *Elenchus* Wallis had suggested the likelihood of Hobbes's plagiarizing those of his results that were actually sound, but he had been forced to conjecture about their possible source. However, his reading of the treatise *Hydraulica* contained in Mersenne's *Cogitata* led him to name Roberval as the source of one particular result concerning the arc length of the curve known as the Archimedean spiral.

The spiral of Archimedes is generated by the compound motion of a point that moves uniformly through a line that itself is rotating uni-

---

49. The treatise is in Roberval 1693, 67–111.

50. See Schumann 1995 for an overview of Hobbes's contributions to Mersenne's *Cogitata.*

formly about one of its endpoints. Thus, in figure 3.13, the spiral *ab-cdefn* arises from the motion of a point through the radius *an* in the same time that *an* completes a single revolution. The second corollary to the twenty-fifth proposition of the *Hydraulica* asserts that the first revolution of the spiral of Archimedes is equal to the arc length of a parabola having a base equal to the radius of the spiral and an axis equal to half its circumference. Mersenne reports that "when I was concerned with this result, a learned man proposed a certain right line that he thought equal to the first revolution of the spiral *abcdefn,* but the revolution of the spiral was greater than this proposed line, and our geometer showed that the spiral was equal to the parabola *GT*" (Mersenne 1644, 129). Roberval was always known by the epithet "our geometer" in Mersenne's text, and Wallis concluded that he must have been the source for Hobbes's result concerning the arc length of the spiral.[51] Elsewhere in Mersenne's *Cogitata,* Roberval is credited with other results with counterparts in *De Corpore,* and Wallis wasted no time in citing these as further instances of Hobbes's presumed plagiarism.[52]

Allegations of plagiarism were common fare in mathematical disputes of the seventeenth century, and there is nothing terribly remarkable about Wallis's use of such a charge against Hobbes. Roberval himself was quick to accuse others of the offense, and his allegations were made so frequently that Hobbes could remark that "Roberval has this peculiar trait: when someone makes public a great theorem he has found out, he immediately sends out letters in which he claims to have found the same result earlier" (*Examinatio 5; OL* 4:188). True to his form, Roberval circulated an open letter in which he accused both Hobbes and Wallis of appropriating various of his results, particularly those concerning the rectification of the spiral.[53] Wallis himself was

51. Hobbes's result and his proof are contained in the appendix.

52. Thus, Wallis claims that the results from chapters 17 and 23 of *De Corpore* concerning centers of gravity and the quadrature of parabolasters can be found in the treatise *Mechanica* among the *Cogitata,* while the method of composition of motions is contained in the *Ballistica,* and other results deriving from Descartes, Galileo, and Fermat are mentioned in the treatise *Mechanica* and in the *Reflexiones Physico-Mathematicae* at the end of the *Cogitata* (*Elenchus* 132–34).

53. The communication from Roberval (now lost) was brought by Thomas White to England, where Brouncker forwarded it to Wallis. The communication (which Wallis refers to as a *"charta"* in a letter to Brouncker that was intended for Roberval) was apparently a broadsheet accusing both Wallis and Hobbes of plagiarism. Wallis agrees that Hobbes is guilty of plagiarism and notes that he himself had attempted to contact Roberval and Gassendi (who had since died) to discern the source of Hobbes's mathematical results. However, he denies that he has ever seen Roberval's work and attempts

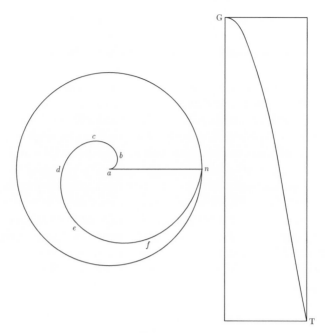

Figure 3.13

certainly no stranger to charges of plagiarism; he had a penchant for grinding self-promotion at the expense of others, and his career was consequently dogged by accusations that he had stolen most of the important work he published.[54]

Although these accusations of plagiarism are significant in the

---

to refute the charge that he is equally guilty of plagiarism. (Wallis to Brouncker for Roberval, 16/26 October 1656; Paris, Bibliothèque Nationale, nouv. acq. fr. 3252, f. 148–52). Wallis's letter to Gassendi requesting further information about the identity of the "vir doctus" mentioned in Mersenne's *Cogitata* is in Gassendi [1658] 1964, 6:540.

54. Wallis's proclivity for appropriating the work of others is detailed in a letter to Hobbes from Aubrey dated June 24/July 4 1675. In it, Aubrey gives an extremely unflattering portrait of a much-despised Wallis. He reports that Robert Hooke "has been as much abused by Dʳ Wallis as any one: he makes it his Trade to be a common-spye. steales from every ingeniose persons discourse, and prints it: viz from Sʳ Ch: Wren God knows how often, from Mʳ Hooke etc. he is a most ill-natured man, an egregious lyar and backbiter, a flatterer, and fawner on my Lᵈ Brouncker & his Miss: that my Lᵈ may keep up his reputation" (*CTH* 2:753). The letter goes on to detail similar charges against Wallis from other quarters. Even taking account of Aubrey's friendship with Hobbes and the latter's wish to hear Wallis denigrated, the letter suggests that Wallis was not a man to be trusted.

course of the Hobbes-Wallis controversy, it is difficult to conclude whether Hobbes actually owed much to Roberval. Because his published output was so small (at least as of the time of *De Corpore*), Roberval could not have had much of an influence on Hobbes by way of publications. The two were acquainted through Mersenne in the 1640s and must have discussed mathematical matters on many occasions (for instance, they both contributed to Pell's campaign against Longomontanus). Nevertheless, Roberval's well-known reticence makes it highly unlikely that he would have shared much with his English acquaintance. Hobbes's close friendship with du Verdus might have been the means of transmission for some of Roberval's ideas, since du Verdus was well versed in Roberval's methods and had prepared several summaries of his results. But there is no evidence that du Verdus met Hobbes before 1651 (which would be quite late in the preparation of *De Corpore*).[55] When pressed, Hobbes acknowledged that he had heard Roberval claim the rectification of the spiral, but denied that he was privy to the demonstration or that he had taken anything of substance from him.[56] In fact, Hobbes later insinuated that Roberval him-

---

55. Malcolm's account of du Verdus's life in the biographical register to Hobbes's *Correspondence* (CTH 2:902–13) is the best source for sorting out the details of the relationship between Hobbes and Roberval. If, as Malcolm suggests, Hobbes and du Verdus did not meet before 1651, this would place their first meeting well after the likely date of composition of a manuscript version of the first seventeen chapters of *De Corpore* (Chatsworth MS. A.10), which probably dates from the 1640s. Although the manuscript differs from the printed *De Corpore* in many respects, the seventeenth chapter (f. 26–30) does contain an analysis of deficient figures.

56. Recounting the matter in the *Six Lessons,* Hobbes declares,

> You say further (you the Geometrician) that I had the Proposition of the Spirall Line equall to a Parabolicall line from Mr. *Robervall.* True. And if I had remembered it, I would have taken also his demonstration, though if I had publisht his, I would have suppressed mine. I was comparing in my thoughts those two Lines, Spirall and Parabolicall, by the Motions wherewith they were described; and considering those Motions as uniform, and the Lines from the Center to the Circumference, not to be little Parallelograms, but little Sectors, I saw that to compound the true Motion of that Point which described the Spirall, I must have one Line equal to half the Perimeter, the other equal to half the Diameter. But of all this I had not one word written. But being with *Mersennus* and Mr. *Robervall* in the Cloister of the Convent, I drew a Figure on the wall, and Mr. *Robervall* perceiving the deduction I made, told me that since the Motions which make the Parabolicall Line, are one uniform, and the other accelerated, the Motions that make the Spirall must be so also; Which I presently acknowledged; and he the next day, from this very method, brought to *Mersennus* the demonstration of their equality. And this is the story mentioned by *Mersennus,* Prop. 25. Corol. 2. of his *Hydraulica.* (SL 6; EW 7:343)

self had first hit upon the solution to the question of the spiral only after Hobbes had put him on the right path.[57]

The problem of finding the arc length of the spiral acquired an especially significant status in the dispute because Wallis himself fell into error in his attempts to solve it. Moreover, Wallis made what is probably the same error as Hobbes had made when he first proposed the rectification of the spiral to Mersenne and Roberval. The difficulty can be summarized as follows, using the techniques of calculus to compute the arc length of the spiral.[58] Let r designate the distance moved along the radius R, s the length of the spiral at r, and φ the angle of rotation at r (see figure 3.14); then (using polar coordinates) we get $r = R\varphi/2\pi$, and $ds = \sqrt{dr^2 + r^2 d\varphi^2}$. Using the general arc length formula $s = \int_0^{2\pi} \sqrt{r^2 + (dr/d\varphi^2)} d\varphi$, the arc length will then be an integral of the form $\int_0^{2\pi} ds$, from which we obtain the result $s = R/2\pi \int_0^{2\pi} \sqrt{1 + \varphi^2} d\varphi$. In his *Arithmetica Infinitorum,* Wallis mistook the problem as one of computing the aggregate of arcs of the spiral's limiting circles, rather than the aggregate of the elements of the spiral. In other words, he computed the integral $\int_0^{2\pi} r d\varphi$, which easily solves to πR. Wallis thereby obtained the false result that the length of the spiral is one-half the circumference of the circumscribing circle. The source of the confu-

---

57. In the fifth dialogue of his *Examinatio,* Hobbes tells essentially the same story about the origin of his rectification of the spiral as he had reported in the *Six Lessons;* however, he adds that there was a fourth person present when the problem was first discussed, and this witness was supposedly prepared to give Hobbes credit for the discovery. An October 1656 letter from Claude Mylon to Hobbes is the basis for this version of the story. Mylon wrote, "M. de Carcavi told me that M. Roberval had demonstrated geometrically the equality between a spiral and a parabola, a proposition which on a previous occasion he had demonstrated by the motion of a point. As you know, you gave him the notion of finding it" (*CTH* 1:315). Writing in the third person, Hobbes declares in the *Examinatio* that "Hobbes sought by letter from the fourth person, whom he did not name, whether or not this sheet [claiming Hobbes and Wallis had plagiarized from Roberval] was Roberval's. He responded that he did not know whose it was, but that he was ready to testify that the inspiration *[lucem]* and the method of demonstration had been taken from Hobbes by Roberval" (*Examinatio* 5; *OL* 4:190). Although he has embellished the account somewhat, Hobbes is clearly referring to Mylon's letter here.

58. The problems involved in this case are by no means trivial, and it seems that Roberval spent a good deal of time unraveling them. See Auger 1962, 72–74, for a summary of the problem and Pedersen 1970, which reproduces two manuscripts containing Roberval's solution, for a more detailed account of the matter. Breidert 1979, 418–20, deals with essentially the same issue, and also in the context of the Hobbes-Wallis dispute.

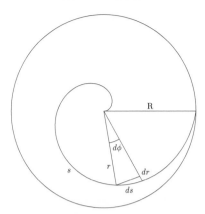

Figure 3.14

sion, as D. T. Whiteside has remarked, is that Wallis's procedure mistakes two curve transformations, one of which is length preserving while the other is area preserving.[59]

When Mersenne reported that the "learned man" Hobbes had erred in solving the problem and proposed a line greater than the true result, he presumably was referring to this very kind of error. Hobbes was apprised of the nature of his mistake by Roberval, who pointed out that the spiral is produced by two motions—one uniform and the other accelerated—while the line whose length Hobbes computed is traced by two uniform motions. Although uniform *circular* motion is included in the definition of the spiral, the point that traces the spiral also passes through the radius of the circular motion; consequently, it lies in the periphery of ever greater circles, which increase with the square of the time. Roberval reasoned that the resulting composition of motions yields the parabola as the arc length of the spiral. In any event, proposition 5 of Wallis's *Arithmetica Infinitorum* originally declared that the length of the Archimedean spiral is one-half the circumference of its limiting circle. It was apparently only after he had seen Roberval's solution to the problem (as reported in Mersenne's *Hydraulica*) that Wallis realized he had made an error, and he hastily attempted to correct it by adding a scholium to the thirteenth proposition that declared that his expression "spiral line" was not intended to

59. See Whiteside's note to a Newtonian treatise on fluxions (Newton 1967–81, 3:308 n. 704).

be the spiral of Archimedes, but rather a collection of arcs of the circles that are circumscribed about the true spiral.[60]

Hobbes jumped at the chance to expose this error, and insisted that "the Aggregate of that infinite Number of infinitely little Arches, is not the Spirall line made by your construction, seeing by your construction the Line you make is manifestly the Spiral of Archimedes; whereas no Number (though infinite) of Arches of Circles (how little soever) is any kind of Spirall at all; and though you call it a Spirall, that is but a patch to cover your fault, and deceiveth no man but yourself" (*SL* 5; *EW* 7:311). In point of fact, when Wallis critiqued Hobbes's comparison of the Archimedean spiral with the parabola (*Elenchus* 125–26) he plainly showed that he had not properly understood the construction of the spiral or its proper rectification. He there faults Hobbes for taking the spiral to be compounded of a uniform and an accelerated motion, and characterizes the Archimedean spiral as a curve generated by the concourse of two uniform motions, writing, "If you had understood the generation of both [the spiral and the parabola], you would instead have concluded that their generation is dissimilar, because the one is generated from the concourse of two motions, of which one is uniform and the other uniformly accelerated, while the other is generated from the concourse of two motions both of which are uniform" (*Elenchus* 126). Hobbes's reply correctly points out that because the point describing the spiral "goes on uniformly in the Semidiameter, it

60. Wallis declares, "It is to be noted that in the preceding propositions concerning the spiral line (some of which I will use in what follows), I have used the term 'spiral line' rather loosely in order to avoid many cumbersome locutions. That is, by the term 'spiral line' (whenever it is compared to the periphery), I wish to be understood (as I indicated in proposition 5) 'the aggregate of all the infinite number of arcs of similar sectors, from which is composed that figure in which the infinite number of sectors of the spiral are inscribed' (which is the meaning I have used in proposition 5 of this work and which Archimedes used in proposition 21 etc. of his treatise on spirals). This spiral taken in the proper sense is always less than this aggregate, most of all near the beginning of the spiral" (1657c, 3:10–11).

The fifth proposition of the *Arithmetica Infinitorum* (which begins the treatment of the spiral and its rectification) was altered in the first volume of Wallis's *Opera Mathematica,* which appeared in 1695. There, he adds a note to the beginning of the proposition, explaining that the spiral he is about to consider is not that of Archimedes (*OM* 1:397). Hobbes's reply to Wallis's original (1656) attempt to mask his error is interesting: "your spiral puts me in mind of what you have underwritten to the diagramme of your prop 5. *The spirall in both figures was to be continued whole to the middle, but by the carelessnesse of the Graver is in one figure* manca, *in the other* intercisa. Truly Sir, you would hardly make your Reader beleeve that a Graver could commit those faults without the help of your own Coppy, nor that it had been in your coppy, if you had known how to describe a spiral line then as now" (Στίγμαι 1; *EW* 7:372).

is impossible it should not pass into greater and greater Circles, proportionally to the Times; and consequently it must have a swifter and swifter Motion circular, to be compounded with the uniform Motion in every Point of the Radius as it turneth about" (*SL* 5; *EW* 7:327).

This episode serves to indicate several important features of the mathematical part of the controversy between Hobbes and Wallis. First, although Hobbes's mathematical work in *De Corpore* is beset with numerous technical and conceptual errors, he pursued problems familiar to the mathematical world of his day and his methods are not completely foreign or unrecognizable. Second, although Wallis is frequently justified in his harsh criticisms, he could clearly overplay his hand in attempting to destroy the entire edifice of Hobbesian mathematics. In particular, Wallis's original misunderstanding of the Archimedean spiral and his clumsy efforts to cover his tracks (particularly in his reply to Roberval's charges of plagiarism) show that Hobbes was not alone in his capacity for mathematical error.[61]

3.5 CHAPTER 20 AND THE FAILED CIRCLE QUADRATURE

Hobbes intended the centerpiece of *De Corpore* to be his quadrature of the circle in the twentieth chapter. This is not to say that he regarded the remainder of his treatise as unworthy of praise or even adulation, but it is clear that he hoped to assert preeminence in the learned world largely on the basis of the solution of the problem of squaring the circle. This intention was not, however, part of the original plan for the work, because the twentieth chapter is a late insertion and not part of the work as originally planned.[62] Indeed, the lack of apparent con-

---

61. This is quite clear in Wallis's letter to Roberval (via Brouncker), which contains an extended but futile argument to show that the passages in the *Elenchus* do not betray a misunderstanding of the case, but instead take Hobbes's spiral to be the same kind as that treated in propositions 5–13 of the *Arithmetica Infinitorum* (Wallis to Brouncker for Roberval, 16/26 October 1656; Bibliothèque Nationale, Paris, nouv. acq. fr. 3252, f. 148–52).

62. It is noteworthy, for example, that the surviving manuscripts of *De Corpore* do not contain the circle quadrature, although they closely follow the printed version in many other respects. Chatsworth MS. A.10 contains drafts of eighteen chapters grouped into two parts. The first part consists of six chapters that correspond very closely to those printed in *De Corpore*. The second part of MS. A.10 consists of twelve chapters, the first six of which closely parallel the structure and wording of the printed *De Corpore*. There is nothing in this manuscript to correspond to the printed twentieth chapter, although the remaining six manuscript chapters deal with a variety of mathematical matters. Moreover, Chatsworth MS. A4 is a fine copy of a nineteenth chapter dealing

nection between Hobbes's twentieth chapter and the surrounding material led Wallis to complain that chapters 18–20 "hang together *like a rope of sand*," because "they have no connexion at all" (*Due Correction* 126). As we have seen, in the *Vindicae Academiarum* Ward challenged Hobbes to make public his geometrical discoveries, and this provoked him to include an attempt at circle quadrature in *De Corpore*. Ward's obvious lack of respect for his mathematical abilities must have enraged the sage of Malmesbury, for he included in chapter 20 several petulant outbursts directed at "Vindex"—his term for the author of the *Vindicae Academiarum*. Thus, in the appendix to the *Vindicae Academiarum* Ward reported that he had "heard that Mr. *Hobbs* hath given out, that he hath found the solution of some Problemes, amounting to no lesse than the Quadrature of the Circle" and promised to "fall in with those that speake loudest in his praise" if Hobbes would publish a successful quadrature ([Ward] 1654, 57).

Confident that he had squared the circle and plainly relishing the chance to humble Ward, Hobbes addressed him contemptuously in the first impression of *De Corpore* with the words:

> But you who are the author of that defamatory book entitled *Vindicae Academiarum,* whatever your concealed name may be, you inept defender [*Vindex*] of the academies which I have never injured; my lying accuser: I absolve you from the payment of those promised praises due to me for solving this problem, judging that to be praised by a man of your manners is considerably less noble than to be censured.[63]

---

with the description and measure of parabolic arcs and spirals; this differs significantly from the published chapter 19 (which deals with angles of incidence and reflection), although some of the material made its way into the published chapters 17 and 18. Furthermore, Sir Charles Cavendish's notes on an early manuscript version of *De Corpore* make no reference to a circle quadrature, although many mathematical topics are discussed. These manuscript versions probably date from the 1640s, and it is certainly possible that Hobbes decided to include a circle quadrature in *De Corpore* at that time, or in the decade or so before the actual appearance of the finished work. Nevertheless, the absence of any quadrature in the manuscript and the fact that Hobbes was drafting and redrafting the twentieth chapter even as *De Corpore* went to press both suggest that it was added hastily. On Sir Charles Cavendish, his connection with Hobbes, and the writing history of *De Corpore,* see Jacquot 1949, 1952a, 1952b, and 1952c.

63. This passage was part of the first impression of *De Corpore,* but the type was reset to tone it down somewhat. The original Latin survives only in Wallis's *Elenchus* at page 121. As first set, the passage read: "Tu vero libelli famosi, qui titulus Vindiciae Academiarum, quicunque celato nomine author es, Academiarum quas nunquam laesi inepte Vindex, Accusator mei mendax, promissas mihi ob soluta haec problemata laudes

Unfortunately for Hobbes, his triumphal outburst was (to say the least) premature. While *De Corpore* was still in press, he showed his attempted quadrature to some friends, who advised him of his error and convinced him to revise the twentieth chapter of his magnum opus. Hobbes reworked the failed quadrature, downgrading it from what had been advertised as the problem "To find a right line equal to the perimeter of a circle," and retitling it "From a false hypothesis, a false quadrature." But he left the construction standing and added a brief demonstration of its falsity. In the published version of the first article of chapter 20, Hobbes admits his error and adds an interesting admission when he comments that he was prepared to declare the problem solved by his initial attempt "except that the insult of malevolent men held me back from this overly hasty conclusion, and made me examine the matter more closely with friends" (Hobbes 1655, 171). It is unclear who these "malevolent men" may have been, and it is also unclear who advised Hobbes of his errors, but it is evident that Hobbes was eager to humble his opponents by making his mathematical powers manifest.[64]

Still convinced that the problem of quadrature could be solved by means no more complex than the simple constructions he had employed earlier, Hobbes attempted another quadrature, but this was ultimately downgraded to a mere approximation. Rather than showing how to find a right line equal to the quadrant of a circle (which, as we have seen, is equivalent to the quadrature of the circle) the second article of chapter 20 was reset to announce only "the finding of a right line as nearly equal to the quadrant of a circle as desired" (Hobbes 1655, 170). This is hardly the magnificent result that Hobbes had in-

---

istas tibi condono: laudari ab homine tuis moribus quam vituperari aliquantò minùs honestem esse judicians." As reset, it reads: "Scio Authorem libelli famosi cui titulus Vindiciae Academiarum, quicunque celato nomine is fuerit, Academiarum quas nunquam laesi ineptum Vindicem, accusatorem meum mendacem, promissas mihi laudes non praestiturum, sed dicturum non deberi. Libenter autem illis careo, vel si debeantur condono laudari ab homine illius moribus quam vituperari aliquanto minùs honestum esse judicans" (Hobbes 1655, 174).

64. It is interesting to note that even as good a friend as du Verdus found Hobbes's decision to publish the "false quadrature from a false hypothesis" rather odd. In a letter to Hobbes of December 1655, he observes that "I might also suspect your motive for putting the section 'A false squaring of the circle from a false hypothesis' in ch. XX, were it not the philosopher's job to give examples of false reasoning just as much as of true reasoning. Besides, it is the action of a generous spirit to admit freely that he has held beliefs which he later learned were false" (*CTH* 1:226). Schumann (1997, xlv n. 65) suggests William Brereton (a friend of Hobbes and student of Pell) as the person who convinced Hobbes of the error in his first attempts at quadrature.

tended, and (as Wallis was happy to point out) it went no further than results that were already well known.[65]

The third quadrature was put forth as an exact result that follows from a method for the arbitrary section of angles. Of course, it too is a failure, and Hobbes admitted that he was made aware of the error only after *De Corpore* was well through the printing process. He was thus forced to add a brief paragraph to the end of chapter 20, declaring:

> Since (after it was written) I have come to think that there are some things that could be objected against this quadrature, it seems better to warn the reader of this than to delay the edition any further. It also seemed proper *[placuit]* to let stand those things that are deservedly directed at Vindex. But the reader should take those things that are said to be found exactly of the dimension of the circle and of angles as instead said problematically. (Hobbes 1655, 181)

Naturally, the net effect of these three failed quadratures is to undermine any claim Hobbes might make for the superiority of his mathematical methodology, at least insofar as such a claim is founded upon his success in the solution of the great classical problems.

Sensitive to the damage that the false quadratures must inevitably do to his reputation, Hobbes attempted to salvage the matter by toning down his harsh words addressed to Ward, and reset his polemic against Vindex to read in the third person:

> I know that the author of that defamatory book entitled *Vindicae Academiarum* (whatever his concealed name may be), that inept defender of the academies which I have never injured, and my lying accuser, will not offer those promised praises to me, but will say they are not to be given. But I freely go without them, or

---

65. Hobbes declares that "geometers can so determine the perimeter of the circle in numbers as to differ from the true value by less than any desired error. Let us then see whether the same can be done by the drawing of right lines" (Hobbes 1655, 172). Wallis responds that "[i]t can most assuredly be done. That is, it is required that there be right lines proportional to these discovered numbers. Or also, that the arithmetical operations by which these numbers are arrived at are carried out in lines. And then whatever can be done in such numbers, the same can be effected in lines, if the same operations are carried out in lines (which is easily done). And thus, if this is what you have determined to do, you will show nothing great, for it is now agreed that this can be done in many ways" (*Elenchus* 119). In fact, Hobbes's construction here adds nothing to what had been shown by Archimedes, Huygens, and Snell. We will see examples of their approximation techniques of the value of π in chapter 6.

if they are due I absolve him from the payment of them, judging that to be praised by a man of his manners is considerably less noble than to be censured. (Hobbes 1655, 174)

He retained, however, another retort to Ward at the end of the third quadrature, in which he announced that "[y]our praises (Vindex) are now owed to me, except that I do not desire them" (Hobbes 1655, 176). But in reality, even Hobbes's more modest estimates of his mathematical success proved vastly overstated and would come back to haunt him.

Making matters much worse for Hobbes was the fact that Wallis had procured an early unbound copy of the first impression *De Corpore,* and was able to reconstruct the unflattering history of Hobbes's attempts to square the circle. This gave Wallis the opportunity to recapitulate Hobbes's earlier efforts and cast an extremely unwelcome light upon the state of his mathematical knowledge as well as his judgment. Much of Wallis's program in the *Elenchus* is devoted to precisely this task, which gave him the chance to taunt Hobbes with the observation that "after privately and publicly, here and abroad, you had boasted that you would bring forth other unknown marvels beyond the quadrature of the circle (which was such a small thing with you); after first, second, and third revisions, you print it a first, second, and third time: in the end you dexterously admit that you have by no means brought forth what you had promised; and then you ask that these things be taken 'as said problematically'" (*Elenchus* 2–3).

Hobbes's disappointment at his failed quadratures was made all the more bitter by the fact that Wallis so publicly paraded the sorry history of chapter 20 of *De Corpore.* Having been provoked by Ward into printing a hastily assembled quadrature and then replacing it with others of no better quality, Hobbes was forced to undergo the humiliation of seeing his efforts dissected and the many layers of error peeled away one by one.[66] So public a failure of his mathematical project would

---

66. Robertson aptly summarized the episode as follows:

One of some copies carelessly issued in the first unamended form having fallen into his hands, [Wallis] was not the man to scorn such a weapon of ridicule, and from it, with his unbound copy, he was able to spell out the whole history of Hobbes's doings from the time of Vindex's challenge. This accordingly he laid bare, showing how, shaken from a brief illusion of triumph by friends concerned for his reputation, again and again during the year had Hobbes persisted in printing loose assumptions as strict truths, and rough and contradictory approximations as exact solutions of an impossible problem; till at last, rather than delay his book longer, he was fain to be content with his lame and impotent conclu-

have discouraged a man of fainter heart than Hobbes. But instead of admitting defeat, Hobbes resolved to vindicate his mathematics and overthrow the entire mathematical corpus of Wallis; the result, as we have already outlined in chapter 1, was that from 1655 until the very end of his life, Hobbes was engaged in his war with Wallis.

sions, made grotesque by the side of the jubilant bursts which he had not the heart to suppress, because he had once had the joy of giving vent to them. (Robertson [1886] 1910, 173–74).

CHAPTER FOUR

Disputed Foundations

Hobbes vs. Wallis on the Philosophy
of Mathematics

It is well for you that they who have the disposing of the
professors places take not upon them to be Judges of Geometry. For if they did, seeing you confesse you have read these
Doctrines in your School, you had been in danger of being
put out of your place.

—Hobbes, Στίγμαι

Wallis's uncompromising critique of Hobbes was not confined to
a purely technical exposition of flaws in his attempted quadratures. He objected to the entire Hobbesian program for mathematics,
and what little he did not denounce as ill founded he dismissed as un-
original, if not plagiarized. Hobbes's efforts in defense of the mathematics of *De Corpore* led him to oppose nearly every point raised by
Wallis and, following the seventeenth-century conventions of polemical literature, the dispute between them became an exchange of point-
by-point rebuttals of the opponent's latest piece, intermixed with the
occasional new charge of folly or ineptitude. Aside from the dubious
virtues of sheer pedantry, it makes no sense to go through all of these
charges and countercharges. However, the two men also debated significant philosophical problems in mathematics, and the object of this
chapter is to outline the most salient of such debates. Again, the disputants' practice of trading rebuttals over previously covered points
means that it is frequently not worthwhile to follow an exchange on a
particular point in complete detail over more than two decades, so I
will refrain from giving an exact chronology of the charges and countercharges. The mathematical themes I find most worth exploring include the role of physical concepts in the philosophy of mathematics,
the nature of ratios, the problem of the "angle of contact" between
curve and tangent, and the admissibility of infinitesimal methods. This

chapter is concerned with outlining these contested points and examining their role in the Hobbes-Wallis controversy. My account requires some excursions into mathematical and historical material that is not often covered, but an adequate understanding of the contested issues requires that we take some relative arcana into consideration. I hope ultimately to show that Hobbes and Wallis adhered to widely divergent traditions in mathematics and that they held completely different conceptions of what mathematics is and how it should be developed.

4.1 MATHEMATICS, PHYSICS, AND THE NATURE OF BODY

Wallis condemned Hobbes's attempt to place mathematics on a thoroughly materialistic foundation, and this aspect of their dispute evolved into a lengthy exchange over the relationship between mathematics and natural philosophy. Wallis was by no means unique in objecting to Hobbes's materialism. As we saw in chapter 2, critiques of materialism were common fare among the many anti-Hobbes polemics in seventeenth-century England, but these tended to focus on the theological and moral dangers alleged to follow from the denial of immaterial substances. Hobbes's critics argued that the rejection of incorporeal substances made it impossible to accept the Christian account of an afterlife, that it reduced God to the status of a material being, and implied that humans have a merely animal (or even mechanical) nature, devoid of moral value or obligations. Ward's *Philosophicall Essay* of 1652 attacked Hobbes's materialism because it threatened the doctrine of immortality, but this was not the end of Ward's campaign against Hobbes; he returns again and again in his 1656 *Exercitatio* to the theme that the denial of immaterial substances is an unsupported dogma inimical to theology and morality.[1] Wallis certainly shared the widespread repugnance over Hobbes's materialism, but his published dispute with Hobbes concentrated on the mathematical issues connected to the Hobbesian conception of body and the foundational role of material concepts in his philosophy of mathematics.

---

1. Thus, Ward insists, "It should first have been shown that there is nothing else contained in the universe except bodies and their motions, and that on that account God, angels, and the eternal souls of men are to be done away with. For this is what he asserts everywhere in *Leviathan,* that there is no spirit, but that it is an absurd connection of words 'spiritual substance' which philosophers are accustomed to have in their ears" (*Exercitatio* 21).

Wallis opposed Hobbes's materialistic foundation for mathematics on two related grounds: first that it fails to accommodate the abstract, immaterial character of pure mathematics, and second that it employs physical principles that are at least doubtful, if not actually false. Although these two objections are clearly related, it is worthwhile to treat them separately because doing so can help illuminate Hobbes's conception of physics and its relationship to mathematics.

The charge that Hobbes degrades mathematics by founding it on the principles of body is an immediate consequence of Wallis's adherence to a traditional distinction between pure and applied mathematics.[2] According to the tradition, pure mathematics is distinguished from its applied counterpart by the fact that it deals with abstract objects not found in nature. Instead, these abstractions inhabit the (presumably immaterial) realm of the intellect, and the theorems of mathematics are thus freed from dependence upon features of the material world. As we saw in chapter 3, Hobbes largely ignored this classical division by developing an account of quantity that makes physical body the sole subject of quantity.

The traditional account of the relationship between mathematical and physical principles derives much of its plausibility from the idea that (purely) mathematical facts do not—indeed must not—depend upon the structure or contents of the material world. Part of what gives mathematics its particularly exalted epistemological status is the intuition that mathematical theorems would remain true even if the world were vastly different than it actually is. We are all inclined to accept that even if there were no bodies at all, or even if the nature of matter were entirely different, 17 would still be a prime number and the Pythagorean theorem would still hold. The traditional distinction between pure and applied mathematics accommodates this idea by locat-

---

2. Although Wallis accepts a traditional division between pure and applied mathematics, he departs from the tradition by claiming that arithmetic is a more abstract and pure science than geometry: "Therefore we say that there are only two pure mathematical disciplines, namely geometry and arithmetic. The first of these deals with discrete quantity or number, the other with continuous quantity or magnitude. But of these indeed one is more and the other less pure: for the subject of arithmetic is so to speak purer and more abstract than the subject of geometry, and on that account it contains more universal speculations that are equally applicable to geometrical matters and to others" (*MU* 1; *OM* 1:18). Nevertheless, he is firmly committed to the view that abstract mathematics deals with objects not found in nature. Moreover, he insists that the truths of mathematics cannot be concerned with the nature of body. Indeed, it is apparently geometry's connection with the nature of space (and hence body) that renders it less pure and abstract in Wallis's estimation.

ing mathematical truth in the relations among abstract objects whose properties are independent of the structure or contents of the actual world. An application of mathematics requires, on this view, additional nonmathematical hypotheses that link the necessary truths of mathematics to contingent features of the physical world.

In his *Elenchus*, Wallis repeatedly castigates Hobbes for introducing physical (rather than mathematical) principles into his account of quantity and thereby contaminating the foundations of mathematics. He asks, for example, "[W]hat need is there for the concepts of body or motion [in Hobbes's definitions of point and line], since the concept of a line can be understood without them?" (*Elenchus* 6). He explains that such physical notions are "plainly accidental, nor do they pertain to their essences, so it is strange to find motion in the definition of a point or line" (*Elenchus* 7). He objects for similar reasons to Hobbes's fundamental dictum that "a body always retains one and the same magnitude, when at rest or moved, but when moved it does not retain the same place" (*DCo* 2.8.5; *OL* 1:93). Wallis holds that this principle is "plainly physical, nor does it regard the present matter, nor is it in any way connected with it, so to what purpose was any mention to be made of it in this mathematical definition of yours?" (*Elenchus* 10). Hobbes's definition of equal ratios in terms of bodies producing equal effects in equal times (*DCo* 2.13.6; *OL* 1:132) meets the same objection. Wallis demands what relevance the definition has to the nature of mathematics and asks "to what end, in order that this notion be understood, need there be a consideration of time or weight, or of any other quantity?" (*Elenchus* 19). In these and other cases Wallis charges that Hobbes takes simple and clear concepts from pure mathematics and renders them complex and obscure by defining them in terms inappropriately borrowed from natural philosophy.

Hobbes's response to such criticisms is to insist that it is only by the introduction of physical concepts that mathematics can be grounded on a firm foundation. Indeed, he holds that "[f]or men that pretend no less to naturall Phylosophy, then to Geometry, to find fault with bringing Motion and Time into a Definition, when there is no effect in nature, which is not produced in Time by Motion, is a shame" (*SL* 3; *EW* 7:242). Thus, where Wallis finds the recourse to time, motion, or body to be the intrusion of nonmathematical concepts into the foundations of mathematics, Hobbes retorts that no genuine demonstration is possible unless it proceeds from the causal principles of body and motion. This aspect of his philosophy of mathematics is summed up in Hobbes's 1666 essay *De Principiis et Ratiocinatione Geometrarum:*

> "But," you will ask, "what need is there for demonstrations of purely geometric theorems to appeal to motion?" I respond: First, all demonstrations are flawed, unless they are scientific, and unless they proceed from causes, they are not scientific. Second, demonstrations are flawed unless their conclusions are demonstrated by construction, that is, by description of figures, that is, by the drawing of lines. For every drawing of a line is motion: and so every demonstration is flawed, whose first principles are not contained in the definitions of motions by which figures are described. (*PRG* 12; *OL* 4:421)

This part of the controversy thus reduces to a conflict over the proper demarcation between mathematics and natural philosophy. As such, it is not readily amenable to resolution, because such fundamental disagreements can rarely be overcome or even profitably debated. Hobbes continued to insist that the only proper foundation for mathematics must be sought in the nature of body, while Wallis viewed this program as an abandonment of the essential features of mathematics. This aspect of the dispute can be illustrated in the exchange drawn from Wallis's *Due Correction,* where he writes: "You adde, *That there is no man (beside such Egregious Geometricians as we are) that inquires the Equality of two bodies, but by measure: And, as for liquid bodies; &c. by putting them one after another into the same vessel, that is to say, into the same place; And, as for hard bodies, they inquire their Equality by weight:* To which I shall reply nothing at all; because you speak therein so like a Geometrician" (*Due Correction* 55–56). Wallis obviously attempts irony in his remark that Hobbes "speaks so like a Geometrician," and his refusal to reply shows that he takes this treatment of measure to be far removed from anything in a genuine science of geometry.

Wallis further faulted Hobbes's materialistic philosophy of mathematics for its dependence upon controversial physical principles, and in particular upon the principle that a body always retains the same quantity. Without declaring his own opinion in the matter, Wallis observed that this tenet had been denied by those who hold that the same body can have lesser or greater quantity as a result of the contrary processes of rarefaction and condensation. In defining equal bodies as those that can occupy the same place (*DCo* 2.8.13; *OL* 1:99–100),

Hobbes identifies the magnitude or quantity of a body with the place it can occupy. But Wallis objects to the definition on the grounds that it "proceeds from the supposition of things that are not true, or that if true must be proved and not supposed" (*Elenchus* 9). In particular, he claims that

> [w]hen you define equality or inequality of one body to another, not by the place it occupies, but by that which it can occupy; I ask what happens if the same body could now occupy a greater and now a lesser space, as for example by rarefaction or condensation? Indeed, if it is conceded that this could be, then the same body could be with respect to another simultaneously equal, greater, and less (which is so absurd that even you must see it to be so); or indeed it could be that (whatever place it may now actually occupy) whenever it then moves a body can occupy the same, as well as a greater or lesser space (successively, just as is necessary in order that your definition be understood). (*Elenchus* 9)

The point at issue here is not whether rarefaction and condensation are processes that actually take place in nature, but whether it is conceivable that there could be such processes. Wallis finds it absurd that Hobbes should adopt an account of mathematics that rules out rarefaction and condensation as necessarily false doctrines, since this makes the science of mathematics (a realm of necessary truths independent of physical matters) contingent upon the status of disputed physical principles.

The theory of rarefaction and condensation has its roots in book 4 of Aristotle's *Physics,* but it was not confined to classical Greek sources. Several seventeenth-century natural philosophers endorsed the theory, among them Sir Kenelm Digby. Digby's natural philosophy was an idiosyncratic attempt to combine the teachings of Aristotle with the mechanistic "new philosophy" of Galileo, Descartes, and other notable modern authors. His *Two Treatises* of 1644 examined the nature of body and the soul, contending for a plenist mechanism in which the human soul is a genuine substance but not a body.[3] Digby proposed rarefaction and condensation as the explanation of various physical phenomena in his *Treatise on the Nature of Body,* and it is interesting to consider his presentation of the doctrine in conjunction with Hobbes's conception of body. According to Digby, the nature of quan-

---

3. For more on Digby's natural philosophy, see Lasswitz 1890, 2:188–207, and Henry 1982.

tity "is nothing else but divisibility; and . . . a thing is bigge, by having a capacity to be divided, or (which is the same) to have partes made of it" (Digby 1644, 9). However, there are obvious cases of bodies that have the same volume but different weights, and the difficulty is to account for this fact. As Digby puts it, "Our measures tell us that their quantities are equall; and reason assureth us, there can not be two bodies in one and the same place; therefore when we see that a pinte of one thing outweigheth a pinte of an other that is thinner, we must conclude that there is more body compacted together in the heavy thing then in the light. . . . But how this comprehension of more body in equal roome is effected, doth not a little trouble Philosophers" (Digby 1644, 16). One proposed solution to this problem is to accept the concept of a vacuum, and explain density as the proportion of matter to empty space in a given body. But Digby accepts Aristotle's argument against the possibility of a vacuum in book 4 of the *Physics,* and regards it as "perfectly demonstrated, that no vacuity is possible in nature; neither great nor little: and consequently, the whole machine raysed upon that supposition, must be ruinous" (Digby 1644, 21).

Digby's solution to the difficulty requires a distinction between quantity and the substance that has quantity, and then defines dense bodies as those having relatively less quantity contained within their substances, while rare bodies contain more quantity in the same amount of substance:

Thus then; remembering how we determined that Quantity is Divisibility: it followeth, that if besides Quantity there be a substance or thing which is divisible; that thing, if it be condistinguished from its Quantity or Divisibility, must of it selfe be indivisible: or (to speake more properly) it must be, not divisible. Putt then such substance to be capable of the Quantity of the whole world or universe: and consequently, you putt it of it selfe indifferent to all, and to any part of Quantity: for in it, by reason of the negation of Divisibility, there is no variety of partes, whereof one should be the subject of one part of Quantity, or another of another; or that one should be a capacity of more, another of lesse.

This then being so, wee have the ground of more or less proportion between substance and quantity: for if the whole quantity of the universe be putt into it, the proportion of Quantity to the capacity of that substance, will bee greater then if but halfe that quantity were imbibed in the same substance. And because proportion changeth on both sides by the single change of onely one side: it followeth that in the latter, the proportion of that substance to its Quantity, is greater; and that in the former, it is lesse; howbeit the substance itselfe be indivisible. (Digby 1644, 22)

If the universe had a perfectly homogeneous distribution of quantity throughout its whole extent, there would be no distinction between rare and dense bodies and the proportion of substance to quantity would be everywhere uniform. However, we observe that natural bodies (air, water, mercury, wood, gold, etc.) differ in their relative density and rarity, and this is accounted for by the fact that substance and quantity are not always in the same proportion throughout the natural world. In its application to the explanation of natural phenomena, this doctrine accounts for the greater divisibility of rare bodies, since they have more quantity (i.e., divisibility) in the same amount of substance as denser bodies. Similarly, the resistance of rare bodies is less than that of dense bodies, by virtue of their greater divisibility. A further virtue of this doctrine (at least in Digby's estimation) is its perfect concurrence with Aristotle's doctrine: "hee telleth us, that that body is rare whose quantity is more, and its substance lesse; that, contrariwise dense, where the substance is more and the quantity lesse" (Digby 1644, 23). With such a distinction between rare and dense bodies, rarefaction can then be defined as a process that compresses more quantity into a body and condensation as one that removes quantity from it.

Hobbes was familiar with Digby's philosophy but regarded it as utter nonsense. In particular, Digby's regard for Aristotle (as well, presumably, as his Catholicism) made him an adherent of the despised "Schools" that figure so prominently in the "Kingdome of Darknesse" denounced in part 4 of *Leviathan*. Digby is clearly one of the targets of Hobbes's denunciation of rarefaction and condensation in *Leviathan*,[4] and although the two had maintained a cordial correspondence

---

4. In summarizing and ridiculing the doctrine of rarefaction and condensation, Hobbes uses language that could have come straight out of Digby:

If we would know why the same Body seems greater (without adding to it) one time, than another they say, when it seems lesse, it is *Condensed* when greater, *Rarefied*. What is that *Condensed*, and *Rarefied*? Condensed, is when there is in the very same Matter, lesse Quantity than before and Rarefied, when more. As if there could be Matter, that had not some determined Quantity; when Quantity is nothing else but the Determination of Matter; that is to say of Body, by which we say one Body is greater, or lesser than another, by this, or thus much. Or as

in the 1630s, Digby later expressed admiration for Ward's and Wallis's campaign against Hobbes.[5]

In any event, Wallis's use of the doctrine of rarefaction and condensation as a weapon against the Hobbesian philosophy of mathematics does not require that such processes be part of the natural order, but merely that they be possible or conceivable. By ruling out rarefaction and condensation as possible explanations of natural phenomena, Hobbes in effect makes his whole philosophy of mathematics depend upon the resolution of a natural-philosophical dispute. Wallis gleefully pointed out the extent to which Hobbes's account of mathematics was enmeshed in controversial physical principles, and he took this as further evidence that Hobbes had mistakenly conflated mathematics with physics:

> But, whether that opinion of Rarefaction and Condensation be true or not: yet since you cannot deny, but that it is at least a considerable controversy, and, by men as wise, and as good Philosophers as M. Hobs, maintained against you: yea and a Controversy not belonging to Mathematicks, but Physicks, or Naturall Philosophy, and there to be determined; it was not wisdome to hang the whole weight of Mathematicks upon so slender a thread, as the decision of that controversy in Naturall Philosophy, which whether way it be determined, is wholly impertinent to a Mathematical Definition. (*Due Correction* 56–57)

Even worse, because Hobbes also denies the possibility of a vacuum, he is apparently left without the resources to explain many natural phenomena (such as evaporation, the freezing of water, etc.) that others had accounted for either by rarefaction or the vacuum.[6] This is

---

if a Body were made without any Quantity at all, and that afterwards more, or lesse were put into it, according as it is intended the body should be more or lesse Dense. (*L* 4.46, 375; *EW* 3:678–79)

5. Four gracious letters from Digby to Hobbes date from 1636 to 1637; Digby's later hostility toward Hobbes is evident in a letter to Wallis from August of 1657, when he writes: "I must not take leave of you, till I have spoken a word or two of your worthy Colleague Doctor Ward. It is some time since I have heard of his booke against M. Hobbes. . . . It is a worthy Triumvirate that you two and Doctor Wilkins do exercise in literature and all that is worthy" (Wallis 1658, 11). Ward makes numerous favorable references to Digby in his *Exercitatio* (see *Exercitatio* 75, 93, 99, 161, and 200), and it is clear that, however idiosyncratic his theological or scientific doctrines, he was regarded favorably by the Oxford scientific establishment.

6. As Wallis observes, "[H]ow to salve these *Phaenomena,* (with many others of the like kind) without either *Vacuum,* which you deny, or *Condensation,* which you laugh

doubly problematic for Hobbes because, in addition to grounding his philosophy of mathematics in disputable (indeed, disputed) physical principles, he appears also to have forsworn the ready means of constructing a mechanistic account of the workings of nature.

Hobbes's defense against these charges was to insist that the disputes over rarefaction and condensation were not, properly speaking, physical debates, but rather verbal quarrels with no relevance to the foundations of mathematics. The empty terms *rarefaction* and *condensation* were used by "Schoole Divines" who did not understand even the first elements of natural philosophy or mathematics, and their speculations were to be left entirely out of any account of the true grounds of mathematics and physics. It is in this sense that Hobbes can claim that

> nature abhorres even empty words, such as are (in the meaning you assign them), *Rarefying* and *Condensing*. And you would be as well understood if you should say (coining words by your own power) that the same Body might take up sometimes a greater, sometimes a lesser place, by *Wallifaction* and *Wardensation,* as by *Rarefaction* and *Condensation.* (*SL* 2; *EW* 7:225)

Schaffer portrays this part of the controversy as a political struggle over the authority and autonomy of mathematicians (or, more particularly, professors of mathematics) to define the fundamental terms of their discipline. He reads Hobbes as holding "that geometers [are] deprived of the right to define their own terms themselves" (1988, 286–87). The Savilian professors' dispute with Hobbes was therefore "a contest over the propriety of the separate authority in civil society of teachers, mathematicians, or divines," in which "Ward and Wallis tried to resist Hobbes's denial of their right to ground geometrical terms" (1988, 288).

Central to this account is Schaffer's interpretation of Hobbes's declaration in the second of the *Six Lessons* that "[i]t is not the work of a Geometrician, as a Geometrician, to Define what is Equality, or Proportion, or any other word he useth, though it be the work of the same man, as a man" (*SL* 2; *EW* 7:222). Schaffer takes this as imposing restrictions upon geometers, which Wallis allegedly opposed by "insist[ing] that geometers properly and necessarily had the right to lay

---

at; (one of which others use to assigne) because you find it too hard a task for you to undertake, (as well you may,) you leave to a *melius inquirendum*" (*Due Correction* 56).

down definitions and then use the privileged principles of geometric reason upon them" (Schaffer 1988, 287).

In reading the dispute this way Schaffer misinterprets the terms of the debate in order to reduce it to an entirely political struggle over the authority to define terms. Hobbes does *not* hold that the geometer (or anyone else) is forbidden to introduce definitions. Rather, his point is that in propounding definitions one is not doing geometry, but engaging in an activity common to all men: the stipulation of how words are to be understood. When Wallis refers to the principle that a body always retains the same quantity as part of a "mathematical definition," Hobbes objects to the use of the term:

> It seems by this, that all this while you think it is a piece of the Geometry of *Euclide,* no less to make the Definitions he useth, then to infer from them the Theorems he demonstrateth. Which is not true. For he that telleth you in what sense you are to take the Appellations of those things which he nameth in his discourse, teacheth you but his Language, that afterwards he may teach you his Art. But teaching of Language is not Mathematick, nor Logick, nor Physick, nor any other Science; and therefore to call a Definition (as you do) Mathematicall, or Physicall, is a mark of Ignorance (in a Professor) unexcusable. (*SL* 2; *EW* 7:225)

Nothing here forbids the geometer from introducing whatever definitions may please him, and this is perfectly in keeping with Hobbes's doctrine that definitions (rather than explicating the essence of a thing) are "truths constituted arbitrarily by the inventors of speech" (*DCo* 1.3.9; *EW* 1:37). Hobbes therefore objects to the concept of a specifically mathematical definition, not because the mathematician lacks the standing to introduce a definition, but because there is nothing essentially mathematical about the act of defining terms.

The degree of arbitrariness inherent in Hobbes's account of definitions does not, however, mean that all definitions are created equal. Some definitions are better than others, and Hobbes insists that the best definitions are "those which declare the cause or generation of that Subject, whereof the proper passions are to be demonstrated" (*SL* 2; *EW* 7:212). Terms such as *rarefaction* lack any proper or coherent definition because (at least according to Hobbes) they could be defined only by joining together words of contradictory signification. Just as he takes the term *immaterial substance* to be a contradiction in terms, Hobbes interprets *rarefied body* as equivalent to "quantity greater in quantity than itself"—a manifest absurdity. According to Hobbes, empty terms of this sort are part of the obfuscatory apparatus of school divinity, and there is no question that he saw his account of quantity as inimical to the designs of Oxford's Savilian professors.[7] Nevertheless, Schaffer errs in depicting this part of the dispute as centering on the authority of geometers to introduce definitions.

The conceptual gulf dividing Hobbes's and Wallis's philosophies of mathematics should be clear from the foregoing discussion. Their differences in the philosophy of mathematics were not confined to questions of the relationship between mathematics and natural philosophy, but I think the other contested points can in large measure be traced back to a fundamental difference over the extent to which mathematics should depend upon principles of body and motion. We can see this more clearly by investigating the two thinkers' polemics over the proper account of geometric ratios, to which I now turn.

4.2 THE THEORY OF RATIOS AND ITS INTERPRETATION

The second fundamental issue raised and debated by Hobbes and Wallis was the thorny question of the nature of ratios. The differences on this point were so great that the interpretation of the theory of (geometrical) ratios occupied a central place in their exchange of polemics for more than twenty years. Of course, Hobbes and Wallis were hardly the only two thinkers to contest this issue. Wallis himself authored a vituperative attack on the *Dialogus de proportionibus* of Marcus Meibom (Wallis 1657c, 3).[8] In fact, the philosophico-mathematical literature of the period contains many disquisitions on the subject.[9] In chap-

---

7. In Στίγμαι Hobbes is quite explicit on this point: "But I beleeve you will by degrees become satisfied that they who say the same Numerical Body may be sometimes greater, sometimes lesse, speak absurdly, and that *Condensation* and *Rarefaction* here, and *definitive* and *circumspective,* and some other of your distinctions elsewhere are but snares, such as School-Divines have invented . . . to entangle shallow wits" (Στίγμαι 1; *EW* 7:384–85).

8. In the dedication to his refutation, Wallis actually notes that he was delayed in publishing by the necessity of answering Hobbes: "[I]n the mean time it happened that Hobbes needed to be punished a second time, now for the errors and bad language he foolishly spewed forth *[effutivit]* against me in an English piece prompted by my *Elenchus,* in which I refuted his geometry. This castigation of Hobbes kept both me and the printer occupied for some time" (*OM* 1:231). Meibom's theory and its connection to these debates is outlined in Sylla 1984, 30–35.

9. Barrow's *Lectiones Mathematicae* are probably among the most important of these. Barrow's principal task in these lectures was to defend the Euclidean theory of

ter 3 I outlined some of the contested issues when I examined Hobbes's account of ratios, and we are now in a position to examine his differences with Wallis. The points at issue in the debate are complex in themselves, and this complexity is augmented by numerous terminological confusions. Nevertheless, the most important aspects of the controversy can be made reasonably clear, and in the process we can gain some understanding of what was at stake, both mathematically and philosophically, in the dispute.

The place to begin unraveling this part of the controversy is with Hobbes's intention of founding the doctrine of ratios on the principles of body and motion. We have already seen that Wallis rejects such an attempt as an unwarranted intrusion of physical principles into mathematics. But Hobbes's program did not simply lead him to place mathematics on a "physicalistic" foundation, for it also encouraged him to propound a strong thesis on the homogeneity of certain magnitudes that had traditionally been deemed heterogeneous. In particular, Hobbes was persuaded by his analysis of motion and the ratios of motions that time is homogeneous to a line. At work here is a principle I mentioned in chapter 3, namely that quantities that do not, at first sight, appear to be captured in the definition of quantity in terms of the three dimensions of body can, nevertheless, be represented or "exposed" by lines. Velocities, times, weights, etc. are all traditionally taken as (one-dimensional) quantities, and Hobbes holds that they can be treated as the quantities of lines by which they are represented (*DCo* 2.12.4–7; *OL* 1:125–26). On the basis of this understanding of quantities and their representation, Hobbes concludes that lines and times are homogeneous quantities because, in the analysis of uniform motion, time is represented by a line and this line is compared to a distance traveled by a moving body:

> Seeing that it has been shown that in uniform motion the lengths covered *[percursum]* are as the parallelograms made by multiplication of the impetus by the time, it will also be, by permutation, as time to length so time to length; and, universally, all the properties and permutations of proportions enumerated in chapter thirteen hold here as well. (*DCo* 2.16.2; *OL* 1:186)

This doctrine permits the "permutation" of terms in a proportion to form two new ratios; thus, if the proportion $T_1:T_2 :: S_1:S_2$ links the

---

ratios, which he does by summarizing and critiquing nearly every alternative account. Another contributor to these disputes was Jacobus Fontialis in his *De idea mirabilis matheseos de Ente* [1660?], in Fontialis 1740, 437–512.

ratio of times elapsed to spaces covered in uniform motion, it can be permuted to form the new proportion $T_1:S_1 :: T_2:S_2$, in which the times and distances are compared directly with each other.

Wallis attacked this portion of Hobbesian doctrine as an intolerable confusion. Where Hobbes had taken the results from the mathematical analysis of motion to show that time and longitude are homogeneous, Wallis observed that the permutation of proportions is admissible only if all the quantities in the proportion are pairwise homogeneous. Thus, because time and length are not capable of direct comparison with one another, they are heterogeneous magnitudes and the terms in the proportion cannot be permuted.[10] Because of his commitment to the numerical conception of ratios Wallis was happy to allow that *ratios* are themselves quantities and, indeed, quantities homogeneous to one another; but he denies that the terms in any two given ratios need be homogeneous. He insists upon precisely this point in chapter 25 of the *Mathesis Universalis,* with the declaration

> All ratios of whatever quantities to one another are homogeneous among themselves. As for example lines and weights are clearly things heterogeneous, nor do they have a ratio to each other; but the ratio of a line to a line and of a weight to a weight are plainly homogeneous. For example, the ratio of a two-foot line to a one-foot line and that of a two-pound weight to a one-pound weight are homogeneous, and even the same ratio, namely double: for if the division is effected the quotient 2 will arise in both cases. (*MU* 25; *OM* 1:36)

It is with this conception of ratios and homogeneity in the background that Wallis can demand of Hobbes: "Tell us then, if you can, what ratio does one hour of time have to two ells of length? There is indeed some ratio of the adjoined numbers, one and two; but there is none of time to length, namely of an hour to an ell [an old English unit of measure, equal to approximately forty-five inches]" (*Elenchus* 41–42). For his part, Hobbes was intent upon maintaining a distinction between different species of magnitudes, and he did not accept that any two magnitudes were necessarily homogeneous. Nevertheless, because he saw the quantity of time as expressible by a line, he was prepared to opt for a general thesis of the homogeneity of time and line. In his

---

10. As Wallis puts it in the *Elenchus,* "And so there can be a ratio of time to time, and also of length to length (which ratios can either be the same or different), because times are homogeneous to times and lengths proportional to lengths, but there can be no ratio of length to time because these are heterogeneous" (*Elenchus* 41).

Six Lessons he grants that "Time and Line are of divers natures" (*SL* 4; *EW* 7:273) but insists that their quantities can be compared by straight lines, and in this sense they may be taken as homogeneous. To Wallis's demand that he declare what ratio an hour bears to an ell, Hobbes replies that it "is the same Proportion that *two Hours* have to *two Ells,*" and by this retort he imagines to have shown Wallis that "your Question was not so subtile as you thought it" (*SL* 4; *EW* 7:273).

The driving force behind this unusual doctrine is Hobbes's insistence that all magnitudes are ultimately derivable from the three dimensions of body. In Hobbes's scheme, there must necessarily be heterogeneity of one-, two-, or three-dimensional magnitudes because lines, surfaces, and solids (the three fundamental dimensions of body) cannot be compared with one another as to quantity. On the other hand, Hobbes reduces such one-dimensional magnitudes as time, mass, density, and temperature to the case of lines, since it is by lines that these magnitudes are exposed and measured. Under criticism from Wallis, Hobbes rephrased this into a declaration of the homogeneity of the *quantities* of times and lines, rather than the homogeneity of times and lines themselves; but he retained the leading idea that all quantities must be traced back to the three dimensions of body. In the *Examinatio,* after declaring it "absurd to say that a stone is quantity, and no less absurd to say that time is quantity" (*Examinatio* 3; *OL* 4:118), Hobbes nevertheless insists that the weight of a stone or the length of time can be measured, and thus have quantity attributed to them. The result is that

> "How much" can be said of natural bodies as well as time, although neither of them can be called quantity abstractly. For all quantity, strictly speaking, is either longitude, or surface, or solid, or as one is accustomed to say, mathematical body. But time, motion, and force, and other things of which it can be asked how much they are, all have quantities, by which it is determined how great they are. And these things surely have one or another of those three dimensions, for the things themselves are measured by them. (*Examinatio* 3; *OL* 4:119)

The result of this doctrine is that "although it is absurd to say that a line is equal to a time, nevertheless it is not absurd to say that a quantity of time is equal to the quantity of a line" (*Examinatio* 3; *OL* 4:119).

Wallis was not alone in objecting to this feature of Hobbes's conception of ratios. Barrow poured scorn on the doctrine, complaining that it was grounded in a confusion of two fundamentally different senses of the word *measure*. As Barrow points out, a quantity can be said to be the measure of another (homogeneous) quantity when it is a part of the measured quantity, as when distances are measured in meters or time in minutes. It is in precisely this sense that Euclid speaks of measure when he declares that "a magnitude is a **part** of a magnitude, the less of the greater, when it measures the greater" (*Elements* 5, def. 1). Hobbes adopts essentially this usage in the *Examinatio* when he defines measure as "one magnitude of another, when it or multiples of it, applied to the other, coincide with it" (*Examinatio* 1; *OL* 4:18). On the other hand, a measure can be, in Barrow's words, "anything that may either conveniently represent or in some way signify another" (*LM* 16, 263), as when time is represented by a line. As Barrow observes, time can be measured in the first sense by minutes or hours (which are equal parts of time itself) or in the second sense by numbers, lengths, or any other method of keeping track of the passage of time. He concludes that if Hobbes "had seen clearly or weighed this ambiguity of the word *measure,* or if he had remained more constant to his own definition, he would not, I think, have so conflated the quantities of all things and been moved by such an argument to pronounce them homogeneous to one another" (*LM* 16, 264).

Barrow is very much in the right here, and his analysis of the problem shows some of the difficulties encountered in Hobbes's insistence that all quantities must ultimately be derived from the three fundamental dimensions of body. It is manifestly absurd for Hobbes to maintain the general thesis of the homogeneity of all one-dimensional magnitudes, since the "permutation" of proportions that he desires can be carried out only after the specification of units to link the otherwise heterogeneous magnitudes. For example, there is no way to connect length and temperature without first specifying a unit (degree) for the measure of temperature, another unit (millimeters, say) for the measure of distance, and a device (such as a mercury thermometer) that is so calibrated that one degree is represented by a specific distance (suppose one degree is represented by one millimeter on the temperature scale). Once such a specification of units has been achieved, it is a trivial matter to establish a correspondence between temperature and length, but the correspondence remains entirely dependent upon the choice of units. In order for the different magnitudes to be truly homogeneous, however, they would have to be capable of direct comparison with one another, and it is on this point that Hobbes's argument reduces either to absurdity or triviality. When he insists that "it is absurd to say that a line is equal to a time, nevertheless it is not absurd to say that a quantity of time is equal to the quantity of a line" (*Examinatio* 3; *OL* 4:119), Hobbes essentially grants his opponents' case. Because time and line cannot be compared directly, they are not homogeneous; but from the fact that their respective quantities can be compared, it follows only that when units have been assigned to one-dimensional magnitudes, these units are capable of comparison with one another.

The general issue of homogeneity of magnitudes, and the relationship between questions of homogeneity and proper mathematical procedures, will occupy us again when we consider the problem of the angle of contact and Hobbes's rejection of analytic methods. For present purposes, however, we can leave the issue aside and return to a more specific consideration of the controversy over the general doctrine of ratios and proportionality.

Hobbes's attempt to define ratios in terms of bodies was intended, in part, to resolve serious questions about the proper interpretation of the theory of ratios. His program of taking ratios as expounded by bodies treats ratios as essentially geometrical, and he resists the "numerical" understanding of ratios in which the ratio of two magnitudes is treated as a quantity and, at least on some presentations of the theory, identified with the quotient of the magnitudes that form the ratio. For his part, Wallis was thoroughly committed to the numerical conception of ratios, and this naturally brought him into conflict with Hobbes's doctrines.

According to Wallis, the numerical theory of ratios is both more easily understood than the relational theory and more general in its application. In particular, he argued that his treatment of ratios as a generalized kind of quotient allows for otherwise complex comparisons of ratios to be reduced to the analysis of quotients. "The quotient of division," he declares in the *Mathesis Universalis,*

> shows the ratio of the dividend to the divisor. So if 12 is divided by 6, it will produce the quotient 2, the ratio commonly known as *double,* which is the ratio that the number 12 has to the number 6. And also a four pound weight is double a two pound weight, because if four pounds are divided by two pounds, the quotient 2 will be produced, or how often [*quoties*] the one weight is contained in the other. And because of this fact, *Where quotients are equal to one another, then also the quantities forming the quotients are in the same ratio.* And indeed the ratio is

estimated by the quotient, and also from the equality of ratios there follows the equality of quotients. (*MU* 25; *OM* 1:135)

As part of his extended case for the priority of arithmetic over geometry, Wallis devoted chapter 35 of the *Mathesis Universalis* to an "arithmetical" presentation of the theory of ratios from book 5 of the Euclidean *Elements*. The theory of ratios and proportions, he explains, "is arithmetical rather than geometrical," providing that arithmetic is understood to encompass more than simply the theory of integers and includes "fractions, surds, the whole of algebra, and (as it is called) the specious or symbolic arithmetic" (*MU* 35; *OM* 1:183). Given this expansive conception of arithmetic, it is no surprise to find Wallis identifying numbers and ratios with the declaration that:

> the whole of arithmetic, if it is regarded more strictly, appears to be scarcely anything other than the doctrine of ratios. And numbers themselves are so many signs of ratios whose common consequent is one, or unity. And here one or unity is taken for an exposed quantity, so that all other numbers (whether integers, fractions, or even surds) are so many signs or exponents of other ratios to an exposed quantity. (*MU* 35; *OM* 1:183)

According to Wallis, the ancients preferred a cumbersome and particularized form of expression when they represented ratios by pairs of lines rather than numerical signs, and in effect they confined the theory of proportions to geometry. This shortcoming was due to an insufficiently general conception of number, which arose principally "because the symbolic method did not then obtain, and even the numerical notation in use everywhere today (and invented by the Indians) had not yet been introduced" (*MU* 35; *OM* 1:183). However, once the resources of symbolic algebra have been brought to bear, a completely general theory of ratios emerges, in which "the quantities of lines, or plane figures, or even solids, or anything else" are represented "by the letters of the alphabet" (*MU* 35; *OM* 1:183).

The scheme proposed by Wallis involves taking the ratio of two magnitudes as a quotient or a number representing how many times the antecedent is contained in the consequent. At first sight, this treatment of ratios as quotients would seem to restrict the theory of proportion to the realm of rational numbers, but Wallis maintained that the generalized understanding of number implicit in the new algebra would expand the concept of quotient to encompass both rational and irrational magnitudes.

This identification of ratios with quotients was not a complete innovation on Wallis's part. As I noted in chapter 3, the numerical theory of ratios had a long history before Wallis. Moreover, the treatment of ratios as quotients is implicit in the numerical theory, since it takes ratios to have "sizes" or "denominations" that are most easily understood as quotients arising from the division of two magnitudes. Before Wallis, the authority of Oughtred's *Clavis Mathematicae,* for example, could be used to underwrite the conception of ratios as quotients, because Oughtred is willing to identify ratios with the result of division. According to his chapter "Of Proportion":

> If of four given numbers the first is to the second as the third to the fourth, these four numbers are said to be proportional. And the relation *[habitudo]* of numbers one to another is found by dividing the antecedent by the consequent: as 31 to 7 is the ratio 4³⁄₇, that is quadruple supertripartient seven. (Oughtred 1693, 15)

This general account of proportion speaks of "numbers" standing in the same ratio when, by division, they are found to have the same relation, which is to say that they form the same quotient. For Oughtred, numbers are not simply positive integers, but any kind of magnitude at all, and he sees no deep difference between numbers and algebraic "species" or letters that can indifferently denote any kind of quantity. As he notes at the beginning of the *Clavis Mathematicae,*

> Magnitudes can be denoted either by numbers signifying their measure, or also by species: as a line seven inches long is designated by 7; or by any one letter or note as *A, B, C,* etc; or by two letters assigned to the termini of the line, as *AB, BC, CD,* etc. And any of these can be taken at will. But you must retain in memory the magnitude for which any species is intended. (Oughtred 1693, 3–4)

By assimilating the theory of ratios into a generalized treatment of number Oughtred, Wallis, and other proponents of the numerical conception of ratios effectively abandoned the classical conception of ratios as a special kind of relation between magnitudes. In their treatment, ratios become quantities ultimately reducible to number.

As we saw in chapter 3, Hobbes rejected this understanding of ratios, and accused Wallis of propagating a serious confusion by identifying ratios with quotients. In the dedicatory epistle to the *Six Lessons,* Hobbes lists among the principles "so void of sense, that a man at the first hearing, whether Geometrician or not Geometrician, must abhor them," the principle that "the Quotient is the Proportion of the Divisor [reading "Division" in Hobbes's original as "Divisor"] to the Dividend" (*EW* 7:186–87). On Hobbes's analysis, a quotient can only be found "in *aliquot parts*," and the general theory of proportions must be phrased in terms that can accommodate incommensurable ratios. He insists that "setting their Symboles one above another doth not make a Quotient," and argues that the presence of incommensurable magnitudes make it "impossible to define Proportion universally, by comparing Quotients" (*SL* 3; *EW* 7:241). Hobbes further objected to the identification of ratios with quotients on the grounds that a quotient or fraction such as ⅔ is an absolute or determinate quantity that cannot be conflated with a ratio, because a ratio is always a relative measure of two quantities that expresses how great one is with respect to another.[11]

Wallis replied to such objections by disingenuously insisting that he had always been scrupulous in observing a distinction between ratios and quotients. One such response to Hobbes's accusations is worth considering at length, for it also gives some idea of the rhetorical standards observed by the two disputants:

> But you adde farther, that *I say, that I make [ratio] to consist in the Quotient.* And is not this abominably false? I neither say so, nor doe so, nor did I give any ground at all for any man (that is in his witts) to believe I did. My words were these, *Videmus igitur Rationis aestimationem esse (secundum Te) penes Residuum, non penes Quotum, & Subductione, non Divisione quaerendam esse.* (And what reason I had to say so, they that consult the place will see.) Now could any man (who had not a great confidence that his English Reader understands no Latine) be so impudent as to say, that in those words, *I say, you make Proportion to consist in the Remainder; and I, in the Quotient?* Can any man, that understands, though but a little Latine, (if he be not either out of his witts, or halfe asleep,) think that these words *Rationis aestimatio est penes Quotum,* (that is, *the Proportion is to be estimated according to the Quotient,* or, to use your own words, *the quotient gives us the measure of the proportion,*) could thus

11. This is the import of Hobbes's declaration at the opening of chapter 13 of *De Corpore,* where he defines ratios after first observing that "great and little are intelligible only by comparison. Now that to which they are compared we have called something exposed; that is, some magnitude either perceived by sense, or so defined by words that it can be comprehended by the mind" (*DCo* 2.13.1; *OL* 1:128).

be Englished *proportion consists in the quotient?* And that then you should raile at us, quite through your Book for saying that *Proportion* is *a certain quotient,* that it is a *number,* that it is an *absolute quantity,* &c. as if we had been so ridiculous as to speak like you. For, that you have so spoken you cannot deny, (and therefore the absurdity what ever it be, lights upon your selfe:) But, to say that *I said so,* or any thing to that purpose, till you can shew where I said it, I take to be, (so farre as a word out of your mouth can be) a *manifest slander.* I neither say so, nor think so. (*Due Correction* 62–63)

This lengthy and vociferous reply repeats an earlier allegation from Wallis's *Elenchus* to the effect that Hobbes absurdly treats ratios as numbers formed by the subtraction of one quantity from another. As I showed in section 2 of chapter 3, this objection depends upon an uncharitable reading of Hobbes's definition of ratio and a failure to take seriously his distinction between arithmetical and geometrical ratios.[12] More interesting is Wallis's heated denial of ever having identified ratios and quotients, because it seems to contradict his other pronouncements on ratios. It may perhaps be true that, strictly speaking, Wallis's works do not contain the explicit formulation "The ratio of two magnitudes is the quotient arising from their division" (or its Latin equivalent). But from what we have seen of his remarks on ratios, it is clear that he does, indeed, take ratios to be quantities falling within the purview of a generalized science of arithmetic. Where earlier authors had confined quotients to the division of integers, Wallis desires to expand the concept of arithmetic to include quotients of irrational

---

12. Where Hobbes had declared in *De Corpore* that "the relation of the antecedent to the consequent according to magnitude, that is to say its equality, excess, or defect is called the ratio and proportion of the antecedent to the consequent; and so ratio is nothing other than the equality or inequality of the antecedent compared to the consequent according to magnitude" (*DCo* 2.11.3; *OL* 1:119), Wallis commented, "But since the quantity, for example, of inequality can be considered either with respect to the remainder, or with respect to how often *[quoad Quotum]* (that is to say, it can be inquired either how much greater this is than that, or how often or how much this is of that; the first of which is made known by subtraction, the second by division) let us see which of these ways of consideration is that which you want to call ratio" (*Elenchus* 14). He then takes Hobbes's remark that "ratio consists in the difference of the antecedent to the consequent, that is in that part of the greater remaining after having taken away the lesser" (*DCo* 2.11.5; *OL* 1:119) to show that Hobbes identifies ratios universally with differences. But, of course, Hobbes distinguishes between arithmetical ratios and geometric ratios in terms not unlike those used by Wallis, and although Hobbes can be sloppy in his use of the term "difference" it is clear that he is not committed to the view to which Wallis objects.

magnitudes, and it is this "larger sense" of quotient that he uses when treating ratios as quotients.[13]

Wallis's insistence that he never actually takes ratios to be quotients is, in effect, a verbal dodge. Not willing to appear to depart too radically from the authority of Euclid and classical geometry, Wallis grants that ratios are "exposed," "determined," "denominated," "designed," or "estimated" by quotients while denying that ratios and quotients are the same thing. But it is difficult to attach much sense to such a distinction between ratios and quotients. All of the relevant facts about ratios are facts about the quotients that "expose" the ratio, and (apart from preserving the appearance of fidelity to the Euclidean tradition) there is no point in insisting upon the distinction. Hobbes took note of this very point, observing that Wallis's pronouncements on his theory of ratios did not sit well with his words in *Arithmetica Infinitorum*.[14] In Wallis's presentation, the ancient theory of ratios has essentially been subsumed into an algebraic theory of numbers, and the numbers corresponding to ratios are in fact quotients.[15] As their dis-

13. Wallis makes this quite explicit in chapter 9 of his *Treatise of Algebra,* where, in discussing the Euclidean definition of ratios, Wallis announces that the "whole Definition of λόγος (Ratio, Rate, or Proportion) . . . is thus rather to be rendered, . . . *Rate* (or Proportion) *is that Relation of two Homogeneous Magnitudes* (or Magnitudes of the same kind,) *how the one stands related to the other, as to the* (Quotient, or) *Quantuplicity*" (*Treatise of Algebra*, 1685, 79). Later, he acknowledges that this requires an expanded conception of the quotient:

But all other Proportions which they call Ineffable, (which are not *ut numerus ad numerum,*) but as Quantities Incommensurable, and for the sake of which, that Scholiast tells us, that Euclide chose to use the word πηλικότητες rather than ποσότης, (for what we commonly call the Quotient in the largest sense) that it might extend to Ineffable as well as Effable Proportion, (as if in Latine he would have said Quantuplum, rather than Quotuplum, lest this should be thought to extend only to Multiples, or but to Effable Proportions;) all these, I say have no peculiar Names allotted; but use to be designed by the Terms themselves, as A to B, or as 1 to √2, or (set Fraction-wise, so as to design a Quotient,) A/B, or 1/√2 &c. (*Treatise of Algebra* 9, 80)

14. In commenting upon the manipulation of ratios in *Arithmetica Infinitorum,* Hobbes remarks, "But first I wonder why you were so angry with me for saying you made proportion to consist in the Quotient, as to tell me it was abominably false, and to justify it, cite your own words *Penes Quotientem;* do not you say here, the proportion is everywhere greater than subtriple, or ⅓? And is not ⅓ the quotient of 1 divided by 3? You cannot say in this place that *Penes* is understood; for if it were expressed you would not be able to proceed" (Στίγμαι 1; *EW* 7:366).

15. Klein, in commenting upon this aspect of Wallis's general theory of number remarks that "the universality of arithmetic as a 'general theory of ratios,' which depends on the homogeneity of all 'numbers,' can be understood only in terms of a symbolic reinterpretation of the ancient 'numbered assemblage,' of the *arithmos.* The object

pute dragged on over the years, Hobbes and Wallis continued to trade accusation and denial on this issue with no significant change in position,[16] and we can leave this part of the quarrel over ratios aside and turn to a consideration of related issues concerning the theory of proportion.

The dispute over the nature of ratios naturally carried over into the thorny issue of the compounding of ratios—a troublesome point that had been the source of debate and confusion well before Hobbes and Wallis. The fundamental difficulty can be illustrated by first considering a pseudo-Euclidean definition that made its way into some editions of book 6 of the *Elements*. The fifth definition of book 6 reads:

> A ratio is said to be compounded of ratios when the sizes (πηλι-κότητες) of the ratios multiplied together make some (?ratio, or size). (*Elements* 6, def. 5)[17]

The remarkable features of this definition are its reference to the "sizes" of the compounded ratios and its application of the arithmetical operation of multiplication to construct a ratio from the sizes of two given ratios. According to the relational theory, a ratio is not a quantity and thus lacks a size, so it is difficult to reconcile this definition with the "official" presentation of the theory of ratios in book 5 of the *Elements*. As Barrow points out, "ratios, as they lack all quantity, can neither be added nor multiplied" (*LM* 20, 326). Moreover, the

---

of arithmetic and logistic in their algebraic expansion is now defined as 'number,' and this means a symbolically conceived ratio—a conception consonant with that of algebra as a general theory of proportions and ratios" (1968, 223).

16. Thus, for example, in the *Examinatio,* one of the interlocutors in the third dialogue reports that "often, as he does here, Wallis says that fractions or quotients are the same thing as ratios. But when he was warned by Hobbes that fractions and quotients are all absolute quantities, but ratios all comparative quantities, he denied that he had said that a quotient is the ratio itself, but that the ratio is according to the quotient. But he attempts to base his whole treatise *Arithmetica Infinitorum* on this foundation: that the quotient itself is the ratio of the dividend to the divisor, so that ⅓ is the ratio of 1 to 3" (*Examinatio* 3; *OL* 4:128). Wallis's rejoinder in *Hobbius Heauton-timorumenos* was to claim, "No, I doe not make *Proportion,* a *Quotient,* or *an absolute Quantity* (that's but his inference, and a weak one.) I say indeed that Proportion *depends upon* the Quotient, is *determined* by the Quotient, *estimated* by the Quotient, and *denominated* by the Quotient; not that it is the Quotient" (1662, 51).

17. The reasons for regarding this definition as a late interpolation are mentioned briefly below. For a more complete discussion, see Heath's commentary in Euclid [1925] 1956, 2:189–90. Sylla 1984, 22–24, also discusses some of the problems with the definition. Even though the definition is not, strictly speaking, part of the Euclidean corpus, it will be more convenient if I refer to it as definition 5 of book 6 of the *Elements,* since that is how it was known to Hobbes, Wallis, and their contemporaries.

definition is never used in the *Elements,* even in the one place where Euclid speaks of compound ratios (*Elements* 6, prop. 23).[18] Henry Savile characterized the definition as one "of the two moles or blemishes on the most beautiful body of geometry" that had engaged the attention of ancient and modern geometers (Savile 1621, 140). The other famous blemish was the parallel postulate, and Savile thought that both should be remedied by showing that they are demonstrable from the remaining Euclidean principles.

For all its theoretical problems in regard to the relational theory of ratios, however, definition 5 makes perfect sense in conjunction with the numerical theory of ratios. If ratios are endowed with sizes, or if they are simply identified with quotients, then it is a simple matter to accept the idea that multiplication can generate a new ratio from two given ratios. Thus, if the ratios 3:7 and 1:4 are compounded, the new ratio that emerges is that found by multiplying the quotients $3/7$ and $1/4$, yielding $3/28$ or 3:28 as the compounded ratio.

Wallis was eager to solve the problems associated with the fifth definition of book 6, and attempted to show that any confusion or obscurity surrounding it could be overcome by, in effect, showing that Euclid himself accepted the numerical theory of ratios. This strategy is clear in his essay "The Geometrical Dispute over the Fifth Postulate and Fifth Definition of Book 6 of Euclid," where Wallis undertakes to rectify the two flaws that Savile had complained of as disfiguring the body of Euclidean geometry. The essay apparently originated as a pair of Savilian lectures in 1651 and 1663, although it was not published until 1693 in volume 2 of Wallis's *Opera Mathematica.*[19] Wallis's treatment of definition 5 proceeds by first claiming that Euclid's definition of ratios as "a sort of relation in respect of size between two magnitudes of the same kind" (*Elements* 5, def. 3) should be understood somewhat differently than the tradition takes it. In particular, he coins the neologism *quantuplicity* and takes the Euclidean definition of ratios to be the following: "A ratio is that relation or habitude of homo-

18. This asserts, "Equiangular parallelograms have to one another the ratio compounded of the ratios of their sides." Euclid does not actually define the term *compound ratio,* but from the context of the proof in *Elements* 6, prop. 23, the meaning is clear enough: If $K{:}L$ and $L{:}M$ are ratios, then the ratio $K{:}M$ is the ratio compounded of the ratio of K to L and that of L to M. In any event, it is noteworthy that the so-called fifth definition from book 6 does not appear in the proof.

19. Prag 1931, 402, reports that "[a]s Savilian Professor at Oxford Wallis was required to work on Euclid's axiomatics, above all on the parallel postulate. In two lectures, February 1651 and July 1663, he fulfilled this obligation." Prag refers here to the essay in Wallis's *Opera* (2:665–78).

geneous magnitudes to one another in which it is shown how the one is to the other, considered according to quantuplicity" (*OM* 2:665). The slippery term *quantuplicity* is Wallis's word for how much one magnitude is in comparison with another, or how many times the one is contained in the other. In particular, he wants to allow relations of quantuplicity that cannot be expressed by ratios of integers, so that the side of a square would be the $1/\sqrt{2}$ quantuple part of the diagonal.[20]

With this (strongly numerical) understanding of ratios in place, Wallis proceeds to interpret the fifth definition of book 6, asserting that the compounding of ratios arises from the multiplication of the quotients that "expose" the compounded ratios. The definition, he says, "is to be understood this way: *A ratio is said to be compounded of ratios when the exponents of the ratios multiplied together make the exponent of that ratio*" (*OM* 2:666). This requires that Euclid's term πηλικότητες (*Elements* 6, def. 5) be taken to designate the quantities or exponents of the compounded ratios, which quantities are therefore nothing other than the quotients of the antecedents and consequents in the ratios. Understood in this way, the definition is no "flaw" to be removed from geometry (as Savile had thought), but simply a stipulation of how the expression *compound ratio* is to be used.[21]

Hobbes naturally resisted such an account of compounded ratios because he saw it as depending too closely on the numerical theory of ratios. In his understanding of the fifth definition in book 6 of the *Elements*, Hobbes holds that Euclid intended the term πηλικότητες to apply, not to the quantities of the compounded ratios, but rather to the quantities that form these ratios. In other words, where Wallis takes

20. The linguistic difficulties surrounding Wallis's pronouncements on this point are sufficiently grave to render all but hopeless any attempt at translation into English. He distinguishes quotuplicity from quantuplicity in much the same way that other authors distinguish aliquot from proportional parts of a magnitude: quotuplicity is a relation of one magnitude to another that can be expressed by a fraction (or, equivalently, a ratio of integers), but quantuplicity is the more general relation between magnitudes that need not be expressible in rational terms. Thus, he contends, "Hoc est, *Quotupla* sit, seu potius *Quantupla*, altera alterius. Quod sic intellectum volo: Nimirum, prout *Pars Aliquanta* distingui solet a *parte Aliquota;* sic ego (quae sunt earum Correlata) *Aliquantuplum* ab *Aliquotuplo* distinguo. Ut *Quantuplum* sit vox Generalis, cujus *Quotuplum* sit un Species, qui *Multiplum* respondeat. . . . Quae Relatio apud Graecos innui solet terminatione πάσιον apud Latinos terminatione *-plum,* apud Nos terminatione *-fold*" (*OM* 2:665).

21. As Wallis puts it, "Therefore there is no reason why this is a demonstrable proposition (any more than are all other definitions), for nothing more is taught here than in what sense he desires this locution to be understood. . . . And therefore there is in this definition no mole to be removed, or blemish to be cleansed away" (*OM* 2:666).

the quantities of the ratios $A{:}B$ and $C{:}D$ to be the quotients A/B and C/D, Hobbes understands the magnitudes A, B, C, D, taken singly, to be the quantities of the ratios.[22] This may seem to be a relatively minor difference of opinion over how best to render a Greek term into mathematical English, but it amounts to nothing less than the contest between the numerical and relational theory of ratios.

This disagreement also set the stage for terminological confusions that would persist throughout the dispute. Following the older tradition and its usages, Hobbes regarded the compounding of ratios as an operation of addition rather than multiplication, since the compounding to two ratios is effected by, so to speak, taking them together. Earlier authors had spoken of the compounding as "addition,"[23] and Hobbes's conception of the matter is not far removed from this earlier view. As it happens, the compound ratio can be found by multiplying the antecedents and consequents of the two compounded ratios, and this fact led to a great deal of the linguistic and conceptual confusion over whether the process of compounding should be characterized as multiplication or addition.[24]

---

22. As Hobbes puts it in *De Corpore*, "Of two ratios, whether arithmetical or geometrical, when the magnitudes compared in both (which are called by Euclid, in the fifth definition of his sixth book, the quantities of ratios) are equal, then one ratio can be neither greater nor less than the other; for one equality is not greater or less than another equality" (*DCo* 2.13.3; *OL* 1:129–30).

23. See Sylla 1984, 11–20, on this usage, which persisted until roughly the time of Newton.

24. Barrow reports that "because when ratios are said to be compounded their denominators are multiplied, it is manifest that ratios are more rightly said to be multiplied than added. Just as when one of the denominators divides another, such an operation will be more justly called the division than the subtraction of a ratio. Nevertheless it has obtained that the former operation is called addition [πρόσθεσις], and the latter subtraction [ἀφαίρεσις]" (*LM* 20, 326–27). In a letter to John Collins from 8/18 September 1668, Wallis commented on the linguistic confusion as follows:

> That which I could wish altered in Mr. Mercator's logarithmotechnica was his manner of expression of the composition of ratios. For composition be a word used by Euclid sometimes for addition, sometimes for multiplication, and there being in him two compositions of ratios, the one mentioned in Def. 14, Lib. 5, which is by addition of the exponents, as when $3/2 + 2/2 = 5/2$, the other by multiplication of the exponents, Def. 5, Lib. 6, as where $3/2 \times 2/2 = 6/4$, which ambiguity hath caused some confusion, especially where the latter is called an addition of ratios. Clavius, and Gregory St. Vincent, and divers others, to avoid this inconvenience, have, for distinction sake, called the former composition by addition, the latter composition by multiplication; with which most writers, who speak distinctly, have used to comply: of which I have spoken at large in what I have writ against Meibomius, and against Mr. Hobbes's fourth dialogue, and elsewhere" (Wallis to Collins in Rigaud [1841] 1965, 2:494–95).

Wallis ridiculed Hobbes's account of the compounding of ratios for its alleged terminological inadequacies, suggesting that Hobbes was ignorant of the most basic operations of arithmetic and their application to the theory of ratios. In Hobbes's presentation, the equivalent of the fifth definition of book 6 of Euclid can be proved as a theorem rather than introduced as a definition. In his parlance, it reads:

> If there are any three magnitudes, or any three things that have some ratio one to another, as three numbers, three times, three degrees, etc., the ratios of the first to the second and of the second to the third, taken together, are equal to the ratio of the first to the third. (*DCo* 2.13.13; *OL* 1:140)

Wallis objected that, instead of speaking of the ratios "taken together," Hobbes should have called the result a "compounded ratio," and in any case the presentation confuses addition with multiplication:

> Let there, for example, be these three quantities: the numbers 6, 3, 1. The ratio of the first to the second is double, of the second to the third triple, and the ratio of the first to the third sextuple: but it is not as you intend, that double and triple taken together equal sextuple (yet these words seem to be understood, when you say that the first and second ratios taken together are equal to the third). For this is completely false, as double and triple taken together are equal (not to sextuple, but) to quintuple. (Although perhaps it could be doubted whether or not you sufficiently understand this.) But by the first and second ratios taken together, you understand the ratio compounded of the first and second, which composition is nevertheless done by the multiplication of the terms (not their addition), as is well known. And this should be called the continuation of ratios rather than their addition (although I do not deny that this appellation can be found in some places). And so the ratio that is compounded, for example, of double and triple is not the double and triple taken together (which is quintuple, because $2 + 3 = 5$) but the triple of the double, that is sextuple, because $2 \times 3 = 6$, in which sense your proposition is true and nearly identical with the fifth definition in the sixth book of Euclid. (*Elenchus* 20)

Hobbes's response to this criticism was to insist that Wallis did not understand the distinction between ratios and numbers. This accusation is, of course, related to his earlier charge that Wallis fails to distinguish between ratios and quotients. Replying to the specific example used by Wallis, Hobbes asks:

> Tell me (egregious professors) how is six to three double Proportion? Is six to three the double of a number, or the double of some Proportion? All men know the number six is double to the number three, and the number three triple to an unity. But is the Question here of compounding numbers, or of compounding proportions? . . . Your instance therefore of six, three, one, is here impertinent, there being in them no doubling, no tripling, no sextupling of Proportions, but of numbers. (*SL* 3; *EW* 7:244–5)

This complaint could obviously have little persuasive effect on Wallis, whose entire approach to the theory of ratios was predicated upon the thesis that there is no deep distinction between numbers and ratios. Wallis takes the number 3 to represent the ratio 3:1, and he simply denies Hobbes's assumption that there is a difference between doubling numbers and doubling proportions.

Aside from its lack of convincing force to a staunch adherent of the numerical theory of ratios, Hobbes's account of ratios had the disadvantage of introducing yet another layer of verbal confusion into the debate. This time, the issue revolves around the question of the difference between such terms as *double* and *duplicate,* and their application to ratios.[25] Classically, in a continued proportion such as $X:Y :: Y:Z$, the magnitudes X and Z are said to be in duplicate ratio, with the idea being that the original ratio of X to Y appears twice, or is duplicated, in the proportion. The ratio $X:Z$ (which, by the way, is compounded of the ratio $X:Y$ and $Y:Z$) can then be said to be the duplicate ratio of the original ratio $X:Y$. On the other hand, a magnitude is double of another when it is twice as great. In exactly the same manner, triplicate ratio can be distinguished from triple quantity, quadruplicate from quadruple, and so forth. The subduplicate of a ratio is obtained by interposing a quantity in continued proportion between the antecedent and consequent of the original ratio; i.e., the subduplicate of the ratio $A:B$ is found by interposing a quantity C such that $A:C :: C:B$. Thus, the ratio $1:\sqrt{3}$ is subduplicate of the ratio 1:3, since $1:\sqrt{3} :: \sqrt{3}:3$.

Hobbes did not follow the established usage, nor did he remain consistent in his own terminology. In the Latin version of *De Corpore* he declared that "the ratio of a greater to a lesser quantity is multiplied by a number when a certain number of ratios equal to it or the same

---

25. See Saito 1993 for an account of duplicate ratios in Euclid.

are added together"; but when the ratio of a lesser to a greater quantity is iterated, "it is not properly said to be multiplied, but submultiplied" (*DCo* 2.13.16; *OL* 1:145). In the English version of *De Corpore* he reverted to the more traditional usage in which "[a] Proportion is said to be multiplied by a Number when it is so often taken as there be Unities in that Number; and if the Proportion be of the Greater to the Less, then shall also the quantity of the Proportion be increased by the Multiplication; but when the Proportion is of the Less to the Greater, then as the Number increaseth, the quantity of the Proportion diminisheth" (*DCo* 2.13.16; *EW* 1:164). This change was presumably due to Wallis's caustic criticism of the usage in the Latin original (*Elenchus* 24), although it is difficult to be sure of this point. In any case, the change in Hobbes's formulation of his doctrine opened a new field for verbal disputes, but these are of little interest to our investigation and can be left out of consideration here.[26]

In the final analysis, this part of the Hobbes-Wallis controversy shows that the two antagonists disagreed over fundamental questions in the philosophy of mathematics, and their disagreement is representative of unresolved tensions in the doctrine of ratios. Others had quarreled over these same issues—although perhaps without the sheer ferocity of Hobbes and Wallis—but the basic questions remained largely unresolved. It is oddly ironic that this debate should have degenerated into exactly the stale wrangling over words that so many proponents of the "new philosophy" found in the debates of their Scholastic predecessors.

4.3 THE ANGLE OF CONTACT

As their dispute progressed Hobbes and Wallis entered into a debate over the problem of the "angle of contact" between a circle and its tangent.[27] This difficulty had been the subject of controversy well before the seventeenth century, and Wallis himself published a *Geometrical Disquisition on the Angle of Contact* in 1656, independently of his

26. The changes in Hobbes's doctrine, as well as much wrangling over the question of what is to count as double or duplicate, are summarized (although not to Hobbes's advantage) by Wallis (*HHT* 70–73).

27. See Maierù 1984 and 1990 for the most comprehensive account of the problems posed by the angle of contact. Thomason 1982 contains a presentation of some of the problems engendered by the angle of contact and their alternative solutions, and Dear 1995b examines the problem in the context of Mersenne's response to Descartes's *Meditations*.

battle with Hobbes. It will be helpful to outline some of the history of this disputed issue before proceeding to an examination of the exchanges between Hobbes and Wallis.

The best place to begin is with Euclid. In the *Elements,* a plane angle is defined as "the inclination to one another of two lines in a plane which meet one another and do not lie in a straight line" (*Elements* 1, def. 8). As I mentioned in chapter 3, this definition permits angles to be formed by curved and straight lines, so the three forms in figure 4.1 are angles in the Euclidean sense: The three kinds of angles here are the familiar rectilinear angle, the curvilinear angle formed by two curves, and the mixtilinear angle formed by a curve and a straight line. It is with certain mixtilinear angles that paradox first appears to threaten.

At proposition 16 of book 3 of the *Elements* Euclid proves that the angle between circle and tangent (which was also known as the "horned angle") is less than any rectilinear angle. The basic idea is as follows: We begin with a circle *ABC* and tangent *BD* (figure 4.2), and then show that any rectilinear angle ∠*DBE* must be greater than the angle of contact ∠*CBD*. Observe that *BE* cuts the circle at *F.* But because ∠*DBF* = ∠*DBE,* the angle of contact must lie within the rectilinear angle ∠*DBE,* and thus ∠*CBD* is less than ∠*DBE.* But because *BE* was chosen arbitrarily, the angle of contact must be less than any rectilinear angle. A similar line of reasoning establishes that the angle ∠*GBC* (known as the "angle in a semicircle") is greater than any acute angle, prompting the suspicion that the angle in the semicircle is equal to the right angle ∠*GBD.*

The angle of contact appears paradoxical because it is a magnitude which seems to violate the principle known as the axiom of Archimedes. In one of many equivalent formulations, it appears as proposition 1 of book 10 of the *Elements* in the form of the claim:

Figure 4.1

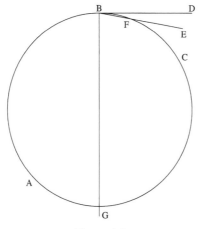

Figure 4.2

Two unequal magnitudes being set out, if from the greater there be subtracted a magnitude greater than its half, and from that which is left a magnitude greater than its half, and if this process be repeated continually, there will be left some magnitude which will be less than the lesser magnitude set out. (*Elements* 10, prop. 1)

The angle of contact appears to violate this condition, since no amount of division of a rectilinear angle can yield an angle less than it. However, angles of contact can be ordered by a less-than relation, because they will be greater or less depending on the radius of the circle from which they are formed. But anything compared by an ordering relation is a magnitude, so the angle of contact seems inconsistent with the Euclidean account of magnitudes.

These difficulties led commentators to speculate about the adequacy of Euclid's definition and to question whether angles were properly quantities at all. Proclus, for example, argued that no angle—whether rectilinear or mixtilinear—is a quantity:

But if [the angle] is a magnitude and all finite homogeneous magnitudes have a ratio to one another, then all homogeneous angles, at least those in planes, will have a ratio to one another, so that a horned angle will have a ratio to a rectilinear. But all quantities that have a ratio to one another can exceed one another by being multiplied; a horned angle, then, may exceed a rectilinear, which is impossible, for it has been proved that a horned angle is less than any rectilinear angle. (Proclus 1970, 98)

Later authors were not prepared to deny that all angles are quantities, but the attempt to fit angles into the traditional theory of magnitudes led to considerable controversy. Most notably, Clavius and the French mathematician Jacques Peletier debated the quantity of the angle of contact in a series of publications beginning with Peletier's *In Euclidis Elementa Geometrica Demonstrationum Libri Sex* (Peletier 1557).[28] In this commentary on the *Elements,* Peletier offered additions, elucidations, and corrections to the Euclidean text, claiming to have "illustrated the principles of geometry by new meditations." In particular he asserted that he had explicated "the nature, formation, and construction of the angle," which had previously not been well understood (Peletier 1557, preface sig. A2r). As part of this undertaking Peletier argued that the so-called angle of contact is not really an angle at all and that it is simply not a magnitude. Clavius opposed this argumentation in his 1574 edition of Euclid. He contended that the angle of contact is a proper angle with a magnitude, although he confessed that it is infinitely small in comparison with a rectilinear angle. The main arguments for each side of the issue are worth summarizing.

Peletier's principal argument proceeds directly from the manifest nature of magnitudes, as exemplified by the Archimedean axiom. He contends that there is "no principle in the whole of geometry that (as I may say) is more naturally true" (Peletier 1557, 75). Thus, any imaginable pair of magnitudes of the same kind must satisfy the condition that the greater of them can, by continued bisection, be made less than the other. This apparently implies that the angle of contact is not a magnitude since no amount of division will ever make a rectilinear angle less than the angle of contact. Peletier also concluded that the angle of the semicircle ($\angle GBC$ in figure 4.2) is a right angle. This follows naturally: the tangent and diameter form a right angle and the angle of the semicircle is the difference between this right angle and the angle of contact, so if the angle of contact is no magnitude, the angle of the semicircle is a right angle (Peletier 1557, 76–77).

Clavius countered this argument by claiming that the angle of contact is divisible (not by a straight line, but by a circular arc), and the classical conception of magnitudes holds that anything divisible is a

---

28. Maierù 1984 and 1990 examine these disputes in detail. A brief overview is also provided by Naux in his study of Clavius (1983, 325–29).

magnitude.[29] So, on Clavius's analysis, the angle of contact is a proper magnitude. He admits that the angle of contact must be smaller than any rectilinear angle, but holds that angles of contact can still be ordered by a less-than relation:

> All angles of contact whatsoever, formed by the tangent line and periphery, are less than any acute angle; but it is not necessary that they are therefore all equal to one another, for one can be greater or less than another, just as we say: any acute angle is less than any right angle, and yet these are not all equal to one another. So also every ant (to take an example from natural things) is less than any man or mountain, and still these can be truly unequal to one another. For this reason Peletier is not correct to conclude that the angle of contact is nothing. (Clavius 1612, 1:120)

Moreover, Clavius argues that the angle of a semicircle is not the same in all circles, but rather is a function of the radius of the circle. Because the angle of the semicircle is the difference between a right angle and the angle of contact, he concludes that as the radius of the circle decreases (and hence the angle of contact increases), the angle in the semicircle must also decrease.[30] But although it may decrease, the angle in the semicircle is still greater than any acute angle, as shown in proposition 16 of book 3 of Euclid.

To Peletier's claim that the angle of contact violates the Archimedean principle, Clavius replied that angles of contact and rectilinear angles are separate kinds of magnitudes, as distinct as surfaces and lines. Both parties to the dispute agreed that the Archimedean axiom applies only to magnitudes of the same kind, since the continued division of a plane figure could never produce a magnitude less than a given line. In asserting that rectilinear angles and the angle of contact are different species of magnitudes, Clavius hoped to make the Archimedean axiom irrelevant to the question of whether the angle of contact is a magnitude and thus evade Peletier's main argument. For Clav-

29. Clavius writes, "On the contrary, I say that any angle of contact can be augmented, and it can be divided infinitely by a curved line, although it cannot be divided by a right line, as Euclid correctly shows" (Clavius 1612, 1:119).

30. Clavius argues, "Although the angle of a semicircle in a greater circle is greater than the angle of a semicircle in a lesser circle, this does not mean that any angle of a semicircle is greater than a right angle. For every right angle will exceed any angle of a semicircle, by the angle of contact, which is made by the periphery and the tangent line" (Clavius 1612, 1:120).

ius, any two angles of contact will satisfy the Archimedean axiom, as will any two rectilinear angles. But because these are separate species of magnitudes, the axiom does nothing for Peletier's case.

This dispute dragged on without resolution as each party issued rebuttals to the claims of the other. Indeed, it is not obvious how the dispute could have been settled, given the conceptual resources available. Peletier and Clavius both accepted Euclid's account of angles, but the Euclidean definition is too vague to resolve the issue between them.[31] Certainly, the circumference and tangent are two lines that "lie in the same plane and meet without lying in a straight line," and there is evidently an "inclination of one to the other," so the angle of contact satisfies the Euclidean definition. But to accept such angles appears to introduce infinitesimal magnitudes into geometry, contrary to the explicitly finitistic standpoint of classical mathematics. Moreover, Clavius's account of the matter would require a substantial reworking of the Euclidean treatment of magnitudes, since declaring angles of contact and rectilinear angles two genuinely distinct species of magnitudes requires a definition of *angle* that would allow them to be distinguished. It is no wonder that the problem of the angle of contact remained unresolved into the seventeenth century.

Wallis argued for Peletier's opinion in some of his Savilian lectures on geometry, which he later collected and published in the form of a *Geometrical Disquisition on the Angle of Contact,* which appeared (along with the more famous *Arithmetica Infinitorum*) as part of the 1656 collection *Operum Mathematicorum Pars Altera* (Wallis 1656b).[32] In this treatise Wallis reviews the origin and course of the

---

31. Peletier comments that the Euclidean definition of angle is "well understood" and "has a tolerable and perspicuous sense." Nevertheless he prefers an alternative: "A plane angle is the cutting *[sectio]* of two lines in a plane" (Peletier 1557, 3–4). Clavius finds the Euclidean definition wholly unproblematic and rejects Peletier's alternative: "But we by no means concede to Peletier that an angle can be made only by two lines that cut one another. In order to form an angle it is sufficient that two lines in a plane incline toward one another and yet do not lie in the same right line, as is clear from the description of the plane angle taken from Euclid, even if these lines do not cut one another if produced. Of this sort is the angle formed by the periphery and the tangent, which is a true angle, as we have shown above" (Clavius 1612, 1:120). The reason for this disagreement should be clear: since the tangent and periphery of the circle touch without cutting, Peletier's alternative definition guarantees that the angle of contact is not an angle at all.

32. The provenance of the *Geometrical Disquisition on the Angle of Contact* as Savilian lectures is established by Wallis's remark that he had earlier undertaken "to show by illuminating demonstrations that truth stood on the side of Peletier in my public lectures, which I then gave over to the printer" (*Elenchus* 34).

controversy, undertaking to prove that "the angle of contact is a non-angle, or a nonquantum; and thus the angle in the semicircle is equal to a rectilinear right angle" (*Angle of Contact* 15; OM 2:630). The crux of Wallis's case is his argument that all angles, whether rectilinear or mixtilinear, are homogeneous. Establishing such a claim would clearly render Clavius's account of the angle of contact untenable. However, before undertaking to prove this, Wallis introduces a crucial assumption concerning the definition of angles. Where Euclid had defined angles in terms of lines that incline toward one another, Wallis suggests that the concept of "inclination" be so understood that only those lines that actually cut have a mutual inclination toward one another.[33] The proof of the homogeneity of all angles then proceeds, although the argumentation is not decisive. The homogeneity of all rectilinear angles is entirely trivial and undisputed, and Wallis thinks that the same can be said for curvilinear angles:

> And further, to any possible rectilinear angle it is also possible to assign a curvilinear angle equal to it: just as Clavius himself demonstrates out of Proclus, at the fifth definition of the fifth book of the *Elements*.[34] For example, the curvilinear angle *DAE* [in figure 4.3] is equal to the rectilinear angle *BAC*; and in the same way to any assignable rectilinear angle there is assignable an equal curvilinear angle; and thus curvilinear angles can bear any given ratio not merely to one another, but also to any rectilinear angle, which ratio rectilinear angles themselves can bear to one another. (*Angle of Contact* 5; OM 2:612)

The weakness of this argument is that it applies only to curvilinear angles where the intersecting lines cut one another. It is certainly true

---

33. Wallis declares: "Peletier preferred this [Euclidean] definition of angle to be slightly altered, so that it is understood of the concourse of lines that cut one another. But no necessity forces this kind of change to be made. Although it is to be said, with Peletier, that only those lines contain an angle that (if produced) will cut one another, nevertheless it is no less true that, according to the same principles, only those lines meet that are inclined to one another. Which shows that the circumference does not make an angle with the tangent right line, and which also will show that these lines are not inclined to one another. So there is no need to change the Euclidean definition on account of this opinion" (*Angle of Contact* 3; OM 2:607).

34. The reference here is to Proclus's discussion of the fourth postulate of Euclid, which asserts that all right angles are equal to one another. Proclus considers the case of a right angle formed by the diameters of two circles, and argues that the angle formed by the intersecting circular peripheries will be equal to a right angle (Proclus 1970, 148–50). Clavius considers the same issue in his remarks on the Euclidean definition of proportionality in book five of the *Elements* (Clavius 1612, 1:208).

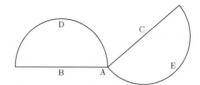

Figure 4.3

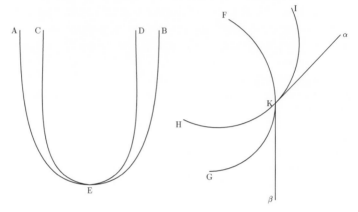

Figure 4.4

that, to any rectilinear angle, there will correspond any number of curvilinear angles (namely, those formed by curves tangent to the legs of the angle at their point of intersection). But this does not show that all curvilinear angles are homogeneous, unless it could also be shown that for every curvilinear angle there is a rectilinear angle equal to it. As it turns out, no rectilinear angle corresponds to curves that touch one another without cutting. In figure 4.4, for example, the curvilinear arcs *AB* and *CD* have a common point at *E,* but no rectilinear angle can be assigned equal to the angle of contact *BED.* Any two curves that cut one another will form angles homogeneous to rectilinear angles; thus the curves *FG* and *HI* in figure 4.4 cut at the point *K,* and the angle *GKI* is equal to the rectilinear angle *αKβ* formed by the tangents *Kα* and *Kβ.*

Nevertheless, Wallis's case is not decisive in showing that all curvilinear angles are homogeneous, unless he can provide an independent reason for restricting the class of curvilinear angles to the case of curves that cut one another. Having satisfied himself that he has shown all rectilinear and curvilinear angles to be homogeneous, Wallis then argues that any mixtilinear angle will likewise be homogeneous to any rectilinear angle. But again, he restricts his attention to cases where a right line cuts a curve. He argues:

> But a mixed angle can also have a ratio to a rectilinear angle, which I think Clavius would not deny. For a right line, if not one that is tangent, then at least one that cuts the periphery (both inside and out) will make a mixed angle with the periphery, which will be greater than at least some rectilinear angle, which then can be multiplied and exceed some assigned rectilinear angle (and on the other hand some rectilinear angle can exceed it). And so, as Clavius himself confesses, it must be said that there is a ratio, in accordance with definition 4 of book 5 of Euclid. (*Angle of Contact* 6; OM 2:613)

This argumentation clearly depends upon the requirement that an angle is formed only by two lines that cut one another. As such, it can do the work Wallis intends for it only if Euclid's definition of angle is taken to imply that when two lines incline toward each other they must cut one another. Wallis was sensitive to this difficulty and tried to overcome it by arguing that lines that are tangent but do not cut (such as those forming the angle of contact) actually do not incline toward each other. Although such lines may *seem* to have mutual inclination in the neighborhood of their concurrence, they do not incline at the point of contact.[35]

Hobbes took a quite different view of the matter. As we have seen, he distinguished between angles of circumlation (or "angles simply so called") from angles of contingence, basing this distinction on "the two ways by which two lines may diverge from one another." Angles of the former kind are generated by the rotation of a line about one of its endpoints, while the latter arise from "continual flexion or curvation in every imaginable point" (*DCo* 2.14.7; OL 1:160). This difference in origin requires that the angles be measured differently. Al-

---

35. Wallis writes, "When Clavius says, arguing from the definition of plane angle, '[t]hat in order to make a plane angle it suffices that two lines in a plane incline to one another,' but it is not required 'that they cut one another,' I admit that this is so; but I deny that the periphery and tangent right line 'are inclined to one another,' at least as far as the point of concourse; they coincide, but they are not inclined. I acknowledge also that (as he elsewhere says) '[t]he angle consists in the inclination of lines at a point that do not lie in a straight line'; but I do not acknowledge that these lines *are inclined*. Rather, with Peletier I affirm that only those lines *incline* (in the point of concourse) that, if produced, will cut one another: nor does Clavius ever show anything to the contrary" (*Angle of Contact* 6; OM 2:614).

though both kinds of angle are magnitudes (since they can be greater or less), they are heterogeneous and cannot be compared with one another as to quantity. Angles of circumlation are measured by the ratio of a comprehended arc to the circumference of a circle; angles of contact are measured in terms of the degree of curvature or flexion with which they are produced.[36]

Hobbes's solution to the problem of the angle of contact thus allows such angles to be construed as magnitudes, but magnitudes heterogeneous to rectilinear angles and measured differently. As he sums up the matter in *De Corpore*:

> An angle of contact, if it is compared with an angle simply so called, no matter how small, stands to it in a ratio of a point to a line; that is, no ratio at all, nor any relation of quantity. For . . . an angle of contact is made by continual flexion: so that in the generation of it there is no circular motion at all, in which consists the nature of an angle simply so called; and therefore it cannot be compared with it according to quantity. (*DCo* 2.14.16; *OL* 1:169–70)

Although Hobbes tended to side with Clavius in this matter, he was not entirely opposed to Peletier's account. In particular, he faulted Clavius for claiming that the angle in the semicircle depends upon the radius of the circle in which it is formed. On Hobbes's interpretation of the matter, the angle in the semicircle is an angle of circulation formed by the divarication of the radius from the periphery of the circle. As such, it is a right angle that must be measured by considering the radius and tangent. Thus Peletier's opinion can be partially vindicated. Or, to put the matter negatively, as Hobbes does in the *Six Lessons,* he disagrees with both authors and with Euclid himself: "[I]n this same Question, I am of opinion that *Peletarius* did not well in denying the Angle of Contingence to be an Angle. And that *Clavius* did not well to say, *the Angle of a Semicircle* was less than *a Right-*

---

36. It is worth noting that this was not always Hobbes's opinion. In his critique of Thomas White's *De Mundo,* Hobbes devotes chapter 23 to a discussion of the angle of contact. His conclusion is that "[i]t is doubtful what measure should be assigned to the angle of contact. If, however, this is an angle and is quantity, then it must possess a measure of the same kind as itself; whence arose the argument between Clavius and Peletier. Here it seems to me that . . . the opinion of Peletier, though true, is [based on] ineffectual arguments; but Clavius, besides putting forward an erroneous view, has contrived further mistaken tenets in order to lend it support" (Hobbes 1976, 262).

lined Angle. And that *Euclide* did not well to leave it so obscure what he meant by *Inclination* in the Definition of a *Plain Angle*" (*SL* 3; *EW* 7:258).

Hobbes's solution to this problem is quite ingenious. By distinguishing two kinds of angles, Hobbes paves the way for separate means of measuring such angles. He thus defends the opinion that the angle of contact is a quantity by showing how to measure it, but in distinguishing the measure of the angle of contact from the measure of more familiar angles he avoids the paradox that threatens if we allow the magnitude of the angle of contact to be compared with that of an ordinary rectilinear angle.

Hobbes had no patience with Wallis's approach to this difficulty. In the third of his *Six Lessons* Hobbes mounted a counterattack against the criticisms of Wallis by reviewing the principal arguments in the *Geometrical Disquisition on the Angle of Contact*. He notes the shortcomings in Wallis's argument for the homogeneity of all angles,[37] and finds fault with his doctrine that there is no mutual inclination between circle and tangent at the point of contact. This latter doctrine poses particularly difficult problems for Wallis, because he is committed to the unusual notion that a curve and its tangent have no inclination at their common point, while a curve and its secant do manifest such an inclination. The idea that points can have inclination to one another is certainly not one with a great deal of intuitive appeal, especially given Wallis's understanding of points as lacking dimension. Hobbes ridicules the doctrine with the remark "I pray you tell me what straddling there is of two coincident Points, especially such Points as you say are nothing. When did you ever see two nothings straddle?" (*SL* 3; *EW* 7:263). Hobbes even puts forward an argument designed to show that lines that cut one another need not form an angle homogeneous to a rectilinear angle.[38] The argument is worth considering briefly for the purposes of clarifying some of the issues in dispute. Hobbes writes:

---

37. Thus, Hobbes notes, "Your first argument therefore is nothing worth, except you make good that which in your second Argument you affirm, namely, That all Plain Angles, not excepting the Angle of Contact, are (*Homogeneous*) of the same kind. You prove it well enough of other Curvilineall Angles; but when you should prove the same of an Angle of Contact, you have nothing to say but . . . *whence should arise that diversity of kind, which [Clavius] dreams of, neither can he at all shew, nor I dream*" (*SL* 3; *EW* 7:259).

38. This example, I must confess, undermines my earlier account of Hobbes's distinction between angles of contingence and angles of circumlation in Jesseph 1993a, where I claim that the two classes of angles are distinguished by whether or not the lines

And why have two straight Lines Inclination before they come to touch, more then a straight Line and an Arch of a Circle? And in the Point of Contact it self, how can it be that there is less Inclination of the two Points of a straight Line and an Arch of a Circle, then of the Points of two straight Lines? But the straight Lines you say will cut; Which is nothing to the Question; and yet this also is not so evident, but that it may receive an objection. Suppose two Circles AGB and CFB to touch in B [figure 4.5], and have a common Tangent through B. Is not the Line CFBGA a crooked line? And is it not cut by the common Tangent DBE? What is the Quantity of the two Angles FBE and GBD, seeing you say neither DBG nor EBF is an Angle? 'Tis not, therefore, the cutting of a crooked Line, and the touching of it, that distinguisheth an Angle simply, from an Angle of Contact. That which makes them differ, and in kind, is, that the one is the Quantity of a *Revolution,* and the other the Quantity of *Flexion.* (*SL* 3; *EW* 7:260–61)

This argument is intriguing for several reasons. First, it highlights Hobbes's methodological tenet that the different species of geometric objects must be distinguished by their causal origin. The angle of contact is formed by the flexion of a line, while ordinary angles are formed by the rotation of lines; thus, the measure of the two kinds of quantity must be adapted to the different manners in which the quantities are produced. Wallis follows the more traditional method of arguing from a definition that is supposed to explicate the essence of the thing defined without reference to the manner in which it is produced. He thus defines the angle in terms of the mutual inclination of two lines, so that the question of whether the circle and its tangent fall under the relevant definition reduces to the question of whether they are inclined to one another. Wallis further sharpens his definition by requiring that lines inclined to one another must also cut, and he takes this criterion to solve the problem. In contrast, Hobbes takes the question of whether two lines cut to be entirely irrelevant to the manner in which they are produced, and hence useless in resolving the issue of the angle of contact.

Beyond this, it should be noted that the example Hobbes gives does

---

that form them cut one another. As the following example shows, this is not a completely accurate representation of Hobbes's views, since he constructs a curve that he claims exemplifies the angle of contact, even though the two lines in it cut one another.

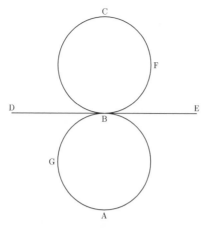

Figure 4.5

pose something of a problem for Wallis: the lines *CFBGA* and *DBE* do, in fact, cut one another, but because *DBE* is also the tangent to the curve at *B,* they do so without forming an angle (or, at least, without forming an angle in the sense that Wallis is prepared to recognize). Thus, Wallis must accept that the angles *DBG* and *FBE* have no magnitude. The case shows that the criterion for formation of an angle must be more complex than simply requiring that two curves cut. To vindicate his account, Wallis must add the requirement that the lines forming an angle cut without being tangent; in those cases where a tangent cuts the curve (such as points of inflexion, as in Hobbes's counterexample), there is no angle.

Hobbes and Wallis traded polemics over the angle of contact for many years, but the fundamental positions did not vary in the least. Hobbes continued to accuse Wallis of misunderstanding the nature of angles; Wallis countered that Hobbes had granted him every contested point. Ultimately, Wallis seems to have been intent upon a willful misunderstanding of Hobbes's solution to the problem, as is evident in his reply to Hobbes's *De Principiis et Ratiocinatione Geometrarum:*

> What he saith here ... concerning the *Angle of Contact;* amounts but to thus much, That, by the *Angle of Contact,* he doth not mean either what *Euclide* calls an *Angle,* or any thing of that kind; (and therefore says nothing to the purpose of what was in controversie between *Clavius* and *Peletarius,* when he says, that *An Angle of Contact hath some magnitude:*) But, that by the *Angle of Contact,* he understands the *Crookedness of the*

Arch; and in saying, *the Angle of Contact hath some magnitude,* his meaning is, that the *Arch of a Circle hath some crookedness,* or is a *crooked line:* and that, of equal Arches, That is the more crooked, whose chord is shortest: which I think none will deny. . . . But, why the *Crookedness of an Arch,* should be called *an Angle of Contact;* I know no other reason, but, because Mr. *Hobs* loves to call that *Chalk,* which others call *Cheese.* (Wallis 1666, 292)

The objection seems quite beside the point. As noted, Hobbes based his distinction between two types of angles on straightforward geometric criteria and introduced two methods for measuring these distinct kinds of magnitudes. Wallis's comments here are revealing because they show that he stubbornly refused to accept Hobbes's case for distinguishing the measure of the angle of contact from the measure of ordinary angles.

Curiously, Wallis eventually adopted a position not radically different from that of Hobbes, although it emerged only in 1684, some five years after Hobbes's death. In his *Defense of the Treatise of the Angle of Contact* Wallis defended his previous treatise on the subject and added some elucidations and clarifications. In the fourth chapter he took up the task of explaining away the apparent "Paradox to sense" that a curve can divaricate from a line without forming an angle. He does this by introducing a distinction between flexion and fraction as the two means by which one line may depart from another:

A streight-line as APp, which we may suppose in a Perpendicular position to AC, may come to change its position, as from Perpendicular to Parallel, (as to some part of it) either by a Break, as at B; or by more such, as at D[,] E; making so many Angles, as there are Breaks; (each part retaining its own streightness as before) or (without any Break) by one continued Bowing, as AF. (1684, 89)

In the former case, the lines are removed from each other by fraction or breaking; in the latter case by flexion or bowing. Fraction produces a genuine angle, while the uniform flexion at every point leaves no angle. It is obvious that the distinction here is the same as Hobbes's distinction between angles of circumlation and angles of contingence. Although Wallis would never have acknowledged it, it is amusing to see that his treatment of the angle of contact ultimately bears a strong resemblance to that of his most bitter antagonist.

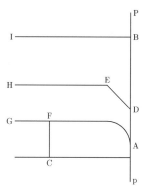

Figure 4.6

4.4 HOBBES, WALLIS, AND THE INFINITE

The topic of the infinite—and particularly the infinitely small—was a source of debate throughout the seventeenth century, and it is no surprise that Hobbes and Wallis should clash over it as well.[39] Their disagreement in this matter focused on Wallis's use of the infinite in his version of the method of indivisibles. The fundamental question at issue is whether a properly developed mathematical theory may postulate the existence of infinitesimal magnitudes, that is to say quantities greater than nothing but less than any assignable finite magnitude.

We saw in chapter 1 that Cavalieri's method of indivisibles was an important tool for the investigation of quadratures, and the method is often portrayed as an example of infinitesimal mathematics. However, Cavalieri himself avoided any direct commitment to the reality of infinitesimal magnitudes. Rather than taking geometric magnitudes as composed of infinite collections of infinitely small parts, Cavalieri tried to fit indivisibles into the classical theory of magnitudes by proposing that "all the lines" of a figure could be a new species of magnitude, on a par with lines, angles, surfaces, and solids. Thus, rather than assert that the area of the figure F is composed of an infinite sum of lines l_i, Cavalieri claims that the collection of lines enclosed in the figure F_1 stands in a determinate ratio to the collection of lines enclosed in another figure F_2. This approach is silent on the issue of the reality of infinitesimal quantities and requires only that one acknowledge that

39. This part of the dispute has been examined by Mancosu and Vailati (1991, 64–70) and Beeley (1996, 272–74).

"all the lines" contained in figures are themselves magnitudes that can be compared to one another in ratios. Later mathematicians were not necessarily as cautious or scrupulous in their approach to the infinite.

Wallis was an enthusiastic proponent of the infinitesimal, and in his presentation of the method he made no scruple of speaking of continuous geometric magnitudes as literally composed of infinite collections of infinitely small parts. Where Cavalieri had tried to avoid any commitment to the reality of infinitesimal magnitudes, Wallis saw no difficulty in taking a line or curve as an aggregate of indivisible points, a surface as an infinite collection of lines, and so forth. He was also quite happy to assume that Cavalieri's presentation of the method had resolved any lingering foundational doubts about the infinite. In his account of the foundation of the method:

> It is understood that any continuum (according to the Geometry of Indivisibles of Cavalieri) consists of an infinite number of indivisibles.
>
> As from an infinity of points, a line; a surface from an infinity of lines; and a solid from an infinite number of surfaces; so also time from an infinity of temporal moments, etc. That is . . . from homogeneous particles, infinitely small and infinite in number, and . . . equal in at least one dimension. (*Mechanica* 5; *OM* 1:645–46)

Notwithstanding the novelty of this approach, at least when judged by the classical standards of rigor, Wallis insists that it is merely a notational variant of the classical "method of exhaustion," which eschews infinitesimal considerations and proceeds by constructing a sequence of approximations converging to the desired result. In Wallis's estimation:

> The Method of Exhaustions (by Inscribing and Circumscribing Figures, till their difference becomes less than any assignable,) is a little disguised, in (what hath been called) Geometria Indivisibilium . . . which is not, as to the substance of it, really different from the Method of Exhaustions, (used both by Ancients and Moderns,) but grounded on it, and demonstrable by it: But is only a shorter way of expressing the same notion in other terms. (*Treatise of Algebra* 74, 285)

Not surprisingly, Wallis admits that the new method must be "applied with due caution" (*Mechanica* 5; *OM* 1:646) in order to avoid familiar paradoxes involving the infinite and the composition of the continuum.

Wallis's procedure rests (in part) on his conviction that arithmetic is the genuine foundation of all mathematics, and that geometry is a science subordinate to that of arithmetic. The true method of solving geometrical problems, according to Wallis, requires that they be represented arithmetically as infinite sums and their solution sought at the level of arithmetical calculation. He portrays this approach as "much more gra[c]eful and agreeable, than the Operose Apagogical Demonstrations . . . which some seem to affect," largely because there is no need for complex reductio ad absurdum arguments in the style of the classical method of exhaustion. Such classical techniques were made necessary "for reasons which now (in great measure) are ceased since the introducing the Numerical Figures, and (much more) since the way of Specious Arithmetic" (*Treatise of Algebra* 78, 298).

Another great advantage to be gained by the new methods is that they show quadrature results to be independent of any specifically geometrical content. The use of infinite series in solving geometric quadratures avoids dependence upon diagrams and represents the results in a manner independent of our intuitions concerning the structure of geometric space. In Wallis's defense of his methods, the lack of diagrams becomes a peculiar strength of his approach:

> If any think [these demonstrations] less valuable, because not set forth with the Pompous ostentation of Lines and Figures: I am quite of another mind. For though such Lines and Figures be necessary where the Truth of a Proposition depends on Local Position: And though they be otherwise of use, sometimes for assisting the Fansy or Imagination (shewing that to the eye, by way of instance, in one particular case, as that of Lines; which is abstractly true in all kinds of Quantity whatever:) Yet where the truth of the Proposition depends merely on the nature of Number or Proportion . . . It is much more natural to prove it abstractly from the nature of Number and Proportion; without such embarrassing the Demonstration. (*Treatise of Algebra* 78, 298)

Wallis characterizes this approach to geometry as a kind of "*Abstractio Mathematica* . . . whereby we separate what is the proper Subject of Inquiry, and upon which the Process proceeds, from the impertinences of the matter (accidental to it,) appertaining to the present case or particular construction" (*Treatise of Algebra* 76, 292). This perspective is evident even in the title of Wallis's main work on the method of indivisibles—*Arithmetica Infinitorum,* or the arithmetic of infinities. His point is that the infinite, and especially the infinitely small, falls within the purview of the general science of quantity, so that the arithmetic of infinities is a legitimate branch of mathematics useful for the investigation of quadratures.

As I mentioned in chapter 1, Wallis's arithmetic of infinities approaches geometric problems of quadrature by first representing a figure as an infinite sum of infinitesimal elements and then calculating the appropriate sum in order to find the area of the figure. This approach requires the evaluation of infinite sums in order to obtain a quadrature. Fundamental to this process is Wallis's method of "induction," which infers the value of an infinite sum from an investigation of the initial cases. As he expresses it:

> The simplest way of investigating this and other problems is to set forth a certain number of cases and observe the resulting ratios, and then compare them with one another in order that the universal proposition can then be known by induction. For example, we have

$$\frac{0+1}{1+1} = \frac{1}{2}$$

$$\frac{0+1+2}{2+2+2} = \frac{3}{6} = \frac{1}{2}$$

$$\frac{0+1+2+3}{3+3+3+3} = \frac{6}{12} = \frac{1}{2}$$

$$\frac{0+1+2+3+4}{4+4+4+4+4} = \frac{10}{20} = \frac{1}{2}$$

$$\frac{0+1+2+3+4+5}{5+5+5+5+5+5} = \frac{15}{30} = \frac{1}{2}$$

$$\frac{0+1+2+3+4+5+6}{6+6+6+6+6+6+6} = \frac{21}{42} = \frac{1}{2}$$

> And in the same way, however far we may proceed, this will always produce a ratio of 1 to 2. (*AI* 1; *OM* 1:365)

The difficulty with such a procedure is obvious enough, since it in effect draws a conclusion about the infinite case from the examination of a few initial cases. In the example cited the ratio of quantities remains constant as the number of terms increases, but Wallis also sought to determine ratios between infinite series where the ratios vary but tend toward a fixed value. In both kinds of reasoning there remains the difficulty of trying to assign a finite value to an infinite sum in the absence of principles that extend to the infinite case. The development of the calculus in the eighteenth and nineteenth centuries eventually led to a rigorous theory of infinite series, but in Wallis's day there were no clearly defined criteria for the summation of infinite series. The rather casual manner in which Wallis approached questions of the infinite has been widely acknowledged, and we need not dwell on it except to note that it was an inviting target for Hobbes's criticism.[40]

Hobbes wasted no time in attacking this weak link in Wallis's mathematics. He ridiculed such "inductions" in his *Six Lessons* with the words: "Egregious Logicians and Geometricians, that think an *Induction* without a *Numeration* of all the particulars sufficient to infer a Conclusion universal, and fit to be received for a Geometricall Demonstration!" (*SL* 5; *EW* 7:308). Lacking a coherent theory of the infinite or a methodology that could rigorously establish results for the evaluation of infinite sums, Wallis tried to defend himself by insisting that his methods comport with the classical standards. In particular, he argued that by including the expression "and the like in other cases," the reasoning "may passe for a proofe, till there be a possibility of giving some instance to the contrary; which, here, you will never be able to doe" (*Due Correction* 42). The alleged classical warrant for this procedure is Euclid's procedure of effecting a construction for a single case, say a given line designated $\alpha\beta$, and then generalizing the result to any other line.[41] The weakness of this pretense is manifest: when a construction

---

40. Scott 1938, 19–21, is representative in regarding Wallis's approach to the infinite as an untoward lapse that can still be interpreted charitably: "Unfortunately, however, he falls into the not uncommon error of regarding infinitely small and zero as synonymous expressions, *i.e.* as though $a - a$ were the same thing as $1/\infty$. . . . Frequently he treats infinity as though the ordinary rules of arithmetic could be applied to it. . . . But this is perhaps understandable. For many years to come the greatest confusion regarding these terms persisted, and even in the next century they continued to be used in what appears to us as an amazingly reckless fashion." Prag 1931, 390, observes that Wallis himself did not use the term "proof" (or its Latin equivalents) in speaking of his methods or their results; this suggests that perhaps the Savilian professor himself did not regard his procedures as completely demonstrative.

41. As Wallis puts it:

[I]f such an induction may not passe for proofe, there is never a proposition in Euclide demonstrated. For all along he takes no other course then such, (or at least grounds his Demonstrations on propositions no otherwise demonstrated.) As for instance; he proposeth it in general . . . *to make an Equilater triangle on a line given.* And then shews you how to doe it upon the line $\alpha\beta$ which he then shews you: and leaves you to supply, *and the same by the like meanes, may be*

is effected on a line or figure "in general position," the construction does not depend upon any specific features of such lines and figures, and hence the result may be generalized to other lines and figures that share the properties appealed to in the construction. Wallis's results for the summation of infinite series lack this kind of generality. Indeed, without an analogue of the schema for mathematical induction, Wallis's results are not rigorously demonstrated. In this particular case Hobbes can embarrass Wallis, precisely because there was no adequate theory of convergence upon which to base the arithmetic of infinities. In point of fact, if the "official" standard of rigor from the seventeenth century is applied to Wallis's principal theorems they must be accorded the status of plausible conjectures rather than genuine demonstrations. Hobbes, of course, never tired of insisting that Wallis's results were undemonstrated, and his continued polemics against the *Arithmetica Infinitorum* are a recurring theme in the dispute.

I showed in chapter 3 that Hobbes was influenced by Cavalieri's method of indivisibles and that he adopted it almost without alteration. Such obvious acceptance of the method might make it surprising that Hobbes should attack Wallis's use of indivisibles, but it is clear that Hobbes understood the method in completely different terms than Wallis. In fact, Hobbes saw Cavalieri as working with a conception of points and lines that differs very little from his own materialistic program for the foundations of mathematics. In particular, Hobbes thought that Cavalieri attributed breadth to lines, and he took this to be part of the foundation of the method of indivisibles. This is brought out in the part of the *Examinatio* that deals with the concept of homogeneous magnitudes, where one of the interlocutors in Hobbes's dialogue declares that:

> Those things that can exceed one another when multiplied are homogeneous, and these are measurable by a measure of the same kind, as lines are measurable by lines, surfaces by surfaces, and solids by solids. However, things heterogeneous are measured by different kinds of measures. Nevertheless, if lines are considered as the most minute parallelograms, as they are considered by those who use the method of demonstration that Bonaventura Cavalieri calls the doctrine of *Indivisibles,* then there will be a ratio between a *right line* and a *plane surface.* And indeed

---

done upon any other streight line; and then inferres his generall conclusion. Yet *I* have not heard any man object, that the induction was not sufficient, because he did not actually performe it in all lines possible. (*Due Correction* 42)

such lines, when multiplied, can exceed any given finite plane surface. (*Examinatio* 2; *OL* 4:75)

It is likely that Cavalieri's deliberate vagueness about the foundations of his method is responsible for Hobbes's unusual interpretation of the method. There is also the possibility that Cavalieri encouraged such an interpretation with some of his less strict statements of the method of indivisibles. In particular, such expressions as the "cloth and book" metaphor in the *Exercitationes* could easily have encouraged Hobbes's misinterpretation. There, Cavalieri writes "it is clear that plane figures are to be conceived by us as like cloth made up of parallel threads, and also solids as like books composed out of parallel pages" (Cavalieri 1647, 3). Despite its apparent affinity to Hobbes's conception of mathematical objects, this passage is really quite misleading. Perhaps not surprisingly, the metaphor was widely known and used to justify various interpretations of Cavalieri's methods.[42] In fact, Cavalieri immediately adds an important qualification: material objects such as cloth and books are composed out of finite collections of finite parts, where each part has some thickness, while the lines and planes considered in the method of indivisibles are "indefinite" in number and are conceived as lacking thickness (Cavalieri 1647, 4). Whatever the source of his interpretation of Cavalieri, Hobbes seems to have seen in him one of his few mathematical allies, notwithstanding the fact that Cavalieri opposed Hobbes's doctrine of the fundamentally material nature of mathematical bodies.[43] Since there is no deep doctrinal affinity between these two thinkers, the latter's reticence on foundational issues

42. De Gandt 1991, 161 n. 12, observes that this passage is not representative of Cavalieri's doctrine, but "unfortunately [it] was more widely known and quoted than his rigorous and technical formulations."

43. The divergence between Hobbes's and Cavalieri's conceptions of mathematics can be understood from Cavalieri's comments on the relationship between mathematics and the study of nature. Cavalieri declares: "Not all things that are considered by geometers are necessarily found in the nature of things just exactly [as they are in mathematical theories], for natural things, frequently sullied with matter and enmeshed in imperfections, are so inconstant that they can hardly be the object of science. For this reason mathematicians are constrained to abstract from this sensible matter and to pursue their studies in the purest and simplest figures, which always remain the same. And thus Archimedes presented the doctrine of spirals, even if perhaps no line of this sort is found in nature. Thus Galileo, and more recently Torricelli, examined the effects produced by local motion according to the same suppositions, regardless how nature in itself may work by local motion. All of which things, even if they were only minimally employed in human affairs, are by no means to be neglected, as they serve as nourishment for our contemplative powers" (Cavalieri 1647, 322). This doctrine is clearly at odds with the Hobbesian conception of mathematics.

is responsible for Hobbes's approval (and reinterpretation) of Cavalieri's doctrines.[44]

Whatever the ultimate source of Hobbes's interpretation of Cavalieri, he insisted that the method of indivisibles could not be properly comprehended without attributing extension to points and breadth to lines. Thus, in the fifth of his *Six Lessons,* he remarks to Wallis: "[Y]ou think it will pass for current, without proof, that a Point is nothing. Which if it do, Geometry also shall pass for nothing, as having no ground nor beginning but in nothing. But I have already in a former Lesson sufficiently shew'd you the consequence of that opinion. To which I may add, that it destroys the method of *Indivisibles,* invented by *Bonaventura;* and upon which, not well understood, you have grounded all your scurvy book of *Arithmetica Infinitorum*" (*SL* 5; *EW* 7:300–301).

Hobbes may well be wrong in thinking that Wallis's methods depend upon the supposition that a point is "nothing," but his complaint can be rephrased as a more cogent objection to the doctrine of infinitesimal parts. If, indeed, a line is "breadthless length," then the addition of lines can never constitute a breadth because it would simply be a sum of zeroes. Wallis himself sometimes preferred to speak of surfaces as composed of infinitely narrow parallelograms rather than breadthless lines, but this minor change does nothing to make his doctrine more coherent. In his *Treatise of Conic Sections* he announces that "I suppose, to begin with (according to the *Geometry of Indivisibles* of Bonaventura Cavalieri), any plane to be made up *[conflari]* so to speak out of an infinity of parallel lines; or (which I prefer) from an infinity of parallelograms of the same altitude. Let the altitude of any one of them be $1/\infty$ of the whole, or an aliquot part infinitely small (the sign ∞ denoting an infinite number), and the altitude of all together being equal to the altitude of the figure" (*Conic Sections* 1; *OM* 1:297). Wallis makes no attempt to distinguish an infinitely narrow parallelogram from a line, except to say that the former is supposed to "be dilatable, or to have so much thickness that by infinite multiplica-

---

44. In this context it is interesting to contrast Hobbes's evaluation of Cavalieri (and his associate Evangelista Torricelli) with the work of later mathematicians, particularly Wallis. In the *Admonitio ad Lectores* at the end of the *Lux Mathematica,* Hobbes remarks that "we had, avid reader, very skillful masters of the human sciences (I speak of geometry and physics) in the most distant ages: above all in geometry Euclid, Archimedes, Apollonius, Pappus, and others from ancient Greece. More recently we have Cavalieri and Torricelli from Italy. . . . But today, I say we are not even staying even, but instead are falling backward" (*Lux* 14; *OL* 5:147–48). Hobbes presumably intends Wallis as one of the principal examples of declining standards.

tion it can acquire a certain altitude or latitude, namely as much as is in the altitude of a figure" (*Conic Sections* 1; *OM* 1:297). However, this alleged difference is immediately rendered worthless when Wallis announces that the altitude of the parallelograms that compose a surface "is supposed to be infinitely small, that is, no altitude, for a quantity infinitely small is not quantity, scarcely differing from a line" (*Conic Sections* 1; *OM* 1:297).

Hobbes found this confused jumble of doctrine the ideal place to launch a counterattack against Wallis. He was particularly emphatic in pressing the point that the infinitely narrow parallelograms that allegedly compose surfaces must each have either some altitude or none. But in either case the doctrine faces incoherence:

> The least Altitude, is Somewhat or Nothing. If Somewhat, then the first character of your Arithmeticall Progression must not be a cypher; and consequently the first eighteen Propositions of this your *Arithmetica Infinitorum* are all naught. If Nothing, then your whole figure is without Altitude, and consequently your Understanding naught. (*SL* 5; *EW* 7:308)

Hobbes found it particularly galling that Wallis should occasionally resort to the expedient of claiming that the altitude of these infinitely small parallelograms need not be considered, and that in such cases the infinitesimal parallelogram is taken for a line.[45] This manner of expression differs very little from Hobbes's doctrine that lines are bodies whose length is considerable but whose breadth need not be considered, and he reminds Wallis that " 'Tis very ugly in one that so bitterly reprehendeth a doctrine in another, to be driven upon the same himself by the force of truth when the thinks not on't" (*SL* 5; *EW* 7:309).

Wallis's departure from Cavalieri's cautious approach to matters of the infinite made him vulnerable to charges that his methods lacked an adequate foundation. Never sparing an opportunity to challenge the competence of his opponent, Hobbes proceeded to level such charges in the first of his *Three Papers Presented to the Royal Society, against Dr. Wallis* and in subsequent publications directed at the Royal Society, including the *Lux Mathematica*. In the *Lux Mathematica* Hobbes

---

45. Wallis, quoted by Hobbes, declares, "*We will sometimes call those Parallelograms rather by the name of Lines then of Parallelograms, at least, when there is no* consideration *of a determinate Altitude; But where there is a* consideration *of a determinate Altitude (which will happen sometimes) there that little Altitude shall be so far* considered, *as that being infinitely multiplied it may be equal to the* Altitude *of the whole Figure*" (*SL* 5; *EW* 7:309).

summarizes his case against Wallis's results by arguing that they all rest upon two fundamentally mistaken principles:

> The first is one that, so he says, comes from Cavalieri, namely this: *that any continuous quantity consists of an infinite number of indivisibles,* or of infinitely small parts. Although I, having read Cavalieri's book, remember nothing of this opinion in it, neither in the axioms, nor in the definitions, nor in the propositions. For it is false. A continuous quantity is by its nature always divisible into divisible parts: nor can there be anything infinitely small, unless there were given a division into nothing. (*Lux* 3; OL 5:109)

This objection is essentially an analogue of the claim that no collection of breadthless lines can be made to compose a surface, except that it is now phrased as the impossibility of dividing a continuous magnitude into indivisible parts. As such, it highlights the grave conceptual difficulties standing in the way of a mathematically sound theory of infinitesimal magnitudes, as well as the extent to which Wallis departs from Cavalieri's statements on the infinite.

The second flawed principle that Hobbes detects at the basis of Wallis's quadratures is one "so absurd that I can scarcely believe it is advanced by a sane man" (*Lux* 3; OL 5:110). The principle, as Hobbes states it, amounts to the assertion that an infinite series of quantities can be understood to have an end, or that the infinite can be taken to be finite. The basis for this charge against Wallis lies in his practice of taking two infinite sums of quantities and concluding that, in the infinite case, a determinate ratio will arise between them. Consider, for example, the thirty-ninth proposition in the *Arithmetica Infinitorum*. We have already examined the proposition briefly in chapter 1 but it is worth another look, particularly because Hobbes himself cites it as a basic example of Wallis's incoherent methods (*Lux* 3; OL 5:109). In the proposition Wallis attempts to establish the ratio between a sum of cubic quantities and a sum of quantities equal to the greatest cubic quantity and equal in number to them. Following his usual "inductive method" Wallis begins by examining the first few finite cases of the sequence, and observing that they tend toward the ratio 1:4. He then concludes that

> if an infinite series is taken of quantities in triplicate ratio to a continually increasing arithmetical progression, beginning with 0 (or, equivalently, if a series of cube numbers is taken) this will

be to the series of numbers equal to the greatest and equal in number as one to four. (*AI* 41; *OM* 1:382–83)

A major difficulty with this reasoning is the supposition that there can be a completed infinite series of quantities "equal in number" to another infinite series. Hobbes complains that Wallis requires that there be a last term to each of the infinite series compared in the ratio, which is in effect the assumption that the infinite can be understood as limited or finite (*Lux* 3; *OL* 5:110; *EW* 7:443). The reasoning is actually doubly problematic because the denominator of the fraction must, in the infinite case, be an infinite sum of infinitely great quantities.

The infinite as understood by Hobbes must always be inexhaustible, essentially incomplete, and beyond our comprehension. In *Leviathan* he declares that "[w]hatsoever we imagine is *Finite*. Therefore there is no Idea, or conception of anything we call *Infinite*. . . . When we say any thing is infinite, we signifie onely, that we are not able to conceive the ends, and bounds of that thing named; having no conception of the thing, but of our own inability" (*L* 1.3, 11; *EW* 3:17). This doctrine has clear precedents in the classical conception of the infinite, and it is linked in Hobbes's scheme of things to a strongly empiricist epistemology in which all ideas must take their origin in sense experience.[46] Because there can be no sensory experience of something infinitely large or small, the most that can be understood by the term "infinite" is the lack of limits. But when Wallis takes an infinite sum of quantities as a given whole, and compares it in a ratio with an infinite sum of quantities (each of which itself is infinite), he in effect treats the infinite as finite, or the limitless as limited.

The issue in dispute here is fundamental to Wallis's whole procedure, and it was debated by numerous others besides Hobbes and Wallis. Indeed, controversy over the nature and status of infinitary methods is a theme that recurs throughout the history of mathematics. In this conflict, Hobbes upholds an essentially classical attitude toward the infinite. He denies that the concept is fully comprehensible and requires that a properly developed mathematical theory make do with a conception of the infinite as inexhaustible and incomplete. This does not mean that Hobbes leaves no room for the infinite in his mathematics. He can grant, for example, that there are infinitely many numbers

---

46. As Hobbes states in *Leviathan*, "The Originall of them all, is that which we call SENSE; (For there is no conception in a mans mind, which hath not at first, totally, or by parts, been begotten upon the organs of Sense.) The rest are derived from that originall" (*L* 1.1, 3; *EW* 3:3).

or an infinite number of triangles on a given base, and he would allow such claims to be stated and proved within a mathematical theory. Nevertheless, he refuses to allow actually infinite collections to be taken as completed wholes.

It is worth remarking that some of Hobbes's procedures in *De Corpore* lend themselves fairly easily to an interpretation that seems at variance with the rejection of the infinite we find in other of his writings. For instance, some of his attempts to square the circle and his quadratures of deficient figures occasionally speak of quantities being divided infinitely, of figures being made up "of indivisible spaces" and of a "last part" of a division of a magnitude being found.[47] It should be remembered, however, that Hobbes's definitions of such terms as *point* and *line* allow for points to have some magnitude (although the magnitude is not considered), and lines to have some breadth (again, this remaining unconsidered). Thus, Hobbes can say that any quantity can be divided into two quantities of the same kind (i.e., there is no least part of a magnitude), but that beyond a certain threshold the resulting magnitudes can be left out of consideration (i.e., there will be some finite part remaining although its magnitude cannot be considered in a demonstration). This permits Hobbes to reject the actual division of magnitudes to infinity while retaining the language of "least parts" and taking the concept of the infinite to be indefinite or essentially incomplete.

Wallis, of course, wants to make the infinite an object of legitimate mathematical inquiry, and in this he represents the "progressive" mathematicians of the seventeenth century whose work culminated in the development of the calculus. But even Wallis did not wish to break completely with the classical tradition. In the fifth chapter of his *Mathesis Universalis,* during a discussion of the nature of number, he remarks: "Since the addition of units can be extended infinitely, nor can there ever be a greatest number because one or more units can be joined to it, it is therefore impossible that a greatest number be assigned, and still less every species of number. And thus although it is impossible that there should be numbers actually infinite, or an infinite number (that is which has no termini but exceeds all limits), nevertheless an infinity of numbers are possible" (*MU* 5; *OM* 1:28). Hobbes was naturally unimpressed by this apparent concession to the more classical conception of the infinite as incomplete, and he remarked that

47. Examples of this sort of language can be found in the appendix, particularly sections 2 and 3.

"even boys know" that there cannot be a greatest number, although he wondered "how boys can know this when they hear geometers themselves speak of a right line divided into parts infinite in number, and read Wallis's book *Arithmetica Infinitorum?*" (*Examinatio* 1; *OL* 4:52). Hobbes's complaints against Wallis's use of the infinite were summarized in the first of his *Three Papers* presented to the Royal Society, which concluded that the procedures employed in the *Arithmetica Infinitorum* were not only ill founded but useless for their intended purpose.

Wallis attempted to deflect Hobbes's criticisms by arguing that he never asserted the existence of an actual infinite. Instead, he claimed to have reasoned hypothetically, showing that if the completion of an infinite process is supposed, then the required result follows. As he put the matter in his reply to Hobbes's *Three Papers:* "Whether those things *Be* or *Be not;* yea, whether they *Can* or *Cannot be;* the Proposition is not at all concerned, (which affirms nothing either way;) but, whether they can be *supposed,* or made the *supposition, in a conditional Proposition.* As when I say, *If* Mr. Hobs *were a Mathematician, he would argue otherwise;* I do not affirm that either *he his,* or ever *was* or *will be* such. I only say (upon such supposition) *If he were,* what he is not; he would not do as he doth" (Wallis 1671, 2241). The stunning ineptitude of this response is worth noting. Wallis seems entirely unconcerned by the possibility that by supposing something inconsistent or impossible, he would deprive his results of an adequate demonstration. Hobbes pointed out exactly the difficulty here, namely that "if the supposition be impossible, then that which follows will either be false or at least undemonstrated" (Hobbes 1671a, 1; *EW* 7:443). Of course, it is the very possibility of an infinite sequence with a last member that is at issue here, and Wallis makes no headway against the objection; his insistence that the completion of an infinite sequence can be harmlessly supposed simply begs the question against those who challenge the coherence of infinitesimal mathematics.

Their opposing views on the infinite crystallized in the two disputants' different evaluations of Torricelli's counterintuitive discovery of the "acute hyperbolic solid" that has an infinite length but a finite volume.[48] This result, first communicated by Torricelli in 1643, involves the solid generated from the hyperbola $xy = a^2$ by first revolving it about one of its asymptotes and cutting it by a plane perpendicular to

48. This theorem and its connection to the Hobbes-Wallis controversy have been studied by Mancosu and Vailati (1991). My treatment of the issue is consequently brief.

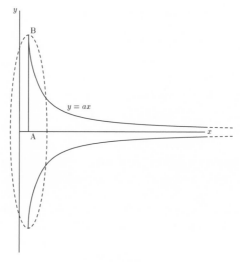

Figure 4.7

the axis of revolution. Thus, in figure 4.7, the hyperbola $y = ax$ is cut by a plane through *AB* and rotated about the x axis, producing a solid of infinite length but finite volume. Wallis cited the result to show that there is no difficulty in supposing an infinite quantity to be completed, bounded, or to have a last member. He points out that "[a] surface, or solid, may be *supposed* to be so constituted, as to be *Infinitely Long,* but *Finitely Great,* (the Breadth continually Decreasing in greater proportion than the Length Increaseth,) and so as to have *no Center of Gravity.* Such is *Toricellio's Solidum Hyperbolicum acutum;* and others innumerable, discovered by Dr. *Wallis,* Monsieur *Fermat,* and others. But to determine this, requires more of *Geometry and Logick* than Mr. *Hobs* is Master of" (Wallis 1671, 2243).

Hobbes's reaction was to insist that Torricelli's result could not be taken to assert the existence of an actually infinite solid of finite volume, for to understand it in that sense "it is not required that a man should be a geometrician or a logician, but that he should be mad" (Hobbes 1671a, 1; *EW* 7:445). Instead, Hobbes tried to show that Torricelli's theorem could be understood by taking the infinite to be indefinite and incomplete. In the final chapter of his *Principia et Problemata aliquot Geometrica,* which bears the title "On the infinite," Hobbes considers the Torricellian result from the standpoint of his doctrine of the infinite. He insists that "the infinite is the same as the unfinished *[imperfectum],* and is neither complete *[finitum],* nor can it ever be finished" (*PPAG* 13; *OL* 5:211). In the case of mathematics, Hobbes claims that the infinite should be understood as the indefinite, or that which can be as great or small as desired. He concludes that "nothing is properly signified by the word *infinite,* except it exceeds every assignable number of given measures *[superet mensurarum datarum numerum omnem assignabilem]*" (*PPAG* 13; *OL* 5:213).

With particular reference to the case of the Torricellian solid, Hobbes considers the objection that the acute hyperbolic solid shows that the infinite can be taken as a fixed and determined quantity, although one greater than any assignable one. He replies that

> the distance which Torricelli supposes *infinite* is to be understood as *indefinite.* Nor could it be understood differently by him, who in quite a lot of demonstrations uses Cavalieri's principles of indivisibles, which are such that their aggregate can be equated to whatsoever given magnitude. Therefore, a proposition as absurd as this, that the infinite is equal to the finite, must not be attributed to Torricelli. In fact, as is clear by the natural light, there can be no solid so subtle that does not infinitely exceed every finite solid. (*PPAG* 13; *OL* 5:213)

This reading of the result runs counter to Torricelli's own understanding of it, and it is not without difficulties of its own. Most obviously, taking the solid as only indefinitely long requires that the equality in the theorem be restated somewhat. Rather than saying that it is of indefinite (finite) length and equal in volume to a given cylinder, Hobbes must say that the solid can be made as long as desired and, in consequence, its volume will approach that of the associated cylinder to within any desired degree of accuracy. Hobbes's own comments are sufficiently brief to make it unclear just how he intends the theorem to be restated, but it is clearly difficult for him to take it at face value.[49]

In the end, this phase of the dispute shows that Hobbes was in some

---

49. On this point Mancosu and Vailati remark that "one could construct an infinite succession of volumes of such solids and claim, by a limit process, that the succession converges to the volume of the cylinder. However, attributing such a view to Hobbes would not only be historically inaccurate, because it would involve forcing our own mathematical notions on him, but would also foist on him conclusions he would hardly tolerate. In fact, although Hobbes would certainly accept the hyperbolic solids whose volumes are elements of this succession, he would have to reject the passage to the limit, since the limit of the succession does not belong to the succession itself, and in order to be equal to the volume of the cylinder the limit must be equal to the volume of an actually infinitely long solid. In sum, Torricelli's result is boldly infinitistic, and Hobbes's attempts at reducing it to a finitistic framework are destined to fail" (1991, 68–69).

important respects more scrupulous and rigorous than his antagonist. Although they almost invariably achieve results that we today recognize as valid, Wallis's methods were shot through with inconsistencies and incoherence, while his pronouncements on the foundations were more obfuscatory than clarificatory. Hobbes rightly pointed out the obscurity of infinitesimal mathematics, and although he did not have a fully developed alternative, his objections were not the ravings of a madman.

CHAPTER FIVE

The "Modern Analytics" and the Nature of Demonstration

> Therefore this analytics is an altogether narrow thing, although it is not completely useless in trigonometry applied to right lines; but because of the great multitude of symbols with which it is burdened today, along with the false opinion that values this method more than it truly merits, it is to be regarded as the plague of geometry.
> —Hobbes, *Examinatio*

The disputes outlined in chapter 4 all concern issues in the philosophy of mathematics, and our study of them shows that there was a deep and apparently irreconcilable conflict between Hobbes's and Wallis's conceptions of mathematics and the principles appropriate to it. These were by no means the only points of contention between Hobbes and Wallis. They also held opposing views on the role of algebra and the status of analytic geometry, and my principal purpose in this chapter is to explore these differences. This topic is clearly related to the disputes we examined in the previous chapter, but its complexity and importance warrant separate treatment. A proper understanding of the issues in dispute requires that we first examine Hobbes's views on the nature of language and demonstration, including his distinction between analytic and synthetic methods. After covering these topics, we can proceed to consider his specific charges against the "modern analytics" he so frequently and vehemently condemned. Along the way we will also be concerned with Wallis's critique of this part of the Hobbesian enterprise, as well as the objections of others, most notably Descartes and Ward.

5.1 HOBBES ON SIGNS, LANGUAGE, AND DEMONSTRATION

Hobbes's philosophy of mathematics is closely linked to his treatment of language and its role in reasoning. Human beings derive many benefits from the possession and use of language, but chief among them is the fact that language makes demonstrative knowledge, or true science, possible. As he argues in *Elements of Law*: "By the advantage of *names* it is that we are capable of *science,* which beasts, for want of them are not; nor man, without the use of them" (*EL* 1.5.4; *EW* 4:21). Hobbes uses the term *science* to convey something quite different from the meaning typically associated with the term today. The sense of the term relevant here derives from the Latin *scientia* and is meant to apply to knowledge that is certain, infallible, and universally applicable. In addition, Hobbes follows the tradition by requiring that scientific knowledge be grounded in an understanding of the causes of things and developed in the form of demonstrations. The locus classicus for such requirements is Aristotle's account of scientific knowledge as presented in the *Posterior Analytics,* a text that dominated Scholastic discussions of scientific methodology. Despite his scorn for the "school philosophy," it is evident that at least in this aspect of his philosophy Hobbes follows the Aristotelian tradition.[1] In fact, Hobbes himself declares that "this saying of Aristotle is true, 'to know is to know through causes'" (*PPAG* 1; *OL* 5:156).

Hobbes's conception of science therefore commits him to the principle that mathematics, politics, or any other body of knowledge that can be cast *more geometrico* by the use of reasoning and demonstration can count as a science. Hobbes also uses the term "philosophy" for such properly established doctrine, and he links such knowledge to

---

1. Aristotle's famous definition of scientific knowledge in *Posterior Analytics* 1.2 reads:

> We think we understand a thing *simpliciter* (and not in the sophistic fashion accidentally) whenever we think we are aware both that the explanation because of which the object is is its explanation, and that it is not possible for this to be otherwise. It is clear, then, that to understand is something of this sort; for both those who do not understand and those who do understand—the former think they are themselves in such a state, and those who do understand actually are. Hence that of which there is understanding simpliciter cannot be otherwise.
>
> Now whether there is also another type of understanding we shall say later; but we say now that we do know through demonstration. By demonstration I mean a scientific deduction; and by scientific I mean one in virtue of which, by having it, we understand something. (71b9–19)

The term *explanation* here translates the Greek αἰτία, which can also be rendered as "cause," provided that causes are construed broadly to include nonmechanical causes. On Hobbes's conception of science, with particular reference to its roots in Aristotelian and Scholastic teachings, see Leijenhorst 1998 and Gargani 1971.

the understanding of causes in a famous declaration in the fourth part of *Leviathan:*

> By PHILOSOPHY, is understood the Knowledge acquired by Reasoning from the Manner of the Generation of any thing, to the Properties; or from the Properties, to some possible Way of Generation of the same; to the end to bee able to produce, as far as matter, and humane force permit, such Effects as humane life requireth. So the Geometrician, from the Construction of Figures, findeth out many Properties thereof; and from the Properties, new Ways of their Construction, by Reasoning; to the end to be able to measure Land, and Water; and for infinite other uses. So the Astronomer, from the Rising, Setting, and Moving of the Sun, and Starres, in divers parts of the Heavens, findeth out the Causes of Day and Night, and of the different Seasons of the Year; whereby he keepeth an account of Time: And the like of other Sciences. (*L* 4.46, 367; *EW* 3:664)

The emphasis here on practical applications as the end for which all scientific activity is undertaken suggests that Hobbes does not see scientific knowledge as valuable for its own sake, but this is not an issue that I will be addressing. His conception of science as concerned with the understanding of causes is, however, an aspect of Hobbes's general methodology that I will discuss later.

Hobbes draws an important distinction between science (thus understood) and prudence, holding that although both kinds of knowledge depend upon the understanding of signs, they are distinguished by the fact that science is certain, infallible, and universal where prudence is probable, conjectural, and particular. Prudence, which Hobbes equates with accumulated experience, involves the interpretation of natural signs and is something we share with the beasts. In contrast, science involves the imposition of arbitrary signs and the construction of syllogisms to draw necessary conclusions. Thus, where prudential considerations show that dark clouds are a likely sign of impending rain, scientific reasoning unerringly demonstrates that a straight line is the shortest line connecting two points. Hobbes draws a memorable distinction between science and prudence in *Leviathan* when he declares:

> As, much Experience, is *Prudence;* so, is much Science, *Sapience.* For though wee usually have one name of Wisedome for them

both; yet the Latines did always distinguish between *Prudentia* and *Sapientia;* ascribing the former to Experience, the latter to Science. But to make their difference appeare more cleerly, let us suppose one man endued with an excellent naturall use, and dexterity in handling his armes; and another to have added to that dexterity, an acquired Science, of where he can offend, or be offended by this adversarie, in every possible posture or guard: The ability of the former, would be to the ability of the later as Prudence to Sapience; both usefull; but the latter infallible. (*L* 1.5, 22; *EW* 3:37)

The certainty and universal applicability of science stem from the fact that it is "knowledge of Consequences, and dependence of one fact upon another" (*L* 1.5, 21; *EW* 3:37), whereas prudence is confined to past experience and depends upon a fallible extrapolation from prior cases. In other words, scientific knowledge unerringly derives its conclusions from the true causes of the phenomenon to be explained, while prudential knowledge amounts to little more than guesswork. Another reason for this difference lies in the fact that the signs upon which prudence depends "are but *conjectural;* and according as they have often or seldom failed, so their *assurance* is more or less; but *never full* and *evident*" (*EL* 1.4.10; *EW* 4:17–18). In contrast, the signs used in demonstrative reasoning carry "full evidence" with them, and this evidence stems from the fact that they are imposed at will rather than arising from chance correlations found in nature.[2] It is in this sense that Hobbes can declare science to be "conditionall Knowledge, or Knowledge of the consequence of words," which arises when "Discourse is put into Speech, and begins with the Definitions of Words, and proceeds by Connexion of the same into generall Affirmations, and of these again into Syllogismes" (*L* 1.7, 30; *EW* 3:52–53).

Because science depends upon the use of signs, it must begin with the creation of a scientific language, a process that Hobbes characterizes as the imposition of names.[3] A name is an arbitrarily chosen sensible thing used to aid the recollection in bringing forth the ideas of things named. Hobbes holds that there can be thought without language, but there can be no science (i.e., philosophy) bereft of signs.

---

2. The contrast between science and prudence is explored further in Barnouw 1990.

3. The literature on Hobbes's theory of language is voluminous. Zarka 1987, pt. 2, is an extensive study of Hobbes's doctrines of meaning, language, and truth, which can

This is because "whatever a man has put together in his mind by ratio-cination without such aid will immediately slip from him," and there-fore "for acquiring philosophy, some sensible reminders are necessary, by which past thoughts can be recalled and each one registered as in its own order" (*DCo* 1.2.1; *OL* 1:12).

The arbitrary character of linguistic signs is the basis of their de-monstrative capacity, and it is a privilege unique to man that he can freely choose the use and meaning of signs and proceed to reason with them. This privilege means that man "can by words reduce the conse-quences he findes to generall Rules," but it is "allayed by another; and that is the priviledge of Absurdity; to which no living creature is sub-ject, but man onely" (*L* 1.5, 20; *EW* 3:33). Absurdity arises when the signification assigned to terms is not kept constant, or when terms of contrary signification are combined. The principal purveyors of absurd speech are school divines, whom Hobbes accuses of cynically mis-leading their students in an effort to increase their own power and hide their ignorance. In *De Homine* he contrasts the epistemic position of humans with that of beasts, concluding that deceptions based in the use of language are a disadvantage peculiar to humans:

> Man, if it should please him (and it will please him as often as it will seem to advance his plans), can teach what he knows to be false in given works; that is, he can lie and make the minds of men hostile to the conditions of society and peace. This cannot happen in the societies of other animals, because they judge what is good and bad for them by their senses and not from the com-plaints of others, the causes of which complaints they cannot un-derstand unless they see them. Moreover, it sometimes happens to those who listen to philosophers and schoolmen that, by the habit of listening, they rashly accept the words they hear even though no sense can be expressed by them (for such are the words that have been invented by the learned to hide their ignorance), and they use them, believing that they are saying something when they say nothing. Finally, because of the ease of speech, the man who really does not think still speaks, and believes that what he says is true, and he can deceive himself. But a beast cannot de-ceive itself. (*DH* 2.10.3; *OL* 2:91–92)

be supplemented with Hungerland and Vick 1973, Isermann 1991, Robinet 1979, and Sacksteder 1981b. Older, but still useful, are Hostetler 1945, Krook 1956, and Robbe 1960.

Thus, language does not "make man better, but only gives him greater possibilities," since the benefits of language are compensated by the potential dangers to which it exposes humans (*DH* 2.10.3; *OL* 2:92).

Although animals undeniably make noises to communicate, Hobbes holds that "the signification that is made by the voice [*vox*] of one animal to another of the same species is not speech [*sermo*], because it is not by their free choice [*arbitrio*] but by the necessity of their nature that their voices signify hope, fear, joy, etc., and are expressed by force of these passions" (*DH* 2.10.1; *OL* 2:88). He concludes that other animals must therefore "lack intellect, for intellect is a certain imagination, but one that arises from the agreed signification of words" (*DH* 2.10.1; *OL* 2:89). The use of speech to signify general truths and the intellectual employment of the imagination thus remain unique parts of the human cognitive apparatus.

The names imposed by human convention can be either proper or general: proper names are "singular to one onely thing," while general names are "*Common* to many things" (*L* 1.4, 13; *EW* 3:21). Once names have been instituted they can function either as marks or signs. A mark is a private name used only to remind the speaker of his previous thoughts, while a sign is a name accepted by others and used to communicate thoughts. Scientific knowledge arises when we employ reason to establish true propositions about the world, a proposition being "*a speech consisting of two names copulated, by which the speaker signifies that he conceives the latter name to be the name of the same thing of which the former is the name;* or (which is the same thing) that the former name is comprehended by the latter" (*DCo* 1.3.2; *OL* 1:27). Thus, if I assert the proposition "London is a city," I indicate that I assign the name 'city' to the same thing that I assign the name 'London', while in asserting "Humans are rational," I show that I assign the name 'rational' to whatever I indicate by the name 'humans'. The truth or falsehood of such propositions then depends only upon the relationship between the names employed in it: "When two Names are joyned together into a Consequence, or Affirmation; as thus, *A man is a living creature;* or thus, *if he be a man, he is a living creature,* If the latter name *Living creature,* signifie all that the former name *Man* signifieth, then the affirmation, or consequence is *true;* otherwise *false.* For *True* and *False* are attributes of Speech, not of things" (*L* 1.4, 14; *EW* 3:23).

This conception of truth as dealing with words rather than things runs against the grain of much seventeenth-century thought. Truth was often reified by thinkers of the period, as when Lord Robert Brooke's treatise *The Nature of Truth* concludes that truth is "the Understanding in its Essence," and undertakes to prove that "the Soule and Truth be One" (Brooke 1641, 33). Even those who resisted such strongly Platonistic accounts of truth held that it involves more than words. Wallis, whose *Truth Tried* is a rebuttal to Brooke, distinguishes a number of different kinds of truth (moral, logical, and metaphysical) and argues that they are "nothing else but an Agreement or Conformity of a *Type* with its *Prototype, Archetypi & Ectypi;* of a Transcript with its Originall; of an *Idea,* or thing representing with that represented; *Signi & Signati*" (Wallis 1642, 2). In this way of thinking about truth it makes sense to speak of "true ideas" as those that agree with their objects; Hobbes rejects such talk as confused holdovers from the Aristotelian notion that "falsity and truth are not in things . . . but in thought" (*Metaphysics* 6.4; 1027b26–27). Hobbes agrees with the Aristotelian tradition that truth and falsity do not reside in things, but rejects the claim that they are in thought. Instead, Hobbes takes truth to be exclusively in words imposed by human convention, so that "the first of all truths had their origin in the arbitrary choice of those who first imposed names on things, or accepted them as imposed by others. So it is true, for example, that man is an animal, for this reason: because it pleased men to impose these two names on the same thing" (*DCo* 1.3.8; *OL* 1:32).

Hobbes conceives of the reasoning by which true propositions are to be established as a kind of arithmetic involving the addition and subtraction of mental contents: "When a man *Reasoneth,* hee does nothing else but conceive a summe totall, from *Addition* of parcels; or conceive a Remainder, from *Substraction* of one summe from another" (*L* 1.5, 18; *EW* 3:29). Such addition or subtraction is performed most easily on signs or words by manipulating them in accordance with purely formal rules, but it is possible to compute without words, in which case our computations will be performed on phantasms. In *De Corpore* (1.1.3; *OL* 1:3–4) Hobbes illustrates this part of his doctrine by an example: suppose a man sees something in the distance, but so indistinctly as not to discern precisely what it is. At this stage he has only the idea of body, for he knows that it must be some kind of visible body. By approaching, he sees that the body moves itself about "now in one place and now in another," and he adds the idea "animated" to his previous idea of body. Upon closer investigation of the animated body he "sees the figure, hears the voice, and perceives other things that are signs of a rational mind, and has a third idea, even though there is as yet no name for it, namely that by reason of which we call anything rational." The three ideas of body, animated, and rational are then drawn into a mental sum by seeing that they all pertain to the same thing; thus arises a new idea (namely, the idea of man) compounded out of the three ideas of body, animation, and rationality. Analyzing the definition of the concept "man" into those of "body," "animation," and "rationality" is a similar kind of calculating process, but one that involves the decomposition of the complex idea into its components.

So described, reasoning or mental arithmetic is performed on particular ideas and without the use of words. The more usual cases of reasoning, however, involve words rather than ideas of particular things named by words. When words are taken as the object of reasoning the process remains one of calculation, but it consists in the drawing of consequences from the manipulation of general names. "Reason, in this sense," Hobbes writes, "is nothing but *Reckoning* (that is, Adding and Substracting) of the Consequences of generall names agreed upon, for the *marking* and *signifying* of our thoughts; I say *marking* them, when we reckon by our selves; and *signifying,* when we demonstrate, or approve our reckonings to other men" (*L* 1.5, 18; *EW* 3:30).

We have already seen that Hobbes's requirements for scientific knowledge include the stipulation that such knowledge be acquired by the manipulation of arbitrary signs, and we can now see that he locates the generality of such knowledge in the universality of general names. To put the matter another way, Hobbes holds that the use of general names in reasoning is essential if the consequences drawn are to apply beyond the range of past experience. Hobbes famously insists that "experience concludeth nothing universally" (*EL* 1.4.10; *EW* 4:18), and he stresses the crucial role of names in the formation of generalized knowledge. It is by the imposition of agreed-upon general names that we can "turn the reckoning of the consequences of things imagined in the mind into a reckoning of the consequences of Appellations" (*L* 1.4, 14; *EW* 3:21), and thereby extend such consequences of appellations beyond the scope of experience to become general or universal.

Hobbes illustrates the universality of reasoning from names with an imaginary geometrical case: suppose that someone entirely ignorant of speech contemplates a particular triangle and two right angles placed beside it; suppose further that this person concludes that, in this particular case, the sum of the interior angles is equal to the sum of the two right angles. Hobbes claims that inability to use words makes it impossible for such a person to generalize this result to cover other cases, while someone who has mastered the appropriate geometric vocabulary will acquire truly universalizable knowledge that the same result holds in all cases:

> But he that hath the use of words, when he observes, that such equality was consequent, not to the length of the sides, nor to any other particular thing in his triangle; but onely to this, that the sides were straight, and the angles three; and that that was all, for which he named it a Triangle; will boldly conclude Universally, that such equality of angles is in all triangles whatsoever; and register his invention in these generall termes, *Every triangle hath its three angles equall to two right angles.* (L 1.4, 14; EW 3:22)

This imaginary case actually echoes Aristotle's example in the *Posterior Analytics,* although it is put to much different purposes. Aristotle had concluded that "since demonstrations are universal, and it is not possible to perceive these, it is evident that it is not possible to understand through perception either; but it is clear that even if one could perceive of the triangle that it has its angles equal to two right angles, we would seek a demonstration and would not, as some say, understand it; for one necessarily perceives particulars, whereas understanding comes by becoming familiar with the universal" (*Posterior Analytics* 1.31; 87b34–39). Hobbes, in contrast, thinks that the universality of demonstrations can be accounted for simply by the use of names and without recourse to universals. I will examine this issue more fully when I consider Hobbes's nominalism and some of its difficulties.

The drawing of consequences from general names is the concern of logic, and Hobbes models his account of scientific inference on the deductive structure of classical syllogistic logic in Aristotle's *Prior Analytics.* A syllogism, in Hobbes's idiom is "speech that consists of three propositions, from two of which the third follows" (*DCo* 1.4.1; *OL* 1:39). In keeping with his account of reasoning as computation, Hobbes treats syllogistic inferences as a kind of mental addition in which the conclusion is drawn as a sum from the two premises (*DCo* 1.4.6; *OL* 1:42). Hobbes observes the standard division of syllogisms into moods and figures, but his treatment of deductive consequence as computation departs rather significantly from the standard view of the matter.[4] Because science aims to "establish universal rules concerning the properties of things," the syllogisms appearing in a scientific demonstration must contain only general names, for it is "superfluous to consider any other mood in direct figure, besides that in which all the propositions are both universal and affirmative" (*DCo* 1.4.7; *OL* 1:44).

This should suffice as a brief outline of the Hobbesian theory of language and demonstration. There are a number of tensions internal to this account, as well as some important objections that opponents raised to it, and we can now proceed to a consideration of such difficulties. In the end, I think that Hobbes's theory can avoid collapse into outright incoherence, but the full scope of his theory and its implications will emerge only after we have investigated his critique of algebra and analytic geometry. For the present, however, we can concern ourselves with issues that do not bear directly on the status of mathematical theories.

5.1.1 *Conventionalism, Causation, and Demonstration*

Hobbes's doctrines certainly seem to result in a conception of science that focuses primarily on such purely linguistic activities as the imposition of names, the analysis of meanings through definitions, and the construction of syllogisms. Indeed, Hobbes often writes as if the principal requirement for the acquisition of scientific knowledge is simply to set purely verbal definitions in order and connect their terms by syllogisms. The universal affirmative propositions appearing in scientific syllogisms would seem to depend for their truth only upon speakers' arbitrary conventions about the meanings of general terms, and this fact suggests that the scientist need tend only to the proper arrangement of terms in his syllogisms. In a summary of his account of science in *Leviathan,* Hobbes comes very close to the implausible claim that science involves nothing more than the correct ordering of names:

> Reason is not as Sense, and Memory, borne with us; nor gotten by Experience onely; as Prudence is; but attayned by Industry; first in apt imposing of Names; and secondly by getting a good and orderly Method in proceeding from the Elements, which are Names, to Assertions made by Connexion of one of them to an-

4. This is not the place for a discussion of the theory of consequence as understood in Hobbes's day. See Ashworth 1974, 120–36, for an overview of postmedieval theories of consequence. For studies of Hobbes's logic see Dal Pra 1962 and De Jong 1986.

other; and so to Syllogismes, which are the Connections of one Assertion to another, till we come to a knowledge of all the Consequences of names appertaining to the subject in hand; and that is it, men call SCIENCE. (*L* 1.5, 21; *EW* 3:35)

Here, Hobbes takes the "Elements" of demonstrative knowledge to be arbitrarily imposed names, and the whole enterprise appears to involve little more than investigating the consequences of such names. This account does have the virtue of making such knowledge completely certain: if demonstrations are grounded in the arbitrary imposition of names and confined to the analysis of names and their interconnections, then (barring inconsistency or ambiguous names) there is no danger that our reasoning might lead to falsehood. On the basis of such pronouncements, it seems that Hobbes is tempted toward an improbable doctrine that secures the truth and certainty of scientific knowledge at the cost of restricting it to the analysis of language. Furthermore, Hobbes's stress on the arbitrariness with which names are "imposed" on the world seems to commit him to a strongly conventionalist conception of truth in which the truth of a proposition amounts to nothing more than speakers' agreement upon the definitions of the terms it contains.[5]

Although Hobbes's stronger statements regarding the role of names may suggest that he conceived of all properly conducted inquiry as involving little more than a manipulation of names, this cannot be the dominant theme in his account of *scientia*. Such an approach clearly conflicts with his insistence that scientific demonstrations must proceed from causes. Hobbes notoriously held that true science must concern itself with the (mechanical) causes of things, whether these be natural phenomena, geometric objects, or the commonwealth. But the search for causes obviously involves more than simply assigning names to things and analyzing definitions. And indeed, Hobbes himself requires that proper scientific definitions contain the causes of the things defined, even to the point of insisting that "names of things that can be understood to have some cause must have this cause or mode of generation in their definitions, as when we define a circle to be a figure arising from the circumlation of a line in a plane, etc." (*DCo* 1.6.13;

5. Sorell 1988, 45–49, argues that Hobbes is not committed to a thoroughly conventionalist theory of scientific truth, although he employs slightly different considerations than those in play here. For an extended study of the role of conventionalism in Hobbes's treatment of science, particularly in the works before *De Corpore,* see Pacchi 1965. Bernhardt 1993 and Meyer 1992, chap. 2, also take up aspects of these questions.

OL 1:72). Elsewhere Hobbes insists that "where there is place for Demonstration, if the first Principles, that is to say, the Definitions contain not the Generation of the Subject; there can be nothing demonstrated as it ought to be" (*SL* epistle; *EW* 7:184). Definitions of this sort cannot be entirely arbitrary or conventional, because it is possible to have a definition that fails to satisfy such a requirement, either by giving no cause of the thing defined or falsely identifying its cause. Thus, we should not take Hobbes's comments on the role of names in science as evidence of a purely conventionalist theory of science or demonstration.[6]

It is admittedly arbitrary and a matter of speakers' convention what word we use to represent any particular thing. But this degree of arbitrariness or conventionality is consistent with there being better or worse definitions of terms, and Hobbes holds that proper definitions are those that reveal the causes of the things defined. For example, speakers of English may use the words *sunrise* or *dawn* interchangeably, and we can agree with Hobbes that it is a quite arbitrary or conventional matter that we use either word to designate a certain type of event. Apart from possible considerations of lyric meter or rhyme scheme, there is really nothing to choose between them. But such arbitrariness in the choice of *words* does not entail that any *definition* is as good as any other. It would be correct (but not fully enlightening) to define either one as "the appearance of the Sun over the horizon." It would be simply incorrect to define either as "the appearance of the Sun over the horizon, as a result of the Sun's revolution about the earth," and indeed the true definition must be "the appearance of the Sun over the horizon, as a result of the earth's circadian rotation." The last of these three definitions is the best precisely because it reveals the true cause of the thing defined, and that it does so is not the result

6. This point has caused some confusion in the literature. Meyer (1992, 95–105) claims that Hobbes's "conventionalist" theory of language and truth runs counter to his insistence that our definitions and reasoning deal with the true natures of things. Pacchi (1965, chap. 8) sees a conflict between Hobbes's "arbitrarism"—the view that truths are arbitrarily imposed by human convention—and his "intuitionism"—the thesis that self-evident first principles can be directly intuited. I agree with Malcolm (1990, 152–53) that "Hobbes's theory of universal truths was a product of his nominalism; and his nominalism was a good deal less extreme than is popularly supposed. He was a nominalist, not an arbitrarist. Hobbes believed that all blue objects, for example, are really similar: our use of the same word to describe them is not a mere freak of human will or fancy. Indeed, his mechanistic theory of sense-perception ensures this, since the nature of the conception in our brains which we connote with the word 'blue' is *caused* directly by the motion of the object which we see. We experience objects as similar because they really do cause similar motions."

of an arbitrary or conventional decision on our part. Hobbes's insistence on the freedom with which humans impose names must therefore be balanced by his recognition that proper definitions express causes. To put the matter another way, when Hobbes insists in *Leviathan* that science must begin with the "apt imposing of Names" we must remember that those names are most aptly imposed when they designate causes, or at least possible causes. Understanding Hobbes's account of definition in this way makes it clear why he can object that "a man may so precisely determine the signification of a word, as not to be mistaken, yet may his Definition be such, as shall never serve for proof of any Theoreme, nor ever enter into any demonstration (such as are some of the Definitions of *Euclide*) and consequently can be no beginnings of Demonstration, that is to say, no Principles" (*SL* 1; *EW* 7:200).

My exposition of Hobbes's program for the philosophy of mathematics in chapter 3 has already considered his view that proper geometric definitions must express the (mechanical) causes of geometric objects. Still, the requirement that demonstrative knowledge proceed from causes may seem to pose a difficulty in the case of mathematics, since we are not in the habit of thinking of this subject as employing causal principles. In fact, however, the insistence upon the causality of mathematical demonstration is hardly a Hobbesian novelty. One notable antecedent is Aristotle's theory of demonstration. Applied to the case of mathematics, the theory of demonstration in the *Posterior Analytics* requires that mathematical proofs proceed from causes.[7] Of course, Aristotelianism has a wider array of causal principles than those we acknowledge today. Because the Aristotelian philosophy takes individual substances as composites of form and matter, its methodology can distinguish between formal, material, efficient, and final causes. Thus, a causal explanation in the Aristotelian tradition can include reference to a substance's form (the formal cause), its matter (the material cause), the process that produced it (the efficient cause), and the end or purpose for which it was produced (the final cause). Given this broad conception of causation, Aristotle required a scientific demonstration be cast in the form of a syllogism whose premises must not only be true, but also express the cause of the conclusion. Demonstra-

---

7. On Aristotle's account of scientific demonstration, see Hankinson 1995. Mancosu (1992a; 1996, chap. 1) has drawn attention to the issues raised by the Aristotelian theory of demonstration in the context of sixteenth- and seventeenth-century mathematics (including the Hobbes-Wallis controversy). Aristotle's philosophy of mathematics is examined in Apostle 1952, Cleary 1982, Jones 1983, Lear 1982, and Mueller 1970.

tions that do not satisfy the causal condition are termed demonstrations τοῦ ὅτι, while those that satisfy it are demonstrations τοῦ διότι; these are standardly called "demonstrations of the fact" and "demonstrations of the reasoned fact," respectively. The point of the distinction is that a demonstration τοῦ ὅτι shows merely that something is the case, while the demonstration τοῦ διότι shows why it is the case by constructing a syllogism whose premises exhibit the cause of the conclusion.

Scholastic philosophy saw extensive debates on the question of whether mathematical demonstrations are true demonstrations, and Hobbes's mathematical and methodological writings should be read against the background of such controversies.[8] One objection raised against the causality of mathematical demonstration argued that the very nature of mathematical objects precluded the possibility of their being the object of demonstrations τοῦ διότι; the Jesuit Benedictus Pereira, for example, held that "mathematical things are abstracted from motion, therefore from all kinds of cause" (Pereira 1576, 70). A similar objection is reported by the Jesuit Martin Smigleckius in his *Logica* of 1634, albeit without his endorsement. He comments that some have objected to the scientific status of mathematics on the grounds that "mathematical objects such as quantities and figures, as they are considered by the mathematician, are not in nature, for indeed there are no lines existing per se, or surfaces abstracted from body, or perfect planes, or bodies perfectly round . . . but it is required for a true and perfect demonstration, that it deals with real beings, not imaginary ones" (Smigleckius 1634, 581). True science, according to this sort of objection, is ultimately a science of natural things and must concern itself with objects in the natural order. To the extent that mathematical objects are regarded as abstracted from the natural world, they cannot be the object of scientific investigation. Another kind of argument against the causality of mathematical demonstration held that the procedures used in Euclidean demonstrations could not express true causes because they introduce factors extrinsic to the object whose properties are being demonstrated; the auxiliary lines or circles constructed in the course of a typical proof seem unrelated to the essence of the original object and are therefore poor candidates for causes of the conclusion.

The standard example of this problem, and one discussed by anyone

8. See Crombie 1977, Jardine 1988, Mancosu 1992a, and Mancosu 1996, chap. 1, for more on this debate.

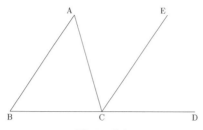

Figure 5.1

who held an opinion on the topic, is Euclid's proof of the theorem that "in any triangle, if one of the sides be produced, the exterior angle is equal to the two interior and opposite angles, and the three interior angles of the triangle are equal to two right angles" (*Elements* 1, prop. 32). The theorem is, obviously enough, of great foundational importance for Euclidean geometry—almost anyone who discusses the nature or essence of the triangle will mention the crucial fact that its interior angles sum to two right angles. The proof, however, makes use of two auxiliary lines that appear unconnected with the intrinsic nature of the triangle. Let ABC (in figure 5.1) be any triangle, and extend BC to D. Draw CE parallel to $BA,$ and there results an equality of angles $\angle ABC$ and $\angle ECD,$ as well as the angles $\angle BAC$ and $\angle ACE.$ Thus $\angle ABC + \angle BCA + \angle CAB = \angle ECD + \angle BCA + \angle ACE,$ and the result is obtained. Pereira and others who disputed the scientific status of mathematics, including Pierre Gassendi in his *Exercitationes Paradoxicae adversus Aristoteleos,* argued that such a proof fails to appeal to proper causes, since the essential parts of the construction are literally external to the triangle. They and others took this as decisive evidence against the claim that Euclidean geometry could satisfy the Aristotelian criteria for scientific demonstration.[9]

Defenders of the scientific status of mathematics (a group that included Wallis, Barrow, and Clavius) argued that, despite appearances, the proofs in Euclid can be interpreted in a manner consistent with Aristotle's theory of demonstration. Details of these defenses are sufficiently remote from my present concerns that they can be left for another day, but it is important to see that this kind of issue is part of the background to Hobbes's project for the philosophy of mathematics.

9. For an account of Gassendi's use of this argument from Pereira, see Mancosu 1996, 19–24. The principal argument for the thesis that geometry fails to meet the Aristotelian standard for science relies on Pereira and can be found in Gassendi [1658] 1964, 3:208–9.

This particular dispute was a matter of great importance in the philosophy of mathematics of the mid–seventeenth century, as is evident from the fact that Barrow devotes the sixth of his *Lectiones Mathematicae* to the topic of "the causality of mathematical demonstration," while Wallis spends the second and third chapters of his *Mathesis Universalis* arguing that the mathematical sciences count as sciences in the strictest sense. One point to stress here is that the causality of mathematical demonstrations was generally regarded as formal causality, in which the form or essence of a geometric object (as expressed in its definition) is causally responsible for and explanatory of the properties demonstrated of it.[10] In this respect, Hobbes's program for mathematics departs from the classical pattern. Hobbes regarded talk of formal causality as part of the empty verbiage of the schools, and his materialistic program for mathematics requires that the definitions of mathematical objects exhibit the kinds of motions by which such objects are produced. Thus, although Hobbes accepts the traditional dictum that "all knowledge is knowledge of causes," he restricts the concept of causality to that of efficient causality, and even this is understood mechanistically, so that it is only by the motion and impact of material bodies that anything can be caused. It is therefore evident that Hobbes's concerns with the causality of mathematical demonstration are anchored in Scholastic and early modern debates over the status of the mathematical sciences, but his solution to the problem departs significantly from the tradition.

His methodological principles also led Hobbes to downplay the distinction between demonstrations τοῦ διότι and τοῦ ὅτι. In his *Mathesis Universalis* Wallis had invoked the standard distinction between τοῦ διότι and τοῦ ὅτι in the course of defending the truly scientific nature of mathematics, arguing that many demonstrations were of the most perfect kind, which he calls "ostensive τοῦ διότι."[11] Hobbes objected

---

10. Barrow, in discussing the causality of mathematical proofs, sums up his opinion on this issue: "Such in truth, and no other, is the causality and mutual dependence of the terms of a mathematical demonstration. That is, a most close and intimate connection of them with one another, which can always be called formal causality, in that from one property first assumed, other attributes *[passiones]* result as from a form. Nor do I think that there is any other causality in the nature of things, in which a necessary consequence may be founded" (*LM* 6, 93).

11. The *Mathesis* states, "But a third sort of demonstration, which is the most perfect of all, is ostensive τὸ διότι, which demonstrates both that something is and why it is. A demonstration is of this sort if someone demonstrates that all the radii of the same circle are equal from the fact that the circle is defined (or at least can be defined) as a plane figure contained within one curve that is everywhere equidistant from the middle

to the classification, after first giving his own definition of demonstration. This reads: "A demonstration is a syllogism, or series of syllogisms, derived from the definitions of names to the final conclusion" (*Examinatio* 1; OL 4:36).[12] He then remarks that

> I was wishing that Wallis had defined what this demonstration τοῦ ὅτι is. For demonstration τοῦ διότι is when someone shows by what cause a subject has such an affection. And so because every demonstration is scientific, and knowledge that such an affection is in the subject comes from the knowledge of the cause that necessarily produces it, there can be no demonstration other than the τοῦ διότι. He rightly says that which is called τὸ ὅτι is not an utterly convincing [κυρίος] demonstration, that is, it is not a demonstration at all. For in the speech of mathematicians, "not to be" and "not to be properly" are the same thing. (*Examinatio* 1; OL 4:38)

The principle that all knowledge and demonstration must arise from causes plays an important role in Hobbes's philosophy of mathematics, and it figures prominently in his rejection of the techniques of algebra and analytic geometry. We will therefore be concerned with it later in this chapter, but we should first turn our attention to some salient features of Hobbes's nominalism.

5.1.2 Hobbes's Nominalism and Its Critics

Hobbes's account of language and reasoning is transparently nominalistic, notwithstanding its reliance upon the thesis that all demonstrative reasoning is from true causes. We can take nominalism to consist of two theses: the negative claim that there are no really existing universals, and the positive claim that the philosophical work traditionally assigned to universals can be done by words or names. To make this definition completely precise, it would help to know what universals are supposed to be, particularly as the doctrine of universals was understood in Hobbes's day. I am in no position to give a complete ac-

---

of the space it comprehends. For if the essence of the circle postulates that its periphery is equidistant from the center, it follows immediately as from a true and proximate cause that all the radii (by which this distance is measured) are also equal. And this is an ostensive demonstration, taken from the proximate and immediate cause" (*MU* 4; *OM* 1:23–24).

12. The very same definition appears in *DCo* 1.6.16: "Demonstratio est syllogismus vel syllogismorum series a nominum definitionibus usque ad conclusionem ultimam derivata" (*OL* 1:76).

count of this enormously complex doctrine, but a few simple points will suffice for the task at hand.[13] Aristotle declares "it is what is always and everywhere that we call universal" (*Posterior Analytics* 1.31; 87b33); and in *De Interpretatione* he remarks that "of actual things some are universal, others particular (I call universal that which is by its nature predicated of a number of things, and particular that which is not; man, for instance, is an universal, Callias a particular)" (*De Interpretatione* 7; 17a38–17b). Although he did not take universals to be separately existing substances, Aristotle held that the universal is present in the particular and can be grasped by the intellect. To understand that the interior angles of a triangle sum to two right angles, the intellect must grasp the universal "triangle" and reason about it; and, in general, to understand that the individual *a* is *F*, one grasps the universal *F*-ness, and sees that it pertains to the individual *a*. Exactly how the intellect goes about grasping these abstract essences or universals—and how these relate to the mundane realm of material things—is a complicated story that can barely be sketched here. Aristotle held that sense perception involved the transmission of "sensible species" from the perceived object to the mind of the perceiver; by analogy, the intellect's grasp of universals involves its reception of "intelligible species" from its proper objects. Like many other proponents of the "mechanical philosophy" Hobbes thought that the doctrine of sensible species was nonsense to be supplanted by a mechanistic account of sensible qualities in terms of the motion and impact of material particles. His scruples extended further, however, and included the abolition of the doctrine of intelligible species in favor of his materialistic theory of the mind.[14]

13. See Largeault 1971 for a study of nominalism and its history, particularly pt. 3, sec. 2, on "the baroque nominalism of Hobbes."

14. Hobbes's attitude is well summarized in *Leviathan,* where, after summarizing his own theory of sensation, he writes:

> But the Philosophy-schooles, through all the Universities of Christendome, grounded upon certain Texts of *Aristotle,* teach another doctrine; and say, For the cause of *Vision,* that the thing seen sendeth forth on every side a *visible species* (in English) a *visible shew, apparition,* or *aspect,* or *a being seen;* the receiving whereof into the Eye is *Seeing;* and for the cause of *Hearing,* that the thing heard, sendeth forth an *Audible species,* that is, an *Audible aspect,* or *Audible being seen;* which entering at the Eare, maketh *Hearing.* Nay for the cause of *Understanding* also, they say the thing Understood sendeth forth *intelligible species,* that is, an *intelligible being seen;* which comming into the Understanding, makes us Understand. I say not this, as disapproving the use of Universities: but because

The nominalistic criticism of the theory of universals typically focuses on its bloated ontology and its attribution of seemingly odd mental powers to quite ordinary human minds. Hobbes is hardly unique in complaining that a world populated by the universal humanity, over and above individual humans, does seem rather crowded; nor is he the only philosopher to have wondered whether a proper theory of human understanding requires the postulation of such mental faculties as intellection or odd objects like intelligible species and the other accoutrements of the traditional theory of universals. In place of such an overcrowded ontology, Hobbes proposes an account of human language and demonstration that makes no commitment to the existence of anything beyond material bodies and their motions.

The universal names in Hobbesian demonstrations are not names of universal things—conceived either as separately existing archetypes to which individuals conform (which we can call real universals) or as abstract concepts in the mind of the demonstrator (which we can term conceptual universals). Hobbes makes clear his rejection of the theory of real universals in the *Elements of Law,* writing that "[t]he universality of *one name* to many things, hath been the cause that men think the *things* themselves are universal; and so seriously contend, that besides Peter and John, and all the rest of the men that are, have been, or shall be in the world, there is yet somewhat else that we call *man,* viz. *man in general,* deceiving themselves, by taking the universal, or general appellation, for the thing it signifieth" (*EL* 1.5.6; *EW* 4:22). His hostility toward a conceptualist treatment of universals is equally clear in *De Corpore* when he lists among the "causes of error" the belief that "the idea of anything is universal; as if there were in the mind a certain image of a man that is not that of some one man, but of man simply, which is impossible; for every idea is both one and of one thing; but they are deceived in that they put the name of the thing for the idea of it" (*DCo* 1.5.8; *OL* 1:53–54). This results in a theory that takes the universal to be a word given general significance by convention, that is to say by its being taken as a representative of a class of objects that resemble one another in some important respect—"One Universall name is imposed on many things, for their similitude in some quality, or other accident: And whereas a Proper Name bringeth

I am to speak hereafter of their office in a Common-wealth, I must let you see on all occasions by the way, what things would be ammended in them; amongst which the frequency of insignificant Speech is one. (*L* 1.1, 4; *EW* 3:3)

to mind one thing onely; Universals recall any one of those many" (*L* 1.4, 13; *EW* 3:21).

Thus stated, Hobbes's nominalism is quite thoroughgoing in its commitment to the principle that only particulars can exist. The denial of both real and conceptual universals, together with the reduction of all universals to arbitrarily imposed signs, goes beyond the famous version of nominalism propounded by William of Ockham.[15] Leibniz, in an early commentary on the work of the Italian humanist Marius Nizolius, remarks that nominalism is to be preferred on grounds of theoretical parsimony, but he has reservations about what he sees as the extremity of Hobbes's nominalism.

> If an astronomer can account for the celestial phenomena with few presuppositions, that is with simple motions only, his hypothesis is certainly to be preferred to that of one who needs many orbs variously intertwined to explain the heavens. From this the nominalists have deduced the rule that everything in the world can be explained without any reference to universals and real forms. Nothing is truer than this opinion, and nothing is more worthy of a philosopher of our own time, so much that, I believe, Ockham himself was not more of a nominalist than is Thomas Hobbes now, although I confess that Hobbes seems to me to be an ultranominalist *[plusquam nominalis]*. Not content to reduce universals to names like the nominalists, he says that the truth of things itself consists in names, and what is more, that truth depends on the human will *[ab arbitrio humano]*, because truth depends on the definitions of terms, and definitions depend on the will. This is the opinion of a man judged among the most profound of our century, and as I said, nothing can be more nominalistic. Yet it cannot stand. As in arithmetic, so also in other disciplines, truths remain the same even if notations are changed, and it does not matter whether a decimal or duodecimal number system is used. (Leibniz 1923–, series 6, vol. 2: 427–29)

Such worries as these, however, confuse Hobbes's nominalism with arbitrarism. As I have already discussed, Hobbes's theory allows for the arbitrary introduction of names, but it does not require that all truths be arbitrary or conventional. Although Hobbes is occasionally sloppy

---

15. For an exposition of Ockham's nominalism and its connection with his philosophy of language and logic, see Loux 1974 in conjunction with Ockham 1974. For the connection between Ockham and Hobbes, see Bernhardt 1985a and 1988.

in expressing his theory, he never says that the truth of, say, an arithmetical proposition depends only upon the specific notation used to express it. He can, after all, hold that the same truths are expressed in both the Latin and English versions of *Leviathan,* even if the expression of such truths must depend upon the appropriate choice of names.[16]

The question of whether Hobbes's nominalism derives from his acquaintance with Ockham's doctrines is difficult to resolve, not least because Hobbes has little to say about Ockham. Bernhardt (1985a, 1988) investigates the Ockham-Hobbes connection and concludes that there is a fundamental difference between the two, viz., Ockham holds to a "semantic" nominalism that admits general concepts as the meanings of universal terms, while Hobbes admits only a "syntactic" nominalism that denies generality to anything but words. Zarka (1985; 1987, 86–88) reaches a similar conclusion and sees Hobbes's nominalism as more thoroughgoing than any of its predecessors. We can grant that Hobbes goes beyond the nominalism of his predecessors without, however, characterizing his theory as the sort of arbitraristic "ultranominalism" denounced by Leibniz.

Even with this qualification, it is evident that Hobbes's doctrine is not without its difficulties. Perhaps the most pressing of these is to account for the fact of resemblance among particulars that share a common universal name in terms that are acceptable within Hobbes's philosophical framework. Hobbes's extreme nominalistic scruples demand a world bereft of abstract concepts as well as universal or general objects. But to grant that speakers can grasp the resemblance between two particulars sorted under the same name seems, at a minimum, to commit the nominalist to the existence of an abstract property of similarity, likeness, or resemblance. Further, it seems quite plausible that such resemblances can be understood without attending to concrete particulars. Moreover, once such resources as these are granted, there appear to be no good grounds for the nominalist to deny that the mind could frame a general concept of a universal without having to confine its attention to the specific particulars. As Zarka has put it, "now the question is to know whether the grasp of this resemblance does not implicitly imply an intellection by which we can form, starting from the representation of individual objects, a general representation, for example that of human nature without reference to individuals" (Zarka 1987, 97). Hobbes seems not to have appreciated this

16. My interpretation of these issues is not original. See Hübener 1977 for a similar take on the matter.

difficulty, and his defense of his view is confined largely to assertions to the effect that there is no need of a faculty of pure intellection or understanding to account for the use of universal names. At best, Hobbes defends his nominalism by claiming that the only alternative is the despised Scholastic language of intelligible species, separated essences, and formal quiddities. Notwithstanding Hobbes's failure to mount a sophisticated argument in defense of his nominalism, it is still worthwhile to investigate contemporary criticisms of his doctrine in order to clarify the philosophical and methodological stakes at issue.

We can begin by contrasting Hobbes's account of reasoning and demonstration with a more traditional theory that takes reason or intellection to be a mental faculty distinct from sensation or imagination. On such a view, sensation is a physical (one might even say physiological) process tied to the body and arising from the stimulation of sense organs; memory is the storage of previous sensory images; and imagination is nothing more than the recollection and recombination of items from memory or sensation. Neither faculty, however, is capable of "pure intellection"—which is a fundamentally nonsensory mode of awareness. Aquinas, in his adaptation of the Aristotelian theory of intellection, held that the senses provide the material upon which the intellect operates; intellectual understanding arises when the "active intellect" abstracts the intelligible species from sensible species. He reasons that

> [b]ecause Aristotle did not allow that forms of natural things subsist apart from matter, and since forms existing in matter are not actually intelligible, it follows that the natures or forms of the sensible things we understand are not actually intelligible. But nothing is reduced from potentiality to actuality except by something in actuality; just as the senses are made actual by what is actually sensible. We must therefore assign on the part of the intellect some power to make things actually intelligible, by the abstraction of the species from material conditions. And such is the necessity for positing an agent intellect. (Aquinas 1964, 11:154)

One consequence of this view is that "on the part of the phantasms, intellectual operations are caused by the senses. But because the phantasms cannot of themselves change the passive intellect, but require to be made actually intelligible by the active intellect, it cannot be said that sensible knowledge is the total and perfect cause of intellectual knowledge, but rather it is in a way the material cause" (Aquinas 1964, 12:36).

We need not be overly concerned with the details of this theory, but it is important to recognize the extent to which it separates the intellect from the senses. Intellectual understanding, although it may make use of sensory ideas as inputs, has as its proper object universals or essences that exceed the grasp of the senses. Barrow phrases his version of this theory by putting intellectual powers in analogy with the senses when he asks: "What if I should assert that the human mind (when rightly constituted and not out of balance, as it is in extreme fools and demented people) has the power of discerning universal propositions by its native faculty, in the same manner as sense discerns particular ones? . . . Such universal propositions the mind directly contemplates and finds to be true by its native power, even without any previous notion and having applied no reasoning; which means of attaining truth is called by the peculiar name intellection [νόησις], and this faculty the intellect [νοῦς]" (*LM* 5, 81). Ward makes a similar case for the immateriality of the intellect when he argues that consideration of such ideas as those of God or the soul "are sufficient to shew the difference betwixt the intellectuall apprehension of things, and the imagination which accompanies our superficiall thoughts, our slight and cursory taking of them to our mindes, and to illustrate that, however in our fancies we may have corporeall representations attending upon these spiritual beings, yet the Ideas whereby the understanding apprehends those simple essences, are incorporeall, and consequently the understanding part of man is incorporeall" (Ward 1652, 57–58).

5.1.2.1 DESCARTES VS. HOBBES ON REASON AND THE INTELLECT The most famous seventeenth-century proponent of this sort of theory is Descartes, who notoriously contrasts the powers of "pure understanding" with those of the imagination in the sixth of his *Meditations*. He there claims that the intellect can frame the idea of a thousand-sided figure—the chiliagon—and even demonstrate necessary properties of it, although the idea of such a figure is too complex to be framed distinctly in imagination (*AT* 7:72). In constructing its purely intellectual idea of the chiliagon the intellect grasps the essence of the figure, and does so without the mediation of sensory ideas. This is not to say that, for Descartes, the intellect lacks all connection to the senses or imagination. In a letter to Princess Elisabeth in June of 1643, Descartes explains that the study of mathematics "exercises mainly the imagination in the consideration of figures and motions" (*AT* 3:691), and the ideas furnished by the imagination provide the intellect with the material from which it can abstract essences. Nevertheless, Descartes maintains that the "perception of universals" is not a matter for the imagination, but (as he explains in a 1640 letter to Regius) it must be attributed "to the intellect alone, which refers an idea that is in itself singular to many things" (*AT* 3:66). It is also significant that Descartes takes the pure understanding to be the source of his idea of himself as an essentially thinking being; the *cogito* delivers an awareness of the self as a fundamentally immaterial substance whose essence is thought, and this awareness comes from the intellect "turning itself towards itself" to grasp such purely intellectual properties as thought, doubt, and selfhood (*AT* 7:73). Hobbes disagrees with this view and holds that talk of pure understanding or intellect is superfluous. As he puts it: "for understanding of the meaning of a universal name, we need no other faculty than the imagination, by which we recollect such names as call to mind sometimes one thing, sometimes another" (*DCo* 1.2.9; *OL* 1:18). This is also the import of his remark that intellect is "a certain imagination, but one that arises from the agreed signification of words" (*DH* 2.10.1; *OL* 2:89).

The conflict between Hobbes and Descartes on this point is evident in the third set of "Objections and Replies" to the *Meditations*. This exchange consists of Hobbes's brief, caustic objections and Descartes's even briefer, more acerbic responses. Fundamental to their differences is the status of reasoning: where Hobbes holds that reasoning involves the manipulation of names, Descartes insists that it is an intellectual process independent of language. Thus, when Descartes claims that the nature of the notorious piece of wax in the second meditation is not revealed by the imagination but is "conceived by the mind alone," Hobbes replies that "there is a great difference between imagining, that is, having some idea, and conceiving in the mind, that is, by reasoning to conclude that something is, or exists. But M. Descartes has not explained to us how they differ." Hobbes goes on to explain that, in his view, "reasoning is simply the joining together and concatenation of names or appellations by means of the verb 'is,'" with the result that "by reasoning we conclude nothing at all about the nature of things, but only about their appellations; that is, whether or not we are combining the names of things in accordance with the conventions we have established by arbitrary choice concerning their significations." Hobbes's conclusion is that "reasoning will depend on names, names on the imagination, and imagination will depend (as I believe it does) on the motions of our bodily organs; and so the mind will be nothing more than motion in various parts of an organic body" (*OL* 5:257–58; *AT* 7:178).

Descartes wholeheartedly opposed nominalism of this sort, and he seems to have taken particular umbrage at the suggestion that reasoning can tell us nothing about things themselves. Cartesian pure thought or understanding grasps the true nature of its objects and does so without relying upon linguistic devices or the manipulation of names. Descartes responded to the objection by insisting that Hobbes had wholly misconstrued the nature of reasoning—it "is not a linking of names, but of the things that are signified by the names." Further, he charged that his English opponent "refutes his own position when he talks of the arbitrary conventions that we have laid down concerning the meaning of the words. For if he admits that the words signify something, why will he not allow that our reasoning deals with this something that is signified, rather than merely with the words?" His frustration with the Hobbesian enterprise ultimately led Descartes to remark that "when [Hobbes] concludes that the mind is motion he could just as well conclude that the earth is the sky, or whatever else he likes" (*OL* 5:258–59; *AT* 2:179).

At its most fundamental level this conflict between Hobbes and Descartes concerns the delicate question of what we think about when we understand a demonstration. For Hobbes, the objects of thought are always concrete particular phantasms (including names, which themselves are "sensible marks"). These phantasms represent the world of external bodies to us; thus, my present greenish visual sensation represents the tree outside my window. Our words do not signify external bodies directly, but rather our thoughts, since "it is manifest that they are not signs of things themselves; for in what sense can the sound of the word 'stone' be a sign of a stone, except that he who hears the word concludes that the speaker thinks of a stone?" (*DCo* 1.2.5; *OL* 1:15). Although its immediate signification is not an external body itself, a name does designate a body indirectly, by signifying a phantasm of it, which in turn represents the body itself. Hobbes therefore often speaks of names as designating the things themselves and omitting the phantasms that forge the link between words and the world. As we have already seen, Hobbes holds that reasoning involves the manipulation of general names (which represent classes of things that resemble one another in some quality). Linking together such names by use of the verb *is* we form syllogisms, whose conclusions yield additional information about the consequences of the names we have imposed. In no case, however, does our reasoning deal either directly with the external objects or with any universal ideas or nonphantasmic intellectual notions.

Descartes holds that in demonstration the intellect deals with its proper object, i.e., concepts or essences that are eternal, immutable, universal, and independent of sense. Except in the case of God, the essence and existence of a thing are distinct, which is to say that from the bare essence of the thing no conclusion follows about whether such a thing exists. Demonstrations therefore need not tell us anything about what the world contains, but they do make manifest the relationship between the essences of things that can exist. Demonstration consequently allows us to gain knowledge about the way things must be structured, if they exist at all. The essence of the semicircle, for example, entails that every triangle inscribed in it is a right triangle, and this result is independent of the structure or contents of the physical world, which may not contain any true circles or triangles. Because "pure thought" or intellect is independent of matter, its objects are also immaterial, and the Cartesian demonstrator, insofar as he attends to his intellectual ideas, is radically separated from the world of bodies. The connection between the demonstrator and the material world is forged by the imagination, which furnishes determinate ideas to the otherwise indeterminate essences of the intellect.[17] Thus, the Cartesian demonstrator may employ visible diagrams in his geometric investigation, and these may assist him in grasping the immaterial essences that are the true objects of his concern. Nevertheless, the business of drawing conclusions and comprehending essences is proper to the intellect alone. It is for this reason that Descartes can hold that in reasoning we deal with the things themselves, not merely names. These "things themselves" turn out to be essences immediately graspable by the intellect and necessarily governing the structure of the world.

Descartes's doctrine of essences found no favor with Hobbes. When Descartes declares that the fact that he can demonstrate properties of the triangle shows that "there is a determinate nature, essence, or form of the triangle, which is immutable, eternal, and not made by me or dependent on my mind," and infers that the nature of the triangle would remain even if there were no triangles outside of his thought,

---

17. Gaukroger 1992, 110, describes this process in the following terms: "It is [in the connection between intellect and the material world] that the necessity for the imagination arises, because the intellect by itself has no relation at all to the world. Entities conceived in the intellect are indeterminate. The imagination is required to render them determinate. When we speak of numbers, for example, the imagination must be employed to represent to ourselves something that can be measured by a multitude of objects. The intellect understands 'fiveness' as something separate from five objects (or line segments, or points, or whatever), and hence the imagination is required if this fiveness is to correspond to something in the world."

Hobbes replies: "If the triangle exists nowhere, I do not understand how it has a nature." In Hobbes's scheme, talk of eternal essences can at best be understood as applying to names. The name *triangle* can be applied to something we have seen, and "if in our thought we have once conceived that the sum of all the angles of a triangle taken together equals that of two right angles, and we give the triangle a second label 'having three angles equal to two right angles,' then even if no angles existed in the world, the name would remain." This allows the truth of the proposition "every triangle has angles summing to two right angles" to be distinguished from the existence of any particular triangle, so that "the truth of the proposition will be eternal."[18] But there can be no immutable essence of a triangle, "if it should happen that every single triangle ceased to exist." Hobbes concludes that any notion of essence "insofar as it is distinct from existence, is nothing more than joining together of names by means of the verb *is;* and so essence without existence is our mental fiction." Needless to say, Descartes was unimpressed with this theory, and in his reply to the objection he simply states that "The distinction between essence and existence is known to everyone; and this talk about eternal names, rather than concepts or ideas of eternal truths, has already been fully refuted" (*OL* 5:271–72; *AT* 7:193).

5.1.2.2 WARD'S CRITIQUE AND THE LIMITS OF MATERIALISM I take it to be unproblematic that there is a significant connection between Hobbes's nominalism and his materialism.[19] By rejecting any theory of pure intellect and insisting upon the adequacy of the imagination to account for our understanding of universal terms, Hobbes rules out the possibility of "higher" mental faculties of the sort that had traditionally been taken as immaterial. Nevertheless, nominalism of this sort does not lead to materialism all on its own. It is only when conjoined with a materialistic account of sensory ideas (and hence of the imagination) that Hobbes's nominalism becomes indissolubly linked with his materialistic theory of the mind. To see this point more clearly it is helpful to recall that George Berkeley champi-

18. Hobbes cannot be too serious about the locution "eternal truth" here, since eternal truth presupposes the existence of eternal names and there is no guarantee that the names forming the proposition will be eternal. Presumably he would hold that when the human race dies out and the world's geometry books crumble to dust, the proposition will vanish with the words in which it was expressed.

19. On the connection between Hobbes's nominalism and materialism, see Zarka 1985 and Zarka 1987, chap. 4.

oned a nominalism nearly identical to that of Hobbes (Jesseph 1993c, chap. 1). Berkeley notoriously had no sympathy for a materialism of any kind. Yet by rejecting a materialistic analysis of ideas of sensation or imagination, Berkeley can accept a Hobbesian brand of nominalism while denying his materialistic conception of the human mind. Berkeley and Hobbes therefore agree that there are no abstract ideas, real universals, or mental faculties of pure intellection; where they disagree is over the role of matter in the origin of ideas of sense. Hobbes's contemporary opponents did not have this option. They all accepted some version of the claim that the senses are, in an important way, "tied" to the body and explicable in mechanistic terms rather than the language of sensible species. As a result, many criticisms of Hobbes's materialism took issue with his nominalism. Ward's *Exercitatio* provides an interesting example of this sort of reaction.

Ward found the Hobbesian theory of reasoning and demonstration as objectionable as Descartes had, and he was particularly concerned to show that the theory undermined the case for the immortality (and immateriality) of the soul.[20] The dangerous nature of Hobbes's opinions is the recurring theme in Ward's *Exercitatio*. He thinks, for example, that the nominalistic pronouncement that "truth and falsehood are attributes of words, not of things" is a "monstrous assertion" connected with Hobbes's political theory, since in the state of nature there is "nothing in itself good or evil." Such views, Ward holds, can only work to reduce truth and right to the arbitrary decrees of whoever holds power, thereby destroying the foundations of morality and religion (*Exercitatio* 46–47).

Ward constantly complains that Hobbes does not mount much of a positive argument for his doctrines, whether metaphysical, scientific, or political. He attributes this to his opponent's "dogmatism" and thinks that one of the best ways to challenge the Hobbesian project is to show the weakness of any possible argument for Hobbes's conclusions. In objecting to Hobbes's nominalistic account of reasoning he therefore takes steps to supply the argumentation that would lead to the conclusion that to understand the meaning of a universal name

20. Ward writes: "Among those who work to prove the immortality of the human soul from the principles of nature, it is usual to adduce as a powerful (or even irrefutable) argument the fact that the human mind in its operations exceeds the whole sphere of corporeal action and accomplishes things that neither the motion of bodies nor the actions of imagination *[phantasia]* (both of which are thought to be common to brutes and men) reach. This philosopher thinks that it is much to his advantage *[sui interesse plurimum]* that the soul is not immortal. This is what he contends in plain chapters

there is no need of any faculty beyond the imagination. This principle, he says, would be sufficiently established if it could be shown that "the concept corresponding to a universal name is the phantasm of a singular thing (although the phantasm of the singular thing does require the imagination)" (*Exercitatio* 30).

Ward takes up the question of what sort of argument might prove that the grasp of a universal name requires only the phantasm of a singular thing. He reports that "either, says Hobbes (for this is the import of his argumentation), the concept corresponding to a universal name is some singular phantasm, or by a universal name is signified that there is something universal in nature that is designated by this name. For example either the concept corresponding to this word 'man' is a phantasm of Peter, or John, etc. or there is some universal man to whom this word corresponds" (*Exercitatio* 30). Since the second alternative is plainly ridiculous, the conclusion follows.

Ward attacks the soundness of the above argument and mounts a challenge to the entire Hobbesian conception of mental faculties. His strategy is to accuse Hobbes of invoking a false dilemma in assuming that the meaning of a universal term must be either a singular phantasm or a universal thing. Instead of these two options, he introduces a third, namely the concept of a "common reason" *[ratio communis]* shared by the particulars that fall under a universal name.

> Who does not see how lax, how fluid, and how completely unsound this type of argument is, which this upstart offers for sale under the name of demonstration? The consequence is this: either the concept corresponding to this word (man) is the concept of Peter, etc. or it is the concept of some common reason, in which Peter, John, and the rest are included, that is, a common and explicable concept of Peter, etc. And this concept is not the phantasm of a singular thing, nor is it the concept of some universal man existing in nature, but the idea of a certain common reason underlying the same species of all the singulars, the formation of which idea can be in no way taken from the imagination, which deals with singulars. (*Exercitatio* 30)

The battle against Hobbes clearly requires more than the statement of the theory of a "common reason," however. Ward promises that "in

---

and considered language *[aperto capite & verbis conceptis]* in *Leviathan*: that the soul dissolves at death, and that those who hold the contrary opinion are not to be tolerated in the commonwealth, on account of the disorder, sedition, and civil wars that arise from this empty opinion of immortality" (*Exercitatio* 29).

order to explain my own opinion and bring Hobbes's tricks to light, I will set forth the method of common human reason in the framing of universals" (*Exercitatio* 30–31).

Indeed, he argues that the inadequacy of nominalism follows from the careful consideration of a case that Hobbes himself had advanced in support of his own theory—that of a person who proves a geometric theorem by attending to the properties of a single triangle. Hobbes had argued that by the imposition of names (such as *triangle* and *equal*) the consideration of a single case could lead to the general result that all triangles have interior angles that sum to two right angles, since "he that hath the use of words, when he observes, that such equality was consequent, not to the length of the side, nor to any other particular thing in his triangle . . . will boldly conclude Universally, that such equality of angles is in all triangles whatsoever" (*L* 1.4, 14; *EW* 3:22). Ward argues that this is an inadequate description of the reasoning involved in a demonstration. He insists that any such particular triangle must be completely specific—it will have a determinate color and position, and its sides will be of specific lengths. The result concerning the equality of the angles can be discerned, but he thinks that there are two ways of doing this: "either by experience (so much as the equality of things can be known by sense), for example by aid of a pair of compasses measuring the angles of the arcs of circles; or it can be found by a deep and attentive *[alta et defixa]* cogitation of the soul" (*Exercitatio* 31).

Ward contends that, in the first case, the reasoning is essentially tied to singular phantasms derived from experience, and there is no chance that the result can be extended to cover other cases. The experience of a particular triangle includes such features as color and situation and is therefore too specific to underwrite a general result. At best such an experience can be taken to show that this individual triangle has angles equal to two right angles, but it says nothing about other cases. To obtain a general result the demonstrator must disregard irrelevant features of this specific triangle and see that the proof depends only upon the "common reason" of all triangles. But such generalization involves higher mental powers than simply imagination. Ward holds that it is manifest that the reasoning involved here "exceeds the power of the imagination, for this does not create ideas, but only receives them from objects, nor can it move beyond the sphere of objects," because anything perceived or imagined must include determinate color, shape, size, etc. He reasons that

it is manifest in this example that the name "having two angles
equal to two right angles" denotes neither this singular triangle,
from which the mind's judgment of equality arose, nor some triangle that is universal (that is to say right-angled and obtuse and
acute all at once), but rather the reason or idea or concept common to all triangles, that is to say one applicable to any triangle.
(*Exercitatio* 31)

Ward therefore concludes that the Hobbesian theory of demonstration
is inadequate precisely because the "phantasms" on which it is based
"cannot exceed those circumstances in which the objects are clothed,
nor can it reach the natures of things, which are made in concepts of
this sort, and expressed by the constitution of universal propositions
or names" (*Exercitatio* 32).

Hobbes, it should be noted, undertook no real response to these
objections. In contrast to his willingness to defend his mathematics at
great length against the objections of Wallis, Hobbes rated Ward's critique of other aspects of his philosophy as beneath notice. At the close
of the fifth of his *Six Lessons*, Hobbes dismissed Ward with the remark
that he will "trust the objections made by you the Astronomer
(wherein there is neither close reasoning, nor good stile, nor sharpness
of wit, to impose upon any man) to the discretion of all sorts of Readers" (*SL* 5; *EW* 7:330).

The sorts of objections raised by Descartes and Ward are not in
themselves decisive against Hobbes's nominalistic account of language
and reasoning, but they raise important questions about the adequacy
of his doctrines. There certainly seems to be more to the process of
inference than the simple manipulation of words or symbols, and we
can at least say that there is little room in Hobbes's scheme for what
we might call the phenomenology of demonstration. In fact, when we
turn to a consideration of his criticisms of algebra and analytic geometry, we will see that Hobbes himself held that proper demonstrations
require more than symbols and rules for their manipulation.

5.1.3 *Mathematics and Natural Philosophy Revisited*

We saw in chapter 4 that Hobbes places physical concepts such as
space, body, and motion at the center of his philosophy of mathematics. Nevertheless, he does think that these two disciplines have important differences. In fact, his criteria for scientific demonstration led
Hobbes to draw a significant methodological distinction between

mathematics (especially geometry) and such natural-philosophical inquiry as physics. The geometer begins his demonstration with definitions that express the true causes of the objects of his inquiry.[21] In contrast, the causes of natural phenomena are generally hidden from us and mathematics consequently has a degree of clarity and certainty exceeding that obtainable in natural philosophy. Because Hobbes holds that our knowledge of nature is confined to "phantasms" or "fancies" in the mind, which in turn are caused by the motion of external bodies, the task of natural philosophy is thus to find a mechanistic explanation of such phantasms. "Your desire," he says to the prospective natural philosopher "is to know the causes of the effects or phenomena of nature; and you confess they are fancies, and consequently, that they are in yourself; so that the causes you seek for only are without you, and now you would know how those external bodies work upon you to produce those phenomena" (*EW* 7:82)

It is axiomatic for Hobbes that "nature does all things by the conflict of bodies pressing each other mutually with their motions" (*DP* epistle; *OL* 4:238). Nevertheless, there are many possible accounts that may satisfy this general requirement of mechanism since "there is no effect in nature which the Author of nature cannot bring to pass by more ways than one" (*EW* 7:88). The result is that natural science must proceed from conjectures about the possible generation of natural phenomena, so there must be an ineradicably conjectural or hypothetical aspect to natural science. The essentially hypothetical nature of natural science leads Hobbes to conclude that—contrary to the tradition—there can be no demonstration τοῦ διότι in physics. Furthermore, since he denies that τοῦ ὅτι "demonstrations" are properly demonstrations at all, Hobbes concludes that there is no hope for a truly demonstrative science of nature. He brings this point out in the *Examinatio,* when he insists that "reasoning that, beginning from true principles, correctly infers a conclusion is properly called demonstration. Nor do I think that Aristotle called reasoning in which there is a paralogism 'demonstration,' not even a τοῦ ὅτι demonstration. And this is how he must have understood reasoning that begins not with definitions but with suppositions (such as physicists use), which are generally uncertain" (*Examinatio* 1; *OL* 4:38). It is in precisely this context that Hobbes made his famous declaration that

21. This should be taken to apply most specifically to Hobbes's own program for properly reformed geometry. Obviously, he does not think that Euclid or those who follow him begin from principles expressing true causes.

[o]f Arts, some are demonstrable, others indemonstrable; and demonstrable are those the construction of the Subject whereof is in the power of the Artist himself; who in his demonstration does no more but deduce the Consequences of his own operation. The reason whereof is this, that the Science of every Subject is derived from a praecognition of the Causes, Generation, and Construction of the same; and consequently where the Causes are known, there is place for Demonstration; but not where the Causes are to seek for. Geometry therefore is demonstrable; for the Lines and Figures from which we reason are drawn and described by ourselves; and Civill Philosophy is demonstrable because, we make the Commonwealth our selves. But because of Naturall Bodies we know not the Construction, but seek it from the Effects, there lyes no demonstration of what the Causes be we seek for, but onely of what they may be. (*SL* epistle; *EW* 7:183–84)

A similar passage in *De Homine* even holds out the hope that the science of geometry can be made complete, in the sense that there could be no unanswerable question regarding figures. Hobbes declares that "many theorems concerning quantity are demonstrable, the science of which is called geometry. Since the causes of the properties that individual figures have are in them because we ourselves draw the lines, and since the generation of the figures depends on our will, to know the properties belonging to any figure whatsoever nothing more is required than that we consider all that follows from the construction that we ourselves make in the figure to be described" (*DH* 2.10.5; *OL* 2:93). It is Hobbes's faith in such definitions that led him to underestimate the difficulty of the classical problems such as the quadrature of the circle and to imagine that their solution would be available to one who had grasped the proper definitions of figures in terms of the motions by which they are produced.

Hobbes's doctrines assimilate mathematics (and demonstrative knowledge generally) into the domain of "maker's knowledge," by grounding its certainty and universality in our construction of the objects known. Some commentators have seen this aspect of Hobbes's philosophy as an inheritance from Francis Bacon, at least to the extent that Bacon saw scientific knowledge as maker's knowledge.[22] Whether

22. On Bacon and the tradition of "maker's knowledge," see Perez-Ramos 1988, esp. 186–93, on the methodological connection between Bacon and Hobbes. Barnouw (1980) argues for a fairly strong Bacon-Hobbes connection, as does Child (1953). Bernhardt (1989, 10) remarks that the relationship between the two is an "obscure point"

such claims of influence ultimately hold up is not a matter I will investigate here, but it is worth noting that Bacon does not hold that mathematical objects are constructions founded in the nature of body, and this part of the Hobbesian program cannot claim a strictly Baconian pedigree.

There is certainly nothing novel or remarkable in Hobbes's claim that mathematics is more certain than natural science. The traditional distinction between pure and applied mathematics, for example, places pure mathematics on a lofty plane of metaphysical certainty, rendering its truths unencumbered by dependence upon the features of the material world. Applied mathematics, such as that in optics or astronomy, introduces physical hypotheses that permit material objects to be treated mathematically, and such hypotheses depend for their truth upon contingent facts about the material world. Thus, the science of optics can be developed "mathematically" by the hypotheses that light is propagated in straight lines and its rays behave in accordance with Euclidean geometry, although it is possible that such hypotheses might fail to be true. Even Descartes, who was not one to regard his physics as so much fallible conjecture, drew a distinction of this sort in a 1638 letter to Mersenne:

> You ask whether I take what I have written about refraction to be a demonstration. I think so, at least insofar as one can be given in these matters without having previously demonstrated the principles of physics by metaphysics (which I hope to do some day, but which has not yet been done) and insofar as any other solution has been demonstrated to a problem of mechanics, optics, astronomy, or anything else that is not pure geometry or arithmetic. But to require me to give geometrical demonstrations in matters that depend on physics is to want me to do the impossible. And if you will not call anything demonstrations except the proofs of geometers, then you must say that Archimedes never demonstrated anything in mechanics, or Vitellio in optics, or Ptolemy in astronomy, which of course nobody says. In such matters it is enough if the authors have supposed things not obviously contrary to experience and if their discussion is coherent

---

in Hobbes's intellectual biography. Tönnies (1925, preface) launches a vigorous polemic against any reading of Hobbes as a successor to Bacon, a view shared by Brandt (1928). Schumann 1984 contains a nuanced overview of the difficult question of Hobbes's relationship to Bacon. My own view is that there is little connection between Bacon and Hobbes on the specifically mathematical issues of concern to us here.

and without paralogism, even though their assumptions may not be strictly true. (*AT* 2:141–42)

Hobbes's contrast between conjectural physics and certain mathematics might seem to comport poorly with his placement of such plainly physical concepts as body, space, time, and motion at the foundation of his mathematics. We have already seen that Hobbes makes these concepts central to his philosophy of mathematics, and one of Wallis's charges against Hobbes is that he makes mathematics overly dependent upon (uncertain) physical principles. How, one might ask, can Hobbes take physical notions as basic to mathematics, and yet claim demonstrative certainty for mathematics while denying it to physics?

The resolution of this apparent contradiction lies in the recognition that Hobbes regards these concepts as basic, not only to mathematics and physics, but to any body of knowledge whatsoever. They are, in fact, the first principles of metaphysics.[23] As such they are known more clearly than other concepts and serve as the basis upon which any science (demonstrative or hypothetical) must be built. The actual derivation of physical phenomena from the first principles is not an option for us. We did not create the world and its phenomena, and in natural philosophy "the most that can be atteyned vnto is to haue such opinions, as no certayne experience can confute, and from wch can be deduced by lawfull argumentation, no absurdity" (Hobbes to William Cavendish, 29 July/8 August 1636; *CTH* 1:33). Mathematical objects, however, are our constructions, and we can have clear, certain, and demonstrative knowledge of their properties because we know how they are generated. In fact, that part of natural philosophy that concerns the general science of motion is capable of demonstration. In *De Homine,* Hobbes distinguishes the a posteriori part of physics (which is purely hypothetical) from the a priori part, which is purely geometrical: "Since one cannot proceed in reasoning about natural things that are brought about by motion from the effects to the causes without a knowledge of those things that follow from every kind of motion; and since one cannot proceed to the consequences of motions without a knowledge of quantity, which is geometry; nothing can be demon-

---

23. It is in this context that Hobbes declares "words understood are but the seed, and no part of the harvest of Philosophy." He then contrasts his fundamental definitions with those of Aristotle, concluding that "this is the Method I have used, defining Place, Magnitude, and the other most generall Appellations in that part [of *De Corpore*] which I intitle *Philosophia prima*" (*SL* 2; *EW* 7:226).

strated by physics without something also being demonstrated a priori" (*DH* 2.10.5; *OL* 2:93). Nevertheless, there is no science of "sensible appearances" because the specific causal mechanisms that produce such appearances remain hidden from our view.[24]

5.2 ANALYSIS, SYNTHESIS, AND MATHEMATICAL METHOD

Hobbes's theory of demonstration provides the background for his version of the distinction between analytic and synthetic methods—a topic of great importance in any study of his mathematical writings. It goes without saying that this distinction did not originate with Hobbes, and indeed some kind of contrast between analysis and synthesis was a commonplace well before the seventeenth century, especially in the philosophy of mathematics.[25] The volume of primary and secondary literature on this topic makes it impossible to treat it in its full complexity, and I will be content to investigate that part of it that bears directly on Hobbes. In point of fact the confusion generated by the many different pronouncements on the nature of analysis and synthesis is so great that one could be forgiven the suspicion that every author who held forth on the subject had his own way of distinguishing the two methods.[26] I begin this section with a quick summary of the

24. Zarka (1996, 73) makes this point as follows:

One could say that the concepts of philosophy are perfectly adapted to grounding the physical sciences, as long as one distinguishes between the science of motion and the science of the sensible world. The science of motion considers in the abstract the effects of one body on another, that is, the laws of impact, and more generally, the laws of the transmission of motion. On the other hand, the science of the sensible world, which Hobbes thinks is physics properly so-called, concerns what appears to the senses and the causes of these appearances. The former science is elaborated a priori from the concepts of first philosophy, while the latter, being concerned with sensible appearance, depends on hypotheses arrived at a posteriori.

25. See Hintikka and Remes 1974 for a study of the method of analysis and its history. More specifically on Hobbes's conception of analysis and synthesis, see Talaska 1988. Hanson 1990 is a very useful study of the role of demonstration in Hobbes's methodology, especially as concerns the issue of analysis and synthesis.

26. Rashed sums up the confusion nicely with the following observation: "Among the problems on the border of philosophy and mathematics, that of analysis and synthesis has occupied a central place for two millennia. Rare indeed are the problems in the philosophy of mathematics which have survived for so long and have given rise to so many writings. Present in shadow in the writings of Aristotle, it is there in person in the works of commentators, of philosophers, and of logicians until the beginning of the last century. It is easy to imagine the diversity of senses and the multiplicity of formulations

classical understanding of analysis and synthesis, then move on to outline how the doctrine was interpreted in the seventeenth century. An exposition of Hobbes's own version of the methods of analysis and synthesis then follows, after which I consider the extent (if any) to which Hobbes's doctrines are an inheritance from the sixteenth-century Paduan school, and specifically the writings of Jacopo Zabarella.

5.2.1 Classical Sources of the Analysis-Synthesis Distinction

The roots of the methods of analysis and synthesis lie in the classical discussions of the philosophy of mathematics, and any discussion of the topic inevitably leads back to a rather small number of classical texts. In fact, it was customary in Hobbes's day to claim a classical pedigree for the distinction. François Viéte's 1591 *In artem analyticam isagoge* is typical. It opens with the declaration that

> there is a certain way of searching for the truth in mathematics that Plato is said first to have discovered; Theon named it analysis, and defined it as the assumption of that which is sought as if it were admitted and working through its consequences to what is admitted as true. This is opposed to synthesis, which is the assuming what is admitted and working through its consequences to arrive at and to understand that which is sought. (Viéte 1646, 1)

Whether Plato can be credited with this methodological discovery is unclear, but there is no doubt that philosophical conceptions had much to do with the framing of the contrast between analysis and synthesis.[27] Aristotle's account of deliberation in the *Nicomachean Ethics* contains an important passage that links analysis with the search for means to a sought end:

> We deliberate not about ends but about what contributes to ends. For a doctor does not deliberate whether he shall heal, nor an orator whether he shall convince, nor a statesman whether he shall produce law and order, nor does any one else deliberate

---

of this question of analysis and synthesis, which then designated a domain vast enough to encompass at once an *ars demonstrandi* and an *ars inveniendi*" (1991b, 131–32).

27. Proclus is presumably the source for attributing the method of analysis to Plato, when he remarks that "there are certain methods [for the discovery of lemmas] that have been handed down, the best being the method of analysis, which traces the desired result back to an acknowledged principle. Plato, it is said, taught this method to Leodamas, who also is reported to have made many discoveries in geometry by means of it" (Proclus 1970, 165–66).

about his end. Having set the end, they consider how and by what means it is to be attained; and if it seems to be produced by several means they consider by which it is most easily and best produced, while if it is achieved by one only they consider how it will be achieved by this and by what means this will be achieved, till they come to the first cause, which in the order of discovery is last. For the person who deliberates seems to inquire and analyze in the way described as though he were analyzing a geometrical construction (not all inquiry appears to be deliberation—for instance mathematical inquiries—but all deliberation is inquiry), and what is last in the order of analysis seems to be first in the order of becoming. And if we come on an impossibility, we give up the search, e.g., if we need money and this cannot be got; but if a thing appears possible we try to do it. (*Nicomachean Ethics* 3.3; 1112b11–26)

This version of the distinction takes analysis to be a kind of "working backwards" from what is sought, with the intention of finding principles from which the desired end can be produced. Synthesis would then appear to be a process of generation, by which the desired result is constructed from first principles (or at least known principles sufficient to generate the thing sought). Proclus, in a rather offhand remark in the first part of his commentary on the *Elements,* connects analysis and synthesis to a broadly Aristotelian theory of demonstration when he says that "[c]ertainly beauty and order are common to all branches of mathematics, as are the method of proceeding from things better known to things we seek to know, and the reverse path from the latter to the former, the methods called analysis and synthesis" (Proclus 1970, 6–7). The epistemic distinction between what is more or less known recalls Aristotle's famous dictum in the *Physics* that the "natural path" of inquiry into principles "is to start from the things which are more knowable and clear to us and proceed towards those which are clearer and more knowable by nature; for the same things are not knowable relative to us and knowable without qualification" (*Physics* 1.1; 184a16–19). The things best known to us turn out to be "inarticulate wholes" that can be analyzed into their elements and principles, thereby leading us back to those things that are better known to nature.

The most complete account of the doctrine of analysis and synthesis in classical literature comes from book 7 of Pappus's *Mathematical Collection.* This book contains a summary of the so-called "treasury of analysis" (ἀναλυόμενος τόπος), which is a compendium of results from different authors that extend beyond the simple elements of the subject.[28] The characterization of analysis and synthesis in Pappus is in terms of the order of exposition:

> [A]nalysis is the path from what one is seeking, as if it were established, by way of its consequences, to something that is established by synthesis. That is to say, in analysis we assume what is sought as if it has been achieved, and look for the thing from which it follows, and again what comes before that, until by regressing in this way we come upon some one principle. . . . In synthesis, by reversal, we assume what was obtained last in the analysis to have been achieved already, and, setting now in natural order, as precedents, what before were following, and fitting them to each other, we attain the end of the construction of what was sought. (Pappus [1875] 1965, 2:635; Heath [1921] 1981, 2:400)

Whatever its obscurities (and they are many) the general drift of this passage is clear enough for our purposes.[29] Analysis starts by assuming the truth of the proposition that is ultimately sought, and deduces consequences from that assumption. Synthesis starts with unchallenged "first principles" and explores their deductive consequences.

The classical tradition also distinguished these two methods by characterizing synthesis as the "method of demonstration" and analysis as the "method of discovery," with the idea that analysis functions as a preliminary to synthesis. Thus conceived, analysis provides a means of exploring the conditions under which a problem can be solved or a theorem proved, and it can thus be used to uncover the fundamental principles that suffice for a genuine synthetic demonstra-

---

28. Pappus declares: "The so-called Ἀναλυόμενος is, to put it shortly, a special body of doctrine provided for the use of those who, after finishing the ordinary Elements, are desirous of acquiring the power of solving problems which may be set them involving (the construction of) lines, and it is useful for this alone. It is the work of the men, Euclid the author of the Elements, Apollonius of Perga, and Aristaeus the elder, and proceeds by way of analysis and synthesis" (Pappus [1875] 1965, 2:635; Heath [1921] 1981, 2:400).

29. Knorr 1986, 354–60, contains a lucid discussion of these difficulties. A central problem is that analysis seems to make synthesis superfluous because the analysis alone would count as a demonstration of the desired result. Pappus and the tradition also distinguish among different kinds of analysis, such as theoretic (for the proof of theorems) and problematic (for the solution of problems). Such refinements are irrelevant for our purposes and will be left aside here.

tion of a proposition. If analytic reasoning leads back to primary truths such as axioms, and if the steps in the process are convertible (in the sense that there is a logical entailment from the consequent to the antecedent), then an analytic procedure can be turned into a synthetic demonstration from the axioms. On the other hand, if the analysis leads to something known to be false the supposition on which the analysis was founded is shown to be false by reductio ad absurdum.

5.2.2 *Analysis and Synthesis in the Seventeenth Century*

Seventeenth-century treatments of analysis and synthesis abounded, as discussion of these methods became a fixture in both the mathematical and philosophical literature (Engfer 1982). Indeed, almost any treatise of the period that deals with method contains a set piece on the distinction between analytic and synthetic procedures. The scope of application of these methods was extended well beyond the realm of pure mathematics, and they were taken as quite general procedures with applications to any subject matter.[30]

Descartes's remarks on analysis and synthesis in the "Second Replies" to the objections to the *Meditations* illustrate the expanded conception of analysis, which applies as well to metaphysics as mathematics.[31] This is one of the more influential seventeenth-century discussions of the matter, although his terminology introduced its share of confusion. He claims, quite traditionally, that "method of demonstration divides into two kinds: the first proceeds by analysis and the second by synthesis." The details of the distinction are less clear. Analysis "shows the true way by means of which the thing in question was discovered methodically and as it were a priori *[tamquam a priori]*, so that if the reader is willing to follow it and give sufficient attention to all points, he will make the thing his own and understand it just as perfectly as if he had discovered the thing for himself" (*AT* 7:155). The cause for confusion here is that a priori reasoning was traditionally taken to proceed from causes to effects and was identified with synthetic presentations, so Descartes has essentially reversed the traditional classification (Lachterman 1989, 158–59). Synthesis, according to Descartes, "employs a directly opposite method where the search is, as it were, a posteriori (though the proof itself is often more a priori

---

30. This is not to say that classical sources regarded analysis and synthesis as confined strictly to mathematics. Still, many seventeenth-century authors are noteworthy for their willingness to extend the methods to include a conception of what Engfer calls "philosophy as analysis."

31. For studies of these objections and replies, see Dear 1995b and Garber 1995.

than it is in the analytic method). It demonstrates the conclusion clearly and employs a long series of definitions, postulates, axioms, theorems, and problems, so that if anyone denies one of the conclusions it can be shown at once that it is contained in what has gone before, and hence the reader, however argumentative or stubborn he may be, is compelled to give his assent" (*AT* 7:156).

In the more specifically mathematical sense of the term *analysis,* Viéte's account is perhaps the most important for the seventeenth century. Viéte distinguishes three kinds of analysis: *zetetics, poristics,* and *rhetics* or (as he also calls it) *exegetics.* He attributes the first two to classical mathematics—zetetics is that "by which is found an equation or proportion of magnitudes between what is sought and what is given," while it is "poristics by which the truth of a proposed theorem is tested by means of an equation or proportion." He characterizes the third branch of analysis—rhetics—as the method "by which the sought magnitude in a proposed equation or proportion is determined." Taking all three of its kinds together, he concludes that analysis "may be called the science of correct discovery in mathematics" (Viéte 1646, 1). Beyond the addition of rhetics or the means of determining equalities or proportions among unknowns, Viéte thinks that the modern analysis is superior to its classical antecedent because "it is no longer confined to numbers, a shortcoming of the old analysts, but works with a newly discovered symbolic logistic that is far more fruitful and powerful than numerical logistic for comparing magnitudes with one another" (Viéte 1646, 1). As we will see, Hobbes regards this "newly discovered logistic" as good for very little, especially in connection with strictly geometric problems.

Descartes seems to have regarded his application of algebra to geometry as an extension of the classical method of analysis, although he voiced his suspicion that the ancient geometers had used powerful methods of discovery that they did not reveal in their written works. Classical geometers' preference for synthetic presentation is therefore not "because they were utterly ignorant of analysis, but because they had such a high regard for it that they kept it to themselves like a sacred mystery" (*AT* 7:156). The idea that the ancients had discovered analytic methods but kept them secret was hardly unique to Descartes and was in fact widespread in the seventeenth century. Wallis, in a letter to Digby from 21 November/1 December 1667, remarks that Archimedes "so concealed the traces of his process of investigation *[inquisitio],* as if he begrudged posterity his art of discovery, from whom he still wanted to extort the approval of his discoveries. Nor was Archimedes the only one; most of the ancients hid their analytics (for it is beyond doubt that they had one) from their posterity" (Wallis 1658, 43).

Wallis also followed the lead of Viéte in seeing the development of algebra as a significant point of superiority for the new methods over the old. In fact, he was prepared to identify algebra with the method of analysis—seeing in it a general technique that can yield the solution to problems of any kind. He frequently refers to "the universal algebra or analytics," or to "the analytic or algebraic method" in the solution of problems (*MU* 11; *OM* 1:53, 59). In the "Inaugural Oration" delivered upon his assumption of the Savilian chair, for example, Wallis gives a brief sketch of the history of mathematics and says of the current era that "beyond the many theorems lately discovered, much is facilitated by this method of discovery. That is to say, algebra, or the analytic practices beyond what were known to the ancients, are now becoming known" (*OM* 1:8). In his *Treatise of Algebra* Wallis announces that his work

> contains an Account of the Original, Progress, and Advancement of (what we now call) *Algebra,* from time to time; shewing its true Antiquity (as far as I have been able to trace it;) and by what Steps it hath attained to the Height at which now it is.
>
> That it was in use among the *Grecians,* we need not doubt; but studiously concealed (by them) as a great Secret.
>
> Examples we have of it in *Euclid,* at least in *Theo,* upon him; who ascribes the invention of it (amongst them) to *Plato.*
>
> Other Examples we have of it in *Pappus,* and the effects of it in *Archimedes, Apollonius,* and others, though obscurely covered and disguised. (*Treatise of Algebra* preface, sig. a2)

Hobbes was particularly annoyed by the identification of analytic methods and algebra, as we will see in section 5. 3. But in order to make sense of his complaints against "the modern analytics" we must proceed to an account of his version of the distinction between analysis and synthesis.

5.2.3 Hobbes on Analytic and Synthetic Methods

Hobbes phrases his version of the analytic-synthetic distinction in terms of causes and effects. Thus, in the Hobbesian scheme, the difference between analysis and synthesis lies in the comparison between the order of reasoning and the order of cause and effect: to reason analytically is to proceed from effects to causes, while the synthetic mode of reasoning follows the natural causal order and moves from causes to effects. As Hobbes himself expresses the contrast in *De Corpore:*

> analysis is ratiocination from the supposed construction or production of a thing to the efficient cause or many coefficient causes of what is produced or constructed. And so synthesis is ratiocination from the first causes of a construction, continued through middle causes to the thing itself that is produced. (*DCo* 3.20.6; *OL* 1:254)

This talk about causes and effects can be interpreted in terms of the premises and conclusions of a demonstration, since proper demonstrations must have premises which express causes. In his discussion of the "analysis of the geometricians" in *De Corpore,* Hobbes clarifies this connection between causes and definitions. He first notes that analytic reasoning "tends ultimately to some equation" so that "there is no end of resolving until at last the very causes of equality and inequality are arrived at, or theorems previously demonstrated from those causes, and enough of them to demonstrate what was sought" (*DCo* 3.20.6; *OL* 1:253). The search for causes must ultimately lead back to definitions:

> And seeing also, that the end of the analytics is either the construction of a possible problem or the detection of the impossibility of a construction; if the problem is possible, the analyst must not stop, until he comes across things in which are contained the efficient cause of what he is to construct. But he must of necessity stop when he comes to first propositions, and these are definitions. In these definitions therefore there must be contained the efficient cause of the construction; I say of the construction, not of the demonstrated conclusion. For the cause of the conclusion is contained in the premised propositions, that is, the truth of the proposition proved is in the propositions that prove it. But the cause of the construction is in these things, and consists in motion or the concourse of motions. Therefore, the propositions in which analysis terminates are definitions, but of such kind as signify the manner in which the thing itself is constructed or generated. (*DCo* 3.20.6; *OL* 1:253)

In order to convert an analysis to a synthesis from such definitions, it is necessary that the steps in the analysis be reversible, in the sense that there is a logical entailment from each result obtained by analysis to its antecedent step. In Hobbes's terminology this requirement means that "the terms of all the propositions should be convertible; or if they are enunciated hypothetically, it is necessary not only that the truth of the consequent follow from the truth of its antecedent, but also that the truth of the antecedent be inferred from the truth of the consequent." Failing this, "when by resolving the principles are arrived at, there is no reverse composition to what is sought" (*DCo* 3.20.6; *OL* 1:252).

Hobbes defines all philosophy (i.e., demonstrative *scientia*) as *"such knowledge of effects or appearances as we acquire by true ratiocination from the knowledge we have first of their causes or generation. And again, of such causes or generations as may be from knowing first their effects"* (*DCo* 1.1.2; *EW* 1:3). Thus, he holds that neither analytic nor synthetic methods are confined to mathematics, but that both can be employed throughout all branches of philosophy. Indeed, Hobbes claims that both modes of reasoning are necessary in any investigation of causes.[32] Mathematics has a place for both analysis and synthesis in the course of proving theorems and solving problems: one proceeds analytically by first assuming what was to be demonstrated and then investigating the conditions necessary for its demonstration. Then, provided that all of the steps in the analysis are reversible, a synthetic demonstration from first principles can be effected. Here the analytic method functions as a preface to synthesis and is intended to aid in uncovering causal principles that can then be used to generate true demonstrations. Should it happen that the analysis leads to an absurdity, then the supposition at the beginning of the analysis is shown to be false.

The differences between mathematics and natural philosophy will be reflected in the kinds of analyses that can be performed in each area. Natural science must begin with phantasms or fancies in the mind whose existence is not in question.[33] Thus, unlike the mathematical

32. I leave aside the question of whether Hobbes sees a true unity of method between natural science and political science. Sorell (1986) argues that there is a methodological disunity between natural and civil science, notwithstanding Hobbes's claims that the methodology of analysis and synthesis is universally applicable. To whatever extent Hobbes is serious about his claim that the business of philosophy is the investigation of causes, it seems reasonable that he would see analytic and synthetic procedures as integral to the philosophical enterprise. Hanson (1990) addresses the question of how far Hobbes's talk of method commits him to the idea that there is a unity of method across disciplines.

33. As Hobbes puts it, "Thus the first principles of all knowledge are the phantasms of sense and imagination, which we know by nature to exist; but to know why they are

case, an analysis in natural science does not run the risk of terminating in absurdity. Analysis in natural science will lead from observed phenomena or phantasms to their possible causes, and synthesis will proceed from causal hypotheses (namely particular motions of bodies) to derivations of their effects (i.e., the phenomena to be explained). But because the causes of natural phenomena can only be the matter for hypothesis or conjecture, the synthetic procedures of the natural scientist will fall short of the demonstrative certainty obtainable in mathematics. In other words, the synthesis available to the natural philosopher will typically be no more than a synthesis from a hypothetical cause, whereas the geometer constructs his synthesis from the true causes of the objects of his investigation.

This conception of analysis as a preface to synthesis is in keeping with the traditional characterization of analysis as the "method of invention or discovery" while synthesis is the true "method of demonstration." Hobbes explicitly endorses this representation of the distinction, holding that it is by analysis that we are led "to the circumstances conducing separately to the production of effects," while synthesis takes us "to that which they effect singly by themselves when compounded together" (*DCo* 1.6.10; *OL* 1:70). Hobbes actually characterizes the synthetic method as the proper method of teaching, as it begins with uncontestable first principles and proceeds by proper syllogisms to the required result: "Therefore the whole method of demonstration is synthetic, consisting in the order of speech that begins from primary or most universal propositions, which are self-evident, and through the continual composition of propositions into syllogisms until the truth of the conclusion sought is understood by the learner" (*DCo* 1.6.12; *OL* 1:71).

Hobbes offers his explanation of light as an example of the proper employment of analysis and synthesis in natural philosophy. First we observe that there is a "principal object" or source of light whenever light is observed; by analysis, we take such an object as causally necessary to the production of light; further analysis shows that a transparent medium and functioning sense organs are required for the phenomenon to present itself. As the analysis continues we infer that a motion in the object is the principal cause of light, and the continuation of this motion through the medium and its subsequent interaction with the

---

or by what causes they are produced there is need of ratiocination, which consists . . . in composition and division" (*DCo* 1.6.1; *OL* 1:59).

"vital motion" of animal spirits in the sense organs are contributing causes. The result is that "in this manner the cause of light can be made up out of motion continued from its origin to the origin of vital motion, the alteration of which vital motion by this incoming motion is light itself" (*DCo* 1.6.10; *OL* 1:69–70). The analysis must stop somewhere, and Hobbes holds that it will terminate in simple, universal things that are self-evidently known and causally sufficient for the explanation of the phenomena. In the case of natural science, the motion of bodies is that self-evident principle upon which the explanation of physical phenomena depends.

Once the analysis has been completed and the phenomenon to be explained is traced back to certain motions, the way is open for a proper synthesis from first principles. This picture of method connects with the account of demonstrations τοῦ ὅτι and τὸ διότι, since "in knowledge of the ὅτι, or that *a thing is,* the search begins with the whole idea. On the other hand, in our knowledge of the διότι, or of the knowledge of causes, that is, in the sciences, the parts of the causes are better known than the whole cause. For the cause of the whole is composed out of the causes of the parts, but it is necessary that the things to be composed are known before the whole compound is known" (*DCo* 1.6.2; *OL* 1:59).

Hobbes discerns three different species of analysis in mathematics, corresponding to the three ways in which he thinks the equality or inequality of geometric figures can be determined, namely analysis by motion, by indivisibles, and by the "powers of lines." The first of these is grounded in the nature of motion and is specifically concerned with the motions by which figures are produced, "for from motion and time the equality and inequality of any quantities can be argued, no less than by congruence; and some motion can be made so that two quantities, whether lines or surfaces, although one is straight and the other curved, are congruent in extent and coincide; which method Archimedes used in his treatise on spirals" (*DCo* 3.20.6; *OL* 1:254). Hobbes identifies the second sort of analysis with the method of indivisibles, since "equality and inequality are often found by the division of two quantities into parts which are considered as indivisible, as Bonaventura Cavalieri has done in our time, and Archimedes in the quadrature of the parabola" (*DCo* 3.20.6; *OL* 1:254). The final style of analysis "is performed by considering the powers of lines, or the roots of powers, by multiplication, division, addition, subtraction, and the extraction of roots from powers, or by finding where right lines terminate in the same ratio" (*DCo* 3.20.6; *OL* 1:254).

The first and third types of analysis are worth investigating somewhat more closely. When he first published *De Corpore,* Hobbes was convinced that he had found a new mathematical method, which he called analysis by motions or the method of motion.[34] He had high hopes for this new type of analysis, having convinced himself that it could deliver results vastly exceeding those available with traditional means. Wallis remarks that in the penultimate impression of chapter 20 Hobbes had claimed in article 6 that the construction used in his quadrature was an example of the "analysis by computation of motions"; this was later abandoned when Hobbes recognized the failure of this attempt to square the circle and the quadrature was amended to be stated "problematically."[35] Evidently confident that this new style of analysis was a fundamental advance, Hobbes had planned to announce:

> From those things which have been said concerning the dimension of the circle, so many others can be deduced, that it seems to me that geometry, which for the longest time has remained stalled in this place as if before the pillars of Hercules, is now thrust forth into the ocean, to navigate the globe of other most beautiful theorems. And indeed I do not doubt that from the known ratio of right lines to the arc of a circle, to the parabola, and to the spiral, knowledge will at last be threshed out *[extitura]* of the ratio of the same right line even to a hyperbolic line, an elliptical line, and even the discovery of any number of continually proportional means between two extremes. (*Elenchus* 129–30)

In the final impression of *De Corpore,* Hobbes's found himself forced to admit the inadequacy of his three circle-squaring efforts and to drop this exultant passage. Instead, he concluded section 20 with the mournful declaration of the quadrature's shortcomings, while justi-

34. He retains the term "method of motion" in the *Six Lessons,* notwithstanding the fact that he had had to give up many of his claimed results. In the fifth lesson he declares: "But because you pretend to the Demonstration of some of these Propositions [in *DCo* 3.17] by another Method in your *Arithmetica Infinitorum,* I shall first try whether you be able to defend those Demonstrations as well as I have done these of mine by the Method of Motion" (*SL* 5; *EW* 7:307).

35. Wallis writes: "Then, after you have further vilified the 'analytics by way of powers,' in order to commend your new 'analytics by the computation of motions' as more proper, you set forth an example of it (at least in the penultimate impression); that is, you show by which motions this analytics had applied to your first quadrature, which analysis we have examined above" (*Elenchus* 129).

fying his decision to leave his bitter remarks aimed at Ward (Vindex) unaltered:

> Since (after this was worked out [*excusa*]) I have come to think that there are some things that could be objected against this quadrature, it seems better to warn the reader of this fact than to delay the edition any further. It also seemed proper to let stand those things which are deservedly directed at Vindex. But the reader should take those things which are said to be found exactly of the dimension of the circle and of angles as instead said problematically. (*DCo* 3.20.8; 1655 edition only)

As I indicated in the third chapter, Hobbes's comments on his powerful new "analysis by motions" invite the conjecture that this method is related to Roberval's method of "composition of motions." Hobbes and Roberval were in contact during Hobbes's stay in Paris in the 1640s, they participated in discussions on mathematical matters at Mersenne's lodgings, and they were both active in the Parisian mathematical community. We also saw that they had a common interest in such problems as the rectification of curvilinear arcs, and particularly in the problem of comparing the arc length of the cycloid with that of the parabola. Roberval's principal innovation in the method of indivisibles was to employ a kinematic analysis in the solution of such problems. This approach involves treating curves as traced by the motions of points and then comparing the infinitesimal motions by which the curve is generated at any given instant. Pierre Costabel has remarked that Roberval's successful solutions to various geometric problems "were connected to a very new conception of curves, in which that construction by a point was given independently of the sort of algebraic analysis that Descartes used. . . . It is this notion that led him to a method of composition of motions that enabled him in 1640 to solve problems of determining tangents" (Costabel and Martinet 1986, 23–24). Whether, in fact, Hobbes's conception of the "analysis by motions" is directly related to Roberval's work remains unclear, since Hobbes's attempted quadrature of the circle contains relatively little reference to the generation of the circle by motion, and does not attempt to compare infinitesimal arcs with rectilinear motions.[36] For his

---

36. Bernhardt observes that "[i]n 1642, during the winter, Hobbes discussed mathematics with the geometer [Roberval] in Mersenne's lodgings. It was the question of the rectification of the arcs of the spiral of Archimedes and the parabola. The surviving documents allow the Roberval scholar Kotiki Hara to conclude that the philosopher without doubt put the geometer on the track of a kinematic demonstration that these

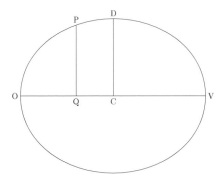

Figure 5.2

part, Wallis was convinced that Hobbes's talk of a new method of "analysis by motions" had been taken from Roberval, misunderstood by Hobbes, and ineptly inserted into the discussion of the circle quadrature.

Hobbes contrasts his new style of analysis to the more traditional analysis "by the power of lines," which is his way of characterizing the classical theory of loci. Consider, for example, the classical definition of the ellipse with major axis OV and center C (figure 5.2). Apollonius and the tradition define it as the locus of points satisfying the condition that the perpendiculars PQ and DC stand in the ratio $PQ^2/CD^2 = (OQ \cdot QV)/(QC \cdot CV)$. In Hobbes's estimation, "that part of analytics that is by powers, although it is regarded by some geometers (not of the first rank) as apt for the solution of all problems, yet it is a thing of no great extent" (DCo 3.20.6; OL 1:256). The limitations of this style of analysis arise from the fact that it "is all contained in the doctrine of rectangles, and rectangled solids. So that although they come to an equation that determines the quantity sought, yet they cannot sometimes by art exhibit that quantity in a plane, but in some conic section; that is, as geometricians say, not geometrically, but mechanically" (DCo 3.20.6; OL 1:256).

Although Hobbes sanctions both analytic and synthetic reasoning in geometry (and is particularly fond of his analysis by motions), he holds that only synthesis can be truly demonstrative. The reason for this should be clear: analysis proceeds hypothetically, but synthesis

discussions left incomplete" (Bernhardt 1989, 107–8). The reference to Kotiki Hara is to a doctoral dissertation defended at Paris in 1965, which I have unfortunately not been able to locate.

leads from acknowledged first principles to their necessary consequences. Synthetic reasoning thus satisfies the traditional (i.e., Aristotelian) requirement that demonstrations proceed from principles better known and more secure than their consequences, and it satisfies Hobbes's requirement that true knowledge be grounded in the investigation of causes. Hobbes thus holds that analytic reasoning is acceptable in mathematics only insofar as it can set the stage for a demonstrative synthesis, and it is in this sense that he claims that "however great anyone is as an analyst, he was earlier as great a geometer; nor do the rules of analysis make a geometer, but only synthesis beginning at the very elements and proceeding by the logical use of them. For the true teaching of geometry is by synthesis, in the method Euclid teaches. And one who has Euclid for a master can be a geometer without Viéte (although Viéte was clearly an admirable geometer); but he who has Viéte for a master, not so without Euclid" (*DCo* 3.20.6; *OL* 1:255–56).

5.2.4 HOBBES, METHOD, AND THE SCHOOL OF PADUA

The picture of method that emerges from this discussion has at least some significant similarity to the methodological doctrine of "resolution and composition" developed in the sixteenth-century by Italian philosophers known as the school of Padua, whose most famous exponent was Jacopo Zabarella (1532–89). On this conception of scientific method, true knowledge comes from first resolving something complex into its constituent parts and then retracing the steps to recompose the complex whole from the simple constituents. Thus, to understand fully the workings of a clock, one first disassembles it into its simpler gears, springs, etc. From knowledge of the simple parts and their mutual connections, one can then reconstruct the clock and acquire insight into its ability to keep time. But the method of resolution and composition is intended to do more than facilitate clock repair. Properly applied, it should extend to all of nature, because the gears and springs of the clock can themselves be resolved into more basic material constituents. Ultimately, the process of resolution should take us to the fundamental causes or first principles from which all of the phenomena of nature can be derived. It is important to stress that the resolution or dissection involved here need not be taken as literal. Especially in the investigation of the most basic aspects of nature, it is sufficient to perform a "thought experiment," in which natural phenomena are resolved only in thought.

Galileo was influenced by the Paduan school, and there is a tradition of Hobbes scholarship that portrays his conception of analysis and synthesis as an inheritance from Padua by way of Galileo.[37] Although I think the evidence for any direct connection between Hobbes and Galileo on this issue is at best inconclusive, there is no doubt that Galileo employed the technique of analysis and synthesis. The most famous example is his explanation of a projectile's path as arising from the composition of several simple motions. Galileo took the complex motion of a projectile such as a cannonball and first resolved its path into combinations of simpler rectilinear motions, including gravitation toward the earth as well as the horizontal and vertical components of the velocity imparted by the cannon. He then applied previous results concerning such rectilinear motions to the determination of the complex motion of the projectile. Finally, he reversed the analysis to show (synthetically) how the parabolic flight path arises from the composition of these different fundamental motions.[38]

Fitting Hobbes into this methodological picture is difficult, notwithstanding his expressed admiration for Galilean science. Prins (1990) has challenged any direct linking of Hobbes to the school of Padua by arguing that Hobbes and Zabarella have entirely different conceptions of science, while Hanson has argued that it is the ancient tradition of geometric analysis, rather than anything in Paduan methodology, that underlies accounts of analysis and synthesis in this period. Hanson writes: "When the new philosophers of the seventeenth century write in a sustained way about method they read very much alike in at least some important respects. That is because their views, on their own accounts, have a common source: not Zabarella and Padua, but the *Mathematical Collection* of Pappus" (1990, 604). Similarly, Malcolm holds that Hobbes's talk of resolution and composition "owed almost nothing to Galileo and very little to the Paduan tradition of commen-

37. A few examples should suffice to show the prevalence of this interpretation. Macpherson declares, "What was needed was a two-part method, which would show how to reach such simple starting propositions, as well as what to do when one had them. Hobbes found it in the method used by Galileo—the 'resolutive-compositive' method" (Macpherson 1968, 25–26). Similarly, Watkins 1965, chaps. 3–4, argues for a Paduan influence on Hobbes, particularly through Galileo and Harvey. Gargani devotes a chapter of his treatment of Hobbes's science to "the technique of resolution and composition in the Paduan commentators on Aristotle" (Gargani 1971, chap. 2). Kersting (1988) argues that Hobbes's unified conception of science is "based on" the Paduan method.

38. The "fourth day" dialogue in Galileo's *Two New Sciences* (Galileo 1974, 217–60) contains the analytic-synthetic approach to projectile motion. See Watkins 1965, 55–63, for an overview of this material and the argument for connecting it with Hobbes's methodological doctrines.

tary on Aristotle. The use of the terms 'resolutio' and 'compositio' was immensely widespread, across a whole range of disciplines: they were simply the Latin equivalents of the Greek terms 'analysis' and 'synthesis', terms used in the Galenist tradition of diagnosis and prognosis, and in the Euclidean tradition of the methodology of mathematical problems" (1990, 153–54).

Although I am persuaded that Hobbes's talk of analysis and synthesis is at best tenuously connected to the Paduan tradition, there is no need to decide the issue here, as nothing crucial depends upon whether Hobbes is indebted to Galileo or the school of Padua for his methodological distinction between analysis and synthesis. Whatever the extent of Hobbes's intellectual debt to Galileo or the school of Padua, his account of analysis and synthesis is certainly fundamental to his conception of mathematical method. It figures prominently in his rejection of algebraic or analytic methods in geometry, to which we must now turn.

5.3 HOBBES AND THE CASE AGAINST ANALYTIC GEOMETRY

Hobbes was contemptuous of the claims made for analytic geometry, and particularly for the claims of increased power and extent of these methods when compared with the classical techniques. He repeatedly argued that the use of algebraic methods can add nothing to classical geometry, and he pursued his claims against the new methods with an almost fanatical intensity. Some have taken this attitude to indicate Hobbes's ignorance of elementary mathematics, as when Peters claims that he "had the temerity to attack Wallis's own work on conic sections treated algebraically, although he knew no algebra himself" (1956, 40). Hobbes was not a mathematical genius, but he was hardly as uninformed as Peters makes him out to be. Still, the vehemence of his attacks on analytic geometry does require some explanation, all the more so because his nominalism would seem to fit conveniently with an approach to mathematics that emphasizes the importance of symbolic algebra.[39]

A large part of Hobbes's hostility to analytic procedures is directed

---

39. Pycior (1997, 143–48) discusses Hobbes's rejection of algebraic methods in the context of his battle with Wallis, reaching the conclusion that "[s]tanding in awe of and deriving personal and intellectual comfort from the certainty of synthetic geometry, Hobbes saw algebra as no more than a 'weapon' in the modern war on geometry" (1997, 148).

against what he takes to be their excessive use of symbols. In *De Corpore* he dismisses the techniques of analytic geometry as "the brachygraphy" (i.e., shorthand) of geometry, a term meant to capture the idea that symbolic methods may abbreviate proofs but cannot fundamentally increase the power of the traditional methods. Such methods, he says, are part of the "analysis by powers," and are not used by geometers "of the first class" (*DCo* 3.20.8; *OL* 1:256). In one especially unrestrained outburst in the *Six Lessons,* he attacks Wallis's use of algebra by asking

> When did you see any man but your selves publish his Demonstrations by signs not generally received, except it were not with intention to demonstrate, but to teach the use of Signes? Had *Pappus* no Analytiques? Or wanted he the wit to shorten his reckoning by Signes? Or has he not proceeded Analytically in a hundred Problems (especially in his seventh Book), and never used Symboles? Symboles are poor unhandsome (though necessary) scaffolds of Demonstration; and ought no more to appear in publique, then the most deformed necessary business which you do in your Chambers. (*SL* 3; *EW* 7:248)

Elsewhere in the *Six Lessons,* he remarks that Wallis's *Treatise of Conic Sections* "is so covered over with the scab of Symboles, that I had not the patience to examine whether it be well or ill demonstrated" (*SL* 5; *EW* 7:316). He ultimately finds in the mathematics of the Savilian professor "no knowledge neither of Quantity, nor of measure, nor of Proportion, nor of Time, nor of Motion, nor of any thing, but only of certain Characters, as if a Hen had been scraping there" (*SL* 5; *EW* 7:330). This much of Hobbes's case against algebraic methods is hardly worth taking seriously, since it amounts only to an "aesthetic" complaint that symbolic methods deface geometric demonstrations.

Hobbes adds a more interesting criticism when he contends that the introduction of algebra cannot add anything to geometry because it distracts the geometer's attention from geometric magnitudes and replaces the contemplation of magnitudes with the manipulation of symbols. This is the import of his accusation that Wallis is misled by empty symbols that have been added on to the old "analysis by powers" but have not produced anything new:

> I verily believe that since the beginning of the world, there has not been, nor ever shall be so much absurdity written in Geometry, as

is to be found in those books of [Wallis]. . . . The cause whereof I imagine to be this, that he mistook the study of *Symboles* for the study of *Geometry,* and thought *Symbolicall* writing to be a new kinde of Method, and other mens Demonstrations set down in *Symboles* new Demonstrations. The way of *Analysis* by Squares, Cubes &c., is very antient, and usefull for the finding out whatsoever is contained in the nature and generation of rect-angled Plains (which also may be found without it) and was at the highest in *Vieta;* but I never saw anything added thereby to the Science of Geometry, as being a way wherein men go round from the Equality of rectangled Plains to the Equality of Pro-portion, and thence again to the Equality of rectangled Plains; wherein the *Symboles* serve only to make men go faster about, as greater Winde to a Winde-mill. (*SL* dedication; *EW* 7:187–88)

Where others had taken algebra to be a way of simplifying demonstra-tions while also enabling geometers to find new results, Hobbes sees algebra as only a new kind of language that has been foisted upon geometry to no purpose.

In a similar vein, he claims that algebra cannot shorten demonstra-tions or make geometry easier to understand. Contrary to the claims of the modern analysts, Hobbes insists that "algebra can yield brevity in the writing of a demonstration, but not brevity of thought. Because it is not the bare characters, or only the words, but the things them-selves that are the objects of thought, and these cannot be abbreviated" (*Examinatio* 3; *OL* 4:97). As Hobbes sees the matter, a proper demon-stration must proceed by way of constructions from causes to effects, but reliance upon algebra simply interposes a collection of symbols between ourselves and the magnitudes we are to construct.

Hobbes's distinction between analytic and synthetic reasoning also accounts, in part, for his rejection of analytic geometry. A typical proof in analytic geometry proceeds "analytically" in the classical sense by first supposing the problem solved and then showing that the solution is algebraically admissible. As Descartes puts it:

If, then, we wish to solve any problem, we first suppose the solu-tion already to have been effected, and give names to all the lines that seem needful for its construction—to those that are un-known as well as to those that are known. Then, making no dis-tinction between known and unknown lines, we must unravel the difficulty in any way that shows most naturally the relations be-tween these lines, until we find it possible to express a single quantity in two ways. This will constitute an equation, since the terms of one of these two equations are together equal to the terms of the other. And we must find as many such equations as we have supposed lines which are unknown. (*AT* 6:372–73)

This haphazard approach is not a proper demonstration by causes and, Hobbes claims, can at best result in a half-finished demonstration that must still be "converted" into a synthesis. True geometry must appeal to causes and constructions, rather than the hypothetical procedure of Descartes and its fortuitous unraveling of a problem by means of equations.

We can thus discern two strands in Hobbes's objection to analytic geometry: he opposes both its excessive use of algebraic symbols and its reliance upon hypothetical procedures. These two strands of Hobbes's critique of analytic geometry can be drawn together and phrased in a single objection: the use of algebraic methods in geometry must be either unscientific or superfluous. For if algebraic techniques have governing principles and do not simply proceed *par hasard,* then these principles must be vindicated by appeal to geometric considerations, thereby making algebra superfluous. Hobbes poses this dilemma in the *Examinatio* as follows:

What else do the great masters of the current symbolics, Oughtred and Descartes, teach, but that for a sought quantity we should take some letter from the alphabet, and then by *right* reasoning we should proceed to the consequence? But if this be an art, it would need to have been shown what this *right* reasoning is. Because they do not do this, the algebraists are known to begin sometimes with one supposition, sometimes with another, and to follow sometimes one path, and sometimes another. . . . Moreover, what proposition discovered by algebra does not depend upon Euclid (6, prop. 16) and (1, prop. 47), and other famous propositions, which one must first know before he can use the rules of algebra? Certainly, algebra needs geometry, but geometry does not need algebra. (*Examinatio* 1; *OL* 4:9–10)

Clearly, the "right" reasoning Hobbes has in mind here will be reasoning from causes to effects, that is to say demonstrative knowledge grounded in a synthetic exposition of the properties of geometric objects. Hobbes holds that we cannot know whether an algebraic procedure is legitimate unless we already know that the geometrical step corresponding to it is an admissible construction. But in such a case there is no need to distract ourselves with the study of algebraic symbols, and we should proceed immediately to the construction.

A final source of Hobbes's suspicion of analytic geometry comes from his thesis that geometry is the fundamental mathematical science upon which all others must be based. Because Hobbes takes arithmetic to depend upon geometry, and because algebra was generally thought to be a kind of "arithmetic of species," he would naturally be suspicious of the idea that an essentially arithmetical method (i.e., algebra) can extend the scope of the foundational science (i.e., geometry).

Although Hobbes's campaign for the elimination of analytic geometry may seem bizarre to the contemporary reader, it was not so terribly far out of the way by the standards of the seventeenth century. Barrow, for example, disparaged the analytic approach and held that the classical methods of synthetic geometry were preferable. And, indeed, the reliance on purely algebraic reasoning in a geometric case can produce a degree of confusion: it is not always clear whether an algebraic transformation corresponds to an admissible geometric construction. Moreover, the authors of the new analysis were not forthcoming with explanations of their techniques that might justify them in terms of more traditional standards. Descartes, for example, occasionally proceeds in a haphazard and seemingly arbitrary manner, with key steps in his reasoning suppressed and justifications of certain moves left unstated. He suggests, with evident irony, that he does not want to deprive his readers of the pleasure of figuring things out for themselves, and we can easily understand how Hobbes could take such procedures as indicative of a complete absence of method.[40]

There is nevertheless a puzzle here, and one that finally leads to a difficulty for Hobbes's whole theory of demonstration. As we have seen, Hobbes's own account of reasoning certainly seems, at first sight, to justify the use of algebra in the study of geometry. Reasoning, as Hobbes understands it, is just the manipulation of arbitrarily imposed signs, and there are no grounds a priori for Hobbes to think that the use of algebraic symbols should be any more suspect than any other kind of signs. Indeed, when Hobbes complains that symbols are an impediment to geometric proof because "it is not the bare characters,

40. Descartes concludes his brief remarks on his method of "unraveling" geometric problems by manipulation of equations in book 1 of the *Géométrie* with the remark that "I will not stop here to explain this in detail, because I would then deny you the pleasure of apprehending it for yourself, as well as the utility of cultivating your mind by working through it, which is in my view the principal benefit to be derived from this science" (*AT* 6:374).

or only the words, but the things themselves that are the objects of thought, and these cannot be abbreviated" (*Examinatio* 3; OL 4:97), his declaration flatly contradicts his pronouncement in the objections to Descartes's *Meditations* that reasoning can tell us nothing at all about the nature of things, but is confined to the manipulation of names arbitrarily imposed by speakers' convention.

I think that the best way out of this difficulty is to rely upon the sort of considerations that help to avoid the conclusion that Hobbes's theory of demonstration is purely conventionalistic. As I argued above, Hobbes thinks that proper demonstrations must be grounded in causes; but where causes are unknown (as in natural science), demonstrations must proceed from causal hypotheses. In the latter case, our assignment of a cause involves the imposition of an arbitrary name for what we take to be the cause of the phenomenon we are explaining. Under these circumstances demonstration need not tell us about the nature of things, since we may have misidentified its cause in our hypotheses. Descartes's claim to uncover and demonstrate the true nature of the piece of wax in the second *Meditation* is obviously inconsistent with such a theory. But in the exceptional case of mathematics (and, politics, as Hobbes understands it), we bring geometric (or political) objects into being and therefore have secure access to the cause of the object whose properties we hope to demonstrate. In such cases, Hobbes thinks that there is room for genuine demonstrations that are not merely hypothetical. Moreover, the very fact that makes such demonstrations nonhypothetical also means that we can deal with "the things themselves" rather than mere names or labels assigned to unknown things. Of course, in order to generalize a result like "all circles have a periphery greater than that of an inscribed polygon" we must still have recourse to language by introducing names like "circle" and "polygon"; but these names are assigned on the basis of true causes and are therefore not arbitrary. The upshot is that the strongly nominalistic themes in Hobbes's theory of reasoning should be taken to apply to the natural sciences, while his complaints against algebra should be read against the background of his doctrine that a properly organized geometry can and should proceed from the true causes of geometric objects.

At the end of the day, Hobbes's rejection of analytic geometry says at least as much about his inability to appreciate the power of new algebraic methods as it does about his criteria for rigorous demonstration. The framing of geometric curves as equations in two unknowns is more than simply the attaching of labels to geometric magnitudes.

Rather, it is a way to place geometric problems in a very general setting and (as Wallis would put it) to remove any specifically geometric content while demonstrating properties of the curves from the algebraic features of their characteristic equations.[41] The fruitfulness of analytic geometry is manifest: with it, mathematicians solved important problems left unsolved by classical methods while opening up whole new classes of problems that could never be attempted within the confines of the classical point of view. Hobbes wanted desperately to make his mark as a mathematician, but his own theory of demonstration encouraged him to toss aside the very tools that might have helped him achieve his dream.

41. This is the "emphasis on the relations rather than the objects" and the "freedom from ontological commitment" that Mahoney (1980) identifies as essential to the rise of algebraic thought and in large measure responsible for the mathematical successes enabled by algebraic methods.

CHAPTER SIX

The Demise of
Hobbesian Geometry

> I do not want to change, confirm, or argue any more about
> the demonstration which is in the press. It is correct; and if
> people burdened with prejudice fail to read it carefully
> enough, that is their fault, not mine. They are a boastful,
> backbiting sort of people; when they have built false con-
> structions on other people's principles (which are either
> false or misunderstood), their minds become filled with
> vanity and will not admit any new truth.
> —Hobbes to Sorbière, 7/17 March 1664

A lthough there may be many disputes in which it is misleading to
speak of winners and losers, there is no question that, at least in
regard to its mathematical aspects, Wallis was the winner in his dispute with Hobbes. He exposed the inadequacy of more than a dozen
Hobbesian circle quadratures and soundly refuted many other of
Hobbes's excursions into the great classical problems. Indeed, he was
so successful in his campaign against Hobbes that the only blemish to
his own reputation was essentially self-inflicted: contemporaries such
as Huygens readily granted the cogency of his arguments but wondered
why he had taken the trouble to expose the vanity of Hobbes's pre-
tenses at such great length.[1] Well before his death in 1679, Hobbes's
once-considerable mathematical reputation had been utterly destroyed
and, although Wallis had hardly enhanced his standing in the course
of the dispute, it was evident to any seventeenth-century observer that
he had defeated his antagonist in their very publicly conducted contest.

---

1. Commenting on the *Elenchus* in a letter to Wallis dated 5/15 March 1656, Huygens remarks that "I was amazed that you judged [Hobbes] worthy of such a lengthy
refutation, although I read your learned and rather sharp *Elenchus* with some pleasure"
(*HOC* 1:392). Oblivious to the suggestion that he might have found a better and more
dignified use for his time, Wallis did not hesitate to publish Huygens's comments in
order to show the world how poorly Hobbes's efforts had fared with the learned public
(*Due Correction* 4–5).

This is not to say that Hobbes had absolutely no allies, but there is hardly anything that one might term a discernibly Hobbesian school of mathematics in seventeenth-century intellectual life.[2]

My purpose in this chapter is to examine Hobbes's increasingly desperate attempts to defend the validity of his mathematical work and thereby to examine the process by which he was led to reject ever-larger portions of geometry in order (so he thought) to shield his claims from refutation. This process went so far that by the end of his life Hobbes found it necessary to condemn essentially all of classical as well as contemporary geometry as ill conceived. Being driven to such an expedient by the continued refutation of his work must certainly have been an unpleasant fate for the man who had once declared classical geometry "the onely Science it hath pleased God hitherto to bestow on mankind" (*L* 1.4, 15; *EW* 3:23). There is much to learn about Hobbes by studying his seemingly forced march toward the abandonment of accepted mathematical practices and results, and we can gain important insights into his philosophy as a whole by attending to some nuances in his changing attitude toward traditional mathematics.

2. Although there was no Hobbesian school of geometry, Hobbes's mathematics was not universally condemned. François du Verdus was generally receptive to Hobbes's mathematical efforts, and Hobbes repaid his mathematical loyalty by dedicating *Examinatio et Emendatio Mathematicae Hodiernae* to him. Eventually du Verdus advised Hobbes to abandon the search for the circle quadrature and admit error, since "you will still have the glory of having penetrated as far as anyone else in that matter—and further than anyone else in every other matter" (Du Verdus to Hobbes 22 December 1656/1 January 1657; *CTH* 1:413). François Pelau, another French admirer of Hobbes, held his mathematical work in even higher esteem. In a letter to Hobbes in May of 1656, Pelau wrote that "it is because of you that I have turned to the truth, having rebelled against it in favor of the Ancients; your section *De Corpore* is a work which will live and be read and admired by the most distant future generations. I was extremely glad to see that you are a great geometer: it was in that capacity, above all, that you touched on my own interests. I have seen the works of Descartes, Gassendi, Galileo, and Mersenne, but they all amount to nothing in comparison with what I learn every day from your book, which I have so thoroughly assimilated that I hardly read anything else: it alone is my entire library, and the subject of all my commentaries" (Pelau to Hobbes, 18/28 May 1656; *CTH* 1:291). Other reports of Hobbes's mathematical supporters are hard to come by. Stubbe reported to Hobbes in early 1657 that admirers in Oxford had composed verses on the battle between Hobbes and Wallis, so that "yo^u may see ye vogue of those youths that pretend to any thing of ingenuity is against D:^{r} Wallis, & yo^u have the good opinion of all who are judges of language, ingenuity or Mathematiques" (Stubbe to Hobbes, 30 January/9 February 1657; *CTH* 1:440). After Hobbes's death, Venterus Mandey translated some of Hobbes's mathematical works into English and evidently found them well suited for teaching such practical subjects as measurement and surveying (Mandey 1682). Still, none of this amounts to a Hobbesian mathematical school.

6.1 THE DESCENT INTO THE ABYSS

I argued in chapter 3 that there is an identifiable core of doctrines that comprise the Hobbesian philosophy of mathematics, the most important of which are the claims that (1) mathematics is a generalized science of body and (2) the first principles of mathematics must express the causes by which mathematical objects are generated. Although Hobbes never abandoned these core principles, his views on the nature of mathematics were not altogether static. In the period after the publication of *De Corpore* (1655) he changed his mind on several significant mathematical issues, and particularly about the extent to which classical geometric methods could be reconciled with his own materialistic program for mathematics. Hobbes was soon forced to face the unpleasant fact that his supposed solutions to classical problems were refuted by Wallis and other European mathematicians. His initial reaction to such refutations was generally to admit error on points of technical detail while insisting that his general conception of mathematics was the only coherent philosophical account of the subject. However, as the refutations mounted and his revised efforts fared as badly as the originals, Hobbes began to challenge the validity of criticisms and to reject the principles upon which they were based. In the end, this process went so far that he abandoned even the most elementary principles of classical geometry and recanted his earlier admissions of error.[3]

I have chosen to illustrate the collapse of Hobbes's geometric program by an examination of his responses to criticisms of his proofs and document the increasingly desperate measures he undertook in defense of his claims. I begin with an account of his replies to technical objections to the geometry of *De Corpore,* proceed to an examination of his unsuccessful efforts to duplicate the cube in the 1660s, then turn to an overview of his *essay De Principiis et Ratiocinatione Geometrarum,* and conclude with a brief study of his attempted quadrature of the circle from 1669. This selection, although necessarily incomplete, gives an adequate picture of the difficulties Hobbes encountered and of the increasing desperation of his responses. Along the way, I

---

3. On Hobbes's recantation of prior admissions of error, Wallis observed with characteristic venom, "How many Quadratures, first and last, Mr *Hobs* hath furnisht us with; I cannot presently tell You. But that they are all true, and all the same, I suppose he would have us beleeve. For though he have formerly confessed some of them to be mistakes; yet he hath now revoked those confessions, and thinks them to be true" (*HHT* 104).

will take the opportunity to study Hobbes's failed campaign for admission to the Royal Society and its connection with his geometric work.

6.1.1 Hobbes in Defense of De Corpore

I showed in chapter 3 that the circle quadrature in section 20 of *De Corpore* has a tortured history that reflects Hobbes's apparent indecision about the soundness of his argumentation. He assembled several different versions of this result, replacing versions one after another as friends made him aware of their shortcomings. In the end, the text of *De Corpore* as Wallis encountered it had a decidedly "pentimento" character that made the refutation and humiliation of Hobbes all the easier. Notwithstanding the acknowledged errors in his first two attempts to square the circle, Hobbes insisted on publishing his original effort under the title "A false quadrature, from a false supposition," and downgraded the second to the status of an approximation, and finally had to admit that the third must be taken "problematically" rather than as a strict demonstration. All of this shows that Hobbes was prepared to recognize potential shortcomings in his geometrical proofs at the time he published *De Corpore,* and it is clear that he was by no means impervious to strictly technical criticisms of his work. He was convinced that he had squared the circle and was more than ready to revel in this imaginary triumph, but he was not irrevocably committed to any one particular solution and he was prepared to reconsider his efforts in the light of objections.

Hobbes's response to Wallis's *Elenchus* shows the same willingness to acknowledge error, despite the fact that his opponent's mocking tone and generally uncivil language gave him reason for resentment. In numerous passages from the *Six Lessons* Hobbes admits the inadequacy of his presentation and argumentation, although he frequently complains about the manner in which Wallis states his objections.[4] We can consequently take it as established that, as of 1656, Hobbes was

4. Thus, in the third of the *Six Lessons* (*EW* 7:267) he admits that the definition of geometric figure at *DCo* 2.14.22 "wants the same word" missing from the earlier definition of parallels at *DCo* 2.14.12 in order to make it completely general. Similarly, at the beginning of the fourth lesson he confesses that *DCo* 2.16 contains "three or four faults, such as any Geometrician may see proceed not from ignorance of the Subject, or from want of the Art of Demonstration, (and such as any man might have mended himself) but from security" (*SL* 4; *EW* 7:269). He further admits that a projected twentieth article to the sixteenth chapter of *DCo* contains "a great error," although the article was left out of the final version (*SL* 4; *EW* 7:297). Finally, he grants that in both of the problems that were supposed to be solved in the eighteenth chapter, "You have truly demonstrated that they are both false; and another hath also Demonstrated the same

ready to accept the authority of classical geometry and to submit his putative demonstrations to the tribunal of prevailing mathematical opinion.

Some examples drawn from the *Six Lessons* illustrate this claim more fully. In reply to Wallis's objections against the (admittedly false) first attempt to square the circle, Hobbes remarks that "seeing you knew I had rejected that Proposition, it was but a poor Ambition to take wing as you thought to do, like Beetles from my egestions" (*SL* 5; *EW* 7:324). Hobbes's point here is that the errors in his initial assault on the problem had already been acknowledged in the final printed text of *De Corpore,* so that the "false quadrature from a false supposition" need not be taken to show mathematical ineptitude on his part. Instead, he thought that by publishing an erroneous result and also showing the error in his procedure he would enable (or at least help promote) a true solution to the problem since "it was likely to give occasion to ingenious men (the practice of it being so accurate to sense) to inquire wherein the Fallacy did consist" (*SL* 5; *EW* 7:324). Wallis was hardly impressed by this display and commented that it was a sign of Hobbes's vanity and self-conceit, since he judged even his miserable failures to merit the attention of the learned world.[5]

Apparently unable to resist the temptation to make yet another foray into the treacherous terrain of circle quadrature, Hobbes even tried to counter Wallis's objections on a particular technical point.[6] But even here his language falls far short of a dogmatic assertion of the validity of his procedures. After attempting to rebut one of Wallis's criticisms of the construction and argumentation in *De Corpore,* Hobbes declares, "And though in this also I should have erred, yet it cannot be denied but that I have used a more natural, a more Geometrical, and a more perspicuous method in the search of this so difficult a Probleme, then you have done in your *Arithmetica Infinitorum*" (*SL* 5; *EW* 7:326). This is obviously not the language of a man who is completely confident of his results. Hobbes claims the superiority of his methods for being more geometrical and perspicuous, but he is

---

another way" (*SL* 5; *EW* 7:319). As we will see, the reference to "another" here is to Claude Mylon.

5. Before commenting on the failed quadratures in chapter 20 of *De Corpore,* Wallis remarked: "You do admit that both [of these quadratures] are false, but you bother to submit them, however false, to our view, obviously having judged that even in your miscarriages there is something of beauty" (*Elenchus* 90).

6. Wallis had argued that Hobbes's construction of a specific line was not equal to a given circular arc (*Elenchus* 108–11). Hobbes's argumentation and its flaws are examined in the appendix, section A.3.1.

willing to accept the possibility of error and does not claim that other approaches are entirely without merit.

This willingness to admit error arises, at least in part, from Hobbes's distinction between two fundamentally different causes of error in demonstration: ignorance and negligence. Errors of ignorance proceed from false or misunderstood principles, while errors of negligence involve the misapplication of correct principles. Errors of negligence are, in Hobbes's estimation, much less reprehensible than those of ignorance. The person "ignorant of that he goes about" commits a serious error "because he was not forced to undergo a greater charge than he could carry through" (*SL* 2; *EW* 7:212). Thus, an error of ignorance in a geometric demonstration shows that the geometer does not really understand the basis of his subject. In contrast, errors of negligence arise "through humane frailty" when true principles are mistakenly applied, the mistake being the consequence of the demonstrator "being less awake, more troubled with other thoughts, or more in haste when he was in writing" (*SL* 2; *EW* 7:212). This kind of error is naturally less reprehensible, first because it is common to all men, but more importantly because it does not reflect a shortcoming in the demonstrator's first principles.

Hobbes's admissions of error in *De Corpore* are confined exclusively to errors of negligence.[7] He remained adamant that his fundamental principles were above reproach, since he regarded his materialistic program for the philosophy of mathematics as the only proper way to understand such key concepts as equality, proportion, angle, point, line, figure, and measure. In other words, Hobbes placed more emphasis on the superiority of his overall philosophy of mathematics than on his ability to solve outstanding problems. He consequently did not regard it as particularly damaging if he should misapply one of his principles in the attempt to solve a problem.[8]

His willingness to acknowledge the validity of purely technical crit-

---

7. As I explain below, there is one apparent exception to this claim, when Hobbes admits that his general definition of parallels is incorrect. This might seem to fit his definition of an error of ignorance, as it involves a false or inadequate definition; but even here Hobbes insists that the difficulty is one of negligence because it is little more than a slip of the pen. Of course, from the fact that Hobbes only admits to errors of negligence, it does not follow that his errors are always such. In fact, as an inspection of his demonstrations readily shows, Hobbes frequently based his claims on false or misunderstood geometric principles.

8. In this context it is worth recalling Hobbes's boast to Wallis that "I am the first that hath made the grounds of geometry firm and coherent. Whether I have added anything to the edifice or not, I leave to be judged by the readers" (*SL* 3; *EW* 7:242).

icisms of his geometric work in *De Corpore* can also be seen in Hobbes's reaction to the objections of the French mathematician Claude Mylon.[9] Through the agency of Hobbes's friend du Verdus, Mylon sent Hobbes several objections to the mathematics of *De Corpore*. In particular, he charged that Hobbes's general definition of parallels (*DCo* 2.14.12; Hobbes 1655, 113) was incomplete, and that a consequence drawn from it did not follow.[10] He further objected to all of section 18 of *De Corpore* on the grounds that "the general problem that he uses to equate straight lines with parabolas would be fine if it were true; but the trouble is, M. Huygens of Zuylichem has demonstrated that it is false" (Mylon to du Verdus, enclosure in du Verdus to Hobbes 20 February/1 March 1656; *CTH* 1:241). Mylon added a demonstration of the falsehood of Hobbes's approach to the rectification of parabolic segments, basing this refutation on Huygens's treatise *De Circuli Magnitudine Inventa*.[11] One odd irony of this encounter between Huygens and Hobbes is that it apparently led Huygens to a

9. For an account of Mylon's life and work, see Noel Malcolm's brief biographical note in *CTH* 2:868–69.

10. The original definition reads, "Parallelarum rectarum definitionem aliquam habemus apud Euclidem, sed Parallelarum in universum definitionem nusquam invenio. Itaque earum Definitio Universalis esto haec. Duae *lineae quaecunque* (sive rectae, sive curvae) atque etiam duae superficies, *Parallelae sunt, in quas duae lineae rectae ubicunque incidenties, facientesque cum utravis ipsarum angulos aequales, sunt inter se æquales.*" The corrected version adds "ad easdem partes" between "ipsarum" and "angulos" (Hobbes 1655, 113; *OL* 1:163). The point is that two lines are parallel when any two right lines that intersect both and make equal angles on corresponding sides also have the same length (*SL* 3; *EW* 7:254). Du Verdus actually suggested the appropriate change to Hobbes, rewriting the original definition to read, "Any two lines whatsoever, straight or curved, and likewise any two surfaces, are parallel, when any two straight lines between them, intersected anywhere, make equal angles and are of equal length" (Du Verdus to Hobbes 20 February/1 March 1656; *CTH* 1:238).

11. The refutation is in the second of two enclosures in a letter from du Verdus to Hobbes, and may have been the work of Huygens himself. We know that Huygens saw and refuted the quadrature in *De Corpore*, since Mylon refers to Huygens's own "examination" of Hobbes's "first construction" in a letter to Huygens (Mylon to Huygens, 13/23 June 1656; *HOC* 1:440). Furthermore, Mylon says in his letter to du Verdus that contains the enclosures that Huygens "is due to write about this from Holland, to Mr Hobbes. If you are curious to see this demonstration of his, I shall send it to you whenever you like" (Mylon to du Verdus 20 February/1 March 1656; *CTH* 1:241). Malcolm reports that both enclosures are in Mylon's hand and attributes both to Mylon (*CTH* 1:240). Whether or not Huygens actually penned the refutation, it remains the case that the real source of the objection are the results in his *De Circuli Magnitudine Inventa*. This treatise can be found in *HOC* 12:91–181. Mylon sent a second version of the argument to Hobbes in April of 1656, again basing his argument on proposition 9 of Huygens's *De Circuli Magnitudine Inventa* (Mylon to Hobbes, 9/19 April 1656; *CTH* 1:272–76).

close examination of the problem of rectifying parabolic arcs, which eventually produced a correct result.[12]

Hobbes's response to Mylon's criticisms was to concede error. With evident reluctance, he admitted that his definition of parallels was inadequate and corrected it in the English version of *De Corpore*. He nevertheless took great pains to deny that his mistaken definition constitutes an error of ignorance, precisely because he could not bring himself to admit error on a point of general principle. In the *Six Lessons* Hobbes tried both to downplay the significance of the erroneous definition of parallels and to contrast Mylon's manners with those of Wallis:

> At the twelfth Article, I confess your exception to my universal definition of Parallels to be just, though insolently set down. For it is no fault of ignorance (though it also infect the demonstration next it) but of too much security. . . . The same was observed also upon this place by one of the prime Geometricians of *Paris,* and noted in a Letter to his friend in these words, *Chap. 14. Art. 12. the Definition of Parallels wanteth somewhat to be supplyed.* And of the Consectary, he says, *it concludeth not, because it is grounded on the Definition of Parallels.* Truely, and severely enough, though without any such words as savour of Arrogance, or of Malice, or of the Clown. (*SL* 3; *EW* 7:255)

It is noteworthy that even in this case Hobbes insists that his error does not involve ignorance, notwithstanding the fact that it concerns a general principle. Given Hobbes's penchant for reprimanding his opponents for their failure to proceed from proper definitions, this is not the kind of error he could cheerfully admit. Thus, rather than acknowledge an error of ignorance on this point, Hobbes characterizes his mistake as a simple slip arising from "too much security"—which is his way of saying that it proceeds from inattention to detail, presumably

---

12. "Occasioned by the refutation of Thomas Hobbes's failed rectification of the parabola, Huygens showed that by employing a system of chords between equidistant parallels to the diameter (which could then be transformed into a system of tangents), the arc of a parabola could be straightened, and he then reduced the rectification to the quadrature of the hyperbola" (Hoffmann 1957, 2:39). Whiteside mentions that Hobbes "tried to show that the general parabola-arc is equal to a rational line-length, not allowing unfortunately for the modifying effect of changing gradient" (1960–62, 328). Huygens first pointed out Hobbes's error in a letter to Wallis from 5/15 March 1656 (*HOC* 1:392), but his solution to the problem was not worked out until 17/27 October 1657 (see *HOC* 14:188–92, 234–36).

due to his conviction that the matter could be resolved easily. In any event, Hobbes simply will not admit that there is a significant flaw in his geometric principles.[13]

Hobbes's response to the second of Mylon's objections was to admit to an error of negligence and to grant that the theorems in Huygens's *De Circuli Magnitudine Inventa* provide a decisive objection to the whole of chapter 18. The details of his response are in a lost letter to Mylon, but we can reconstruct their general outlines from Mylon's testimony.[14] In a letter to Huygens, Mylon reports that Hobbes replied to his objections in a letter, and that "Mr. Hobbes told me a month ago that it is true that he erred in the equation he used to rectify the parabola, but that his error was not one of method, since he had only taken one line for another in his construction. He thus says that he has corrected it" (Mylon to Huygens, 18/28 June 1656; *HOC* 1:439–40). Inevitably, the new version fared no better and fell victim to the very same kind of refutation.[15]

In sum, Hobbes's reactions to the criticisms of the mathematics in

---

13. Another example of this appears in Hobbes's response to refutations of his treatment of accelerated motion (*DCo* 3.16.4, corollary 1; *OL* 1:190–91). He replies to Wallis that "your exception I confess is just, and (which I wonder at) without any incivility. But this argues not Ignorance, but Security. For who is there that ever read any thing in the Coniques, that knows not that the parts of a Parabola cut off by Lines Parallel to the Base, are in Triplicate Proportion to their Bases? But having hitherto designed the Time by the Diameter, and the *Impetus* by the Base; and in the next Chapter (where I was to calculate the proportion of the Parabola, to the Parallelogram) intending to design the Time by the Base, I mistook and put the Diameter again for the Time; which any man but you might as easily have corrected as reprehended" (*SL* 4; *EW* 7:279).

14. Mylon's account is confirmed by Hobbes's remarks in the *Six Lessons,* where he reports that "the fault [with the treatment of parabolic arcs in chapter 18] was not in my method, but in a mistake of one Line for another and such as was not hard to correct; and is now so corrected in the English as you shall not be able (if you can sufficiently imagine Motions) to reprehend" (*SL* 5; EW 7:319). This is almost exactly what Mylon reports as Hobbes's reply to his own objections.

15. Mylon reports to Huygens: "I sent [Hobbes] a demonstration of the falsity of this new construction, having examined it in nearly the same manner that you did the first construction. Your ninth proposition in *De Circuli Magnitudine Inventa* seems to have been made expressly to be the touchstone *[piere de touche]* of such propositions" (Mylon to Huygens, 18/28 June 1656; *HOC* 1:440). The proposition to which Mylon refers asserts that the whole circumference of a circle is less than two-thirds of an inscribed equilateral polygon plus one-third of a similar polygon circumscribed about the circle (*HOC* 12:136). The letter from Mylon to Hobbes outlining this objection is presumably that from 9/19 April 1656 (*CTH* 1:272–76). Hobbes's reply no longer survives, but it is possible that the revised version of chapter 18 that he appended to the *Examinatio* and subsequently printed in the 1668 *Opera* (Hobbes 1660, 182–85) is the same as the one reported by Mylon.

De Corpore show him to have been generally engaged in the mathematics of his day in the 1650s. Although his grand designs did not meet with the approval for which he had hoped, Hobbes still accepted the same basic framework as his critics and he was ready to have his work judged by the standards of demonstration that prevailed throughout the European mathematical community. There is consequently no reason to imagine some kind of "radical incommensurability" between Hobbes and his critics at this point in his career, precisely because Hobbes both understood the objections raised to his work and attempted to make his proofs conform to the recognized standards of geometric demonstration. As it happens, even his amended efforts were unsuccessful, but even this does not show that there is some kind of barrier to mutual understanding between Hobbes on the one side and Wallis, Mylon, or Huygens on the other. All the players in the dispute understood each other's claims and agreed upon the standards by which they should be judged. However, Hobbes's acceptance of his opponents' standards of proof was not permanent. As we will see, his defense of his attempted duplication of the cube led him to reject essentially all of classical mathematics.

6.1.2 The Beginning of the End:
Hobbes on the Duplication of the Cube

Hobbes's confidence in his mathematical abilities remained high in the early 1660s, but he was less willing to accept the authority of traditional geometric opinion in determining the validity of his results. This fact is best illustrated by an analysis of his 1661 duplication of the cube and his response to criticisms of it. Where he had earlier been prepared to admit error, Hobbes had by this time become less willing to acknowledge even technical shortcomings in his demonstration, and he contested the applicability of algebraic arguments against his geometric constructions. He pursued this resistance even to the point of rejecting the most basic principles of mathematics, finally abandoning the Pythagorean theorem and the entire classical theory of magnitudes in order to preserve his alleged results from algebraic refutations. It is worthwhile to follow this process in some detail. By considering his replies to various criticisms of his mathematics, we can see that Hobbes did not make a sharp break with classical geometry, but instead saw himself forced to adopt ever more desperate measures to defend the integrity and validity of his (alleged) solutions to the great classical problems.

Hobbes published *La Duplication du Cube par V.A.Q.R.* anonymously in Paris in 1661.[16] The reasons for his choice of anonymous publication in a foreign language and in a foreign country are unclear. Robertson claims that he intended "to put Wallis and other critics off the scent, and extort the judgement that apparently was now to be denied to any mathematical work of his" ([1806] 1910, 179), and Scott repeats this opinion almost verbatim (1938, 167). Although Hobbes would later make no attempt to hide his authorship of this or any other of his mathematical pieces, his own remarks on the cube duplication lend some support to Robertson's opinion.[17] Whatever the reason for the peculiar manner of its publication, Wallis immediately recognized Hobbes's handiwork and refuted it in a letter dated 23 June/3 July 1661.[18] To my knowledge the only surviving copy of the piece is item 85 among the Hobbes manuscripts in the library at Chatsworth House. This is a single sheet printed on both sides, although it

16. The expression "V.A.Q.R.," Hobbes later tells us, was meant to stand for "Un Autre Que Roberval" (*DP; OL* 4:295).

17. In particular, Hobbes's *Seven Philosophical Problems* contains a dialogue on the problem, in which the interlocutor A asks, "Have you seen a printed paper sent from Paris, containing the duplication of the cube, written in French?"; to which the intrepid B replies, "Yes. It was I that writ it, and sent it thither to be printed, on purpose to see what objections would be made to it by our professors of algebra" (*SPP* 8; *EW* 7:59). Wallis was the first to conjecture that Hobbes's choice of a French publication was a ruse designed to confuse his opponents:

> Observing that *Mr Hobs's Geometry* (whether by reason of others Envy, or for what other cause, *I* will not now dispute) was not now in any great Repute; and, fearing least that *Odium Hobii,* which he so much complaines of, as so prejudicial to Man-kind in hindring the reception of his Notions, without which it is impossible to make any progresse in the Search of Nature; . . . might be prejudicial also to this of the *Cube,* (and, thereby, not onely deprive him of the Credit, but all man-kind of the Benefit, of his New Discovery:) To obviate those evils; he caused his *Probleme* of *Doubling the Cube,* to be printed in *French;* . . . and divers papers of it to be given abroad, which were pretended to be brought from *Paris;* (For had it been in *English,* or thought to be done at home, the Matter would presently have betrayed the Author:) Not doubting, but that, the *Odium* would cease to operate when the Person was concealed; and, no *Prejudice* obstructing an impartial Estimate, his Demonstration would presently find Reception and Approbation: Which could not afterwards be withdrawn, when He should appear to be the Author. (*HHT* 127–28)

18. This letter is contained among Hobbes's papers in at Chatsworth House (Hobbes MSS., letter 51). Malcolm has shown that is unlikely to have been addressed to Hobbes because it refers to him in the third person and uses terms of address more suited to a nobleman, probably Viscount Brouncker (*CTH* 1:xlvi). In any event, the letter must have been forwarded to Hobbes.

must originally have been issued with an accompanying diagram; no place or date of publication is listed, but Hobbes's later comments show that it was published in Paris. The date of publication is fixed by Wallis's letter reporting that the piece had been recently brought over from France. Hobbes responded by including the duplication of the cube at the end of his 1661 *Dialogus Physicus,* together with Wallis's objections, those of Laurence Rooke (professor of geometry at Gresham College), and his replies to them.[19] I propose to work through Hobbes's original attempt, the objections of both Wallis and Rooke, and Hobbes's response to these objections, in order to bring Hobbes's conflict with traditional geometry more clearly into focus.

We saw in chapter 1 that the classic problem of doubling the volume of the cube is stated as follows: given a cube with edge of length r and volume r^3, construct a cube with volume $2r^3$. In the geometric parlance of the day, the problem reduces to that of constructing two mean proportionals between the lines r and $2r$, since if we are given quantities x and y such that $r{:}x :: x{:}y :: y{:}2r$, we have (by eliminating the term y) $2r^3 = x^3$. Hobbes attacks the problem with construction in figure 6. 1. Given the line segment AB, construct the square $ABCD$, whose sides AB and AD are bisected at E and H by line segments EF and HG. The problem is solved if we can find two lines in continued proportion between AD and DF. Produce BA and CD to points O and P, respectively, such that $AB = AO = DP$, and describe the semicircle BDO. In the segment BC, mark off BR equal to the semidiagonal BI and drop the perpendicular RS, joining AD in S. Bisect SD at T and extend AD to V, such that $DV = DF = AD/2$. With center T and radius TV, draw the circle $VXYZ$, to cut DC at X, AD at Y, and RS produced at Z. Hobbes asserts that the line segments DY, DX are two mean proportionals between AD and DF; i.e., $AD{:}DY :: DY{:}DX :: DX{:}DF$. Since $DF = AD/2$, this would suffice to duplicate the cube. This construction, with some simplifications and cosmetic alterations, re-

19. These are printed as *OL* 4:288–96; the duplication is printed there essentially unaltered in a translation from French to Latin, while Wallis's objections are transcribed without significant alteration from the letter. The original duplication is reproduced in the appendix. Here, my presentation of Hobbes's solution and Wallis's refutation is not a literal translation from either source, because differences in notation and minor variants in the construction make it impractical to reproduce each version exactly. Wallis reprinted the whole exchange, together with a new set of replies to Hobbes's responses, in *HHT* 129–47. Another account of Hobbes's efforts at cube duplication can be found in Grant 1996, 120–23. Hobbes identifies Rooke only as the professor of mathematics at Gresham College, and this only on the occasion of his second defense of the duplication of the cube in the *Problemata Physica* of 1662 (*OL* 4:382).

mained Hobbes's claimed solution to the problem for the rest of his life.[20]

The details of Hobbes's argument can be found in section A.2 of the appendix and will be left aside here. It is important for the present context to note that Hobbes's construction succeeds only if he can establish that the line YZ, when extended, meets DP at P. It should come as no surprise to learn that the construction fails to solve the problem, notwithstanding the detailed construction and convoluted argument he adduces for the result.[21] The inadequacy of the argument can be gleaned from the objections raised by Wallis in his letter of 23 June/ 3 July 1661. Wallis's refutation of the *Duplication du Cube* comes in three parts: first a purely geometric argument to show that YZ extended does not meet DP at P; second the observation that, because the demonstration makes no use of the length of the line DP, it could be assigned any length whatever and hence the demonstration fails to determine that YZ extended intersects DP at P; and third an essentially algebraic reductio ad absurdum argument showing that the original construction cannot yield lines DY and DX exhibiting the desired proportionality. The first two arguments are not sufficiently interesting to detain us,[22] but it is worthwhile to consider the third, as Hobbes's reaction to it is quite significant.

20. It can be found with inessential variants in the *Dialogus Physicus* of 1661 and the English translation of that work in *Seven Philosophical Problems* from the same year (although this was not actually published until 1682, after Hobbes's death). One characteristic of Hobbes's original construction that I have left out of this summary is the superfluity of many of the lines and arcs he constructs. As can be easily seen from an examination of the original version given in the appendix, many of the lines Hobbes inserts in the construction are irrelevant to his argument. Wallis drew attention to this feature of Hobbesian mathematics when he remarked, "You may perhaps wonder (and so did I till I knew Mr *Hobs* was the Author of that Paper) why he should clog his *Figure,* and the *Construction of it,* with such a Multitude of superfluous Lines and Letters, whereof he makes not use at all either in the *Construction of the Probleme* or the *Demonstration* of that Construction" (*HHT* 132). As his cube duplication went through several versions, Hobbes did delete some of these extraneous lines, but the essentials of the construction did not change.

21. This part of Hobbes's cube duplication underwent a number of changes, as he sought arguments to show that the prior construction did, in fact, yield a line YZ that, when extended, meets DP at P.

22. In the first of these objections, Wallis shows that the construction and demonstration used by Hobbes only license the inference that a particular figure in the construction is a rectangle, not a square as Hobbes believed. Wallis examines the matter at greater length in *HHT* 134–44. The second objection (although obviously fatal to Hobbes's demonstration) seems to have had no effect on Hobbes, who simply dismisses it as the work of a man with no grasp of logic or demonstration.

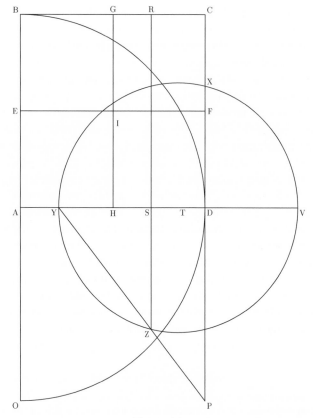

Figure 6.1

Wallis shows that the proportion $AD{:}DY :: DY{:}DX :: DX{:}DF$ cannot obtain, reasoning by reductio ad absurdum.[23] Assign DV the value 1, and then $DA = 2$. By the construction of the semicircle with radius AD, we have $AS = AI = \sqrt{2}$, $SD = AD - AS = 2 - \sqrt{2}$, and $TD = 1 - \frac{1}{2}\sqrt{2}$. Similarly, by construction, we have $TV = 2 - \frac{1}{2}\sqrt{2}$, while $DY = 3 - \sqrt{2}$ and $DX = \sqrt{DY} = \sqrt{3 - \sqrt{2}}$. Then, to the three lines

23. Wallis's objection appears in the letter of 1661, is quoted verbatim by Hobbes in the *Dialogus Physicus* and is re-stated in *HHT* 140. Rooke's objection is known to us only through Hobbes's quotation of it. Wallis indicates that a third version of the same objection was circulated: "And I could tell him of a Third; from a *Noble Hand.* But this Third was in *Symbols,* and therefore he did not think fit to understand it, or take any notice of it" (*HHT* 144). The "Noble Hand" is Viscount Brouncker, as is evident from the letter of Moray to Huygens in September of 1661 (*HOC* 3:339–43).

DV, DX, DY, a fourth proportional will be $(3 - \sqrt{2}) \times \left(\sqrt{3 - \sqrt{2}} \right)$, which works out to approximately 1.997, a result less than the desired value of 2. Hobbes is close, but that does not count. In fact, he is off by the difference between $\sqrt{1681}$ and $\sqrt{1682}$.

Rooke's refutation follows essentially the same line of reasoning. He begins by letting the quantity *AB* or *AD* be taken equal to 2. Then *DF* or *DV* will be equal to 1. Therefore *AV* = 3. *BR* or *AS* is then √2. Therefore *SV* or *YD* is 3 − √2. By stipulation, $AD^3 = 8$, so $DY^3 = 45 - \sqrt{1682}$, or nearly 4. But $45 - \sqrt{1681} = 4$. Therefore, *DY* is slightly less than the greater of two means between *AB* and *DV*.

As I mentioned earlier, Hobbes's replies to these criticisms were published in a brief addition at the end of his 1661 *Dialogus Physicus.* His remarks here contrast sharply with his earlier responses to critiques of the mathematical efforts in *De Corpore.* Rather than admit the cogency of his opponents' technical objections or attempt to repair an acknowledged error of negligence, Hobbes remained adamant. To Wallis's purely geometric objections to the "proof" that the line YZ extended intersects *P,* Hobbes simply comments that "it is surely to be marveled that there is anyone who dares to respond when he does not understand such an easy and perspicuous demonstration," and then asks "who does not see?" that the original construction and argument are conclusive (*DP; OL* 4:289). Hobbes brings a similar rebuttal against Wallis's argument that the failure to determine the length of line *DP* renders the demonstration useless. First he claims that such an objection could be made "by someone who would dare to pronounce, so by chance and without demonstration, only if he could not understand so perspicuous a demonstration and is bereft not only of art but even of intellect." He then alludes to Wallis's service of the Parliamentary cause in the 1640s by adding that the author of the objection "was still so ingenious that he interpreted, or (as they say) deciphered, letters sent to the king and others who stood with the king in the Civil War" (*DP; OL* 4:290).[24]

---

24. Robertson credits this mention of Wallis's cryptanalytical activities with reigniting the dispute: "Wallis, who had deftly steered his course amid all the political changes of the time, managing ever to be on the side of the ruling power, was now apparently stung to fury by a wanton allusion in the 'Dialogus' to his old achievement of deciphering the defeated king's papers, whereof he had boasted in his 'Inaugural Oration,' as Savilian professor, in 1649, but after the Restoration could not speak or hear too little. The revenge he took [by the publication of *Hobbius Heauton-timorumenos*] was crushing" (Robertson [1886] 1910, 181).

Hobbes's reaction to the reductio arguments of Wallis and Rooke is more detailed and far-reaching. In his strategy of reply to these objections Hobbes parts company with the geometry of his contemporaries and adopts the incoherent doctrine that algebraic results can have no bearing on geometric constructions. Commenting on Wallis's version of the objection, the interlocutors in the *Dialogus Physicus* conclude that the argument depends upon a confusion of a rectangle with a right line:

B: But what is it that sent him astray in such a brief calculation?

A: He went astray because even if he is an adequate symbol monger *[symbolicus]* he is nevertheless a poor geometer.

B: I was not asking about that, but rather the source of the error in his calculation.

A: It is none other than this: he thinks that DX is equal to $\sqrt{3 - \sqrt{2}}$.

B: But isn't DX a mean proportional between DV and DY, that is between 1 and $3 - \sqrt{2}$? And the product of 1 and $3 - \sqrt{2}$, that is $3 - \sqrt{2}$ (since multiplying by 1 changes nothing) is equal to a square on DX, and DX itself is equal to $\sqrt{3 - \sqrt{2}}$?

A: The author of the refutation *[refutator]* certainly calculated in this way, but did so wrongly. Although $3 - \sqrt{2}$ multiplied by *one* simply makes $3 - \sqrt{2}$, changing nothing. Yet if it is multiplied by *one line,* here of course DV, it makes the rectangle contained in DV and DY. But the rectangle contained in DV and DY cannot be equal to its side DY. You see therefore this whole error arises from the fact that he used a rectangle for a right line.

B: Certainly. And the cause of the error was, as you said, an ignorance of geometry. (*DP; OL* 4:292)

Hobbes tried to bolster this criticism by addressing Rooke's refutation, intending to show that the algebraic calculations could indeed be made to yield inconsistent results. He presents two different computations of the quantity $(3 - \sqrt{2})^3$, concluding that "these two calculations, although done according to the rules of algebra, nevertheless do not agree, either among themselves or with the geometric calculation. So it is most certain that DY is the greater of two means between AD and DV; and thus the cube of DY is 4. Thus the examination of geometric problems by algebra is generally useless" (*DP; OL* 4:294). The source of the discrepancy in this instance is, however, entirely on Hobbes's part, since the difference in the calculations arises from his taking $(\sqrt{2})^3 = 2$ at one point.[25]

The point of view set forth here is a fundamental departure from the standard conception of mathematics, for it essentially denies that arithmetical or algebraic calculation can be applied to test the correctness of a geometric construction. Indeed, as Hobbes pursued the consequences of this doctrine, its inconsistency with all of traditional mathematics became all the clearer, and he soon found himself questioning the validity of such basic propositions as the Pythagorean theorem. Before we proceed it may be useful to try to elucidate this doctrine somewhat, as it plays a central role in the collapse of Hobbes's geometrical program.

The main issue here concerns the proper interpretation of the doctrine of roots. Classically, the nth root of a number k is a number r such that $r^n = k$; interpreted geometrically, the square root of a number k is a line of length r that forms the side of a square whose area is k square units. Hobbes denies that the geometric understanding of roots can be subsumed into the general treatment of roots as numbers. He holds that geometric "multiplication" is a process of "drawing lines into lines," by which a rectangle is produced, or the drawing of a line into a square to produce a solid; this process generates a geometric object of higher dimension from two objects of lower dimension and cannot proceed beyond the third dimension. Algebraic multiplication, on the other hand, combines numbers together to produce other numbers, but it never generates objects of a different type. For this reason the process of algebraic multiplication can also extend to fourth and higher powers, unlike its geometric counterpart. Insistence upon this supposedly great difference between two kinds of multiplication ulti-

---

25. This error in calculation is probably the incident reported by Pope in his biography of Ward, when (speaking of Rooke, who was a close associate of Ward's) he relates the following story. "Mr. *Hobbs* published a little Treatise concerning Mathematics, wherein, amongst other things, he pretends to give the Square of a Circle; which when Mr. *Rooke* read and considered, he found it false, and went to Mr. *Hobbs* to acquaint him with it, but he had no patience to hear him; therefore when he went next to visit Mr. *Hobbs,* he carried with him a Confutation of his Quadrature, and left it behind him at his departure. Mr. *Hobbs* finds and reads it, and by want of attention, casts it up wrong, for it was accurately Calculated, and truly written, and thence insultingly concludes, since that Learned Persons Confutation was false, his own Quadrature must of necessity be true" (1697, 126). It seems, however, that this is a garbled account of the exchange between Rooke and Hobbes on the duplication of the cube rather than the quadrature of the circle.

mately led Hobbes to his most radical (indeed, incoherent) dissent from the mathematics of the seventeenth century.

Hobbes links this doctrine to his general philosophy of mathematics, arguing that because the "algebraists" are accustomed to treat lines as lacking breadth, their calculations must fail to account for the small overlap between two lines in a geometric figure. In the *Dialogus Physicus,* one of the interlocutors asks "although the arithmetical calculation differs from the geometric one, why does it differ by so little, namely as much as the difference between √1681 and √1682?" The answer is quite remarkable: "Because whoever multiplies lines considered as without latitude does not make a plane, but a number of lines. But whoever draws one right line into another right line does not make a number of lines, but a plane surface. Thence it necessarily happens that in the sides of the planes the points, which are in the common angles of two right lines, are counted twice, and in the sides of cubes three times" (*DP; OL* 4:294–95).

At this stage in his career (i.e., the early 1660s), Hobbes was clearly opposed to the prevailing view of geometry, but his resistance to the authority of classical mathematics was not yet absolute. He continued to seek the opinions of recognized mathematicians (excepting Wallis) in confirmation of his duplication of the cube, and his attitude could hardly be characterized as a dogmatic dismissal of the mathematical principles and practices of his day. John Pell, whom Hobbes had assisted in the campaign against the circle-squaring efforts of Longomontanus nearly twenty years earlier, recorded a meeting with Hobbes on 31 March/10 April 1662, as Hobbes was preparing his *Problemata Physica* for the press. Pell represents the aged sage of Malmesbury as convinced of the validity of his result but still concerned to gain the approval of the mathematical world:

> This morning M^r. Thomas Hobbes met me in the Strand, & led me back to Salisbury house, where he brought me into his chamber, and there shewed me his Construction of that Probleme, which he said he had solved, namely *The Doubling of a Cube.* He then told me, that Viscount Brounker was writing against him. But, said he, I have written a Confirmation & Illustration of my Demonstration; and to morrow I intend to send it to the presse, that with the next opportunity I may send printed coppies to transmarine Mathematicians, craving their censure of it. On this side of the sea, said he, I shall hope to have your approbation of it. I answered, that I was then busy, and could not per-

swade my selfe to pronounce of any such question, before I had very thoroughly considered it, at leysure, in my owne chamber. Where-upon he gave me these two papers, bidding me take as much time as I pleased. Well, said I, if your work seeme true to mee, I shall not be afraid to tell the *world* so: But if I finde it false, you will be content that I tell *you* so: But privately, seeing you have onely thus privately desired my opinion of it. Yes, said he, I shall be content, and thanke you too. But, I pray you, doe not dispute against my Construction, but shew me the fault of my Demonstration if you finde any. Thus we then parted, I leaving him at Salisbury house, and returning home. (British Library MS. Add. 4425, f. 238r)

The "two papers" to which Pell refers in this account are presumably a Latin version of Hobbes's circle quadrature and its accompanying diagram (British Library MS. Add. 4225, f. 215, f. 217). Pell obliged by refuting the duplication and leaving a copy of the refutation with Hobbes, to which Hobbes evidently responded by offering a new argument to show that his construction succeeded.[26]

The significance of this account is great, for it shows something of the state of Hobbes's mathematical methodology in the spring of 1662. He was then still prepared to submit his work to the judgment of recognized mathematicians, but he was clearly less open to criticism than he had been. Hobbes's insistence here that Pell "not dispute against my Construction, but shew me the fault of my Demonstration" is an interesting restriction. The demand shows that Hobbes was not completely prepared to accept all of the principles of traditional mathematics, although he certainly saw value in some criticisms from the classical standpoint. A geometric proof (or attempted proof) such as Hobbes's *Duplication du Cube* traditionally contains two sections: the construction and the demonstration.[27] The construction is typically stated in the imperative mood and bids the reader effect certain constructions, such as "take a line of length AB in the diagonal CD," or "divide the arc SQ into two equal parts." The demonstrative phase argues that the lines and figures thus constructed have certain properties. This section of the proposition thus asserts, for example, that "the line

26. A copy of Pell's refutation survives as British Library MS. Add. 4425, f. 216, which contains Pell's note: "I left a copy of this with Mr Hobbes May 5." Hobbes's reply is MS. Add. 4425, f. 236 (a diagram) and f. 237 (a revised argument, with a note from Hobbes to Pell).

27. See Heath's discussion of the formal divisions of a geometric proposition in Euclid [1925] 1956, 1:129.

BV is a mean proportional between the lines AD and DF," basing the assertion on specific features of the construction. There are, obviously enough, constraints on what can be constructed or demonstrated. The demonstration must avoid errors of logic and the construction must contain only operations permitted by the postulates of the geometric system. In the case we are considering, the construction must use only the familiar "compass and rule" operations characteristic of Euclidean geometry. By insisting that Pell confine his criticisms to the demonstration in the duplication of the cube, Hobbes in effect declared that his construction was immune to challenge and that he was prepared only to consider challenges to his efforts to show that the construction is a successful duplication of the cube. In part, this is due to Hobbes's insistence upon the claim that a properly developed geometry will use definitions that show how to construct the relevant objects. If the construction is in error, then there is reason to fear that it proceeds from false or inadequately understood first principles, which would be a much more serious sort of error.[28]

From the evidence of Pell's report, it is clear that Hobbes had high hopes for his duplication of the cube and its accompanying "confirmation and illustration." In particular, he had convinced himself that Continental mathematicians would welcome his latest excursion into the great classical problems, and he appears to have sent copies of this effort to his friend Samuel Sorbière in Paris with instructions to show them to leading mathematicians and gather their responses.[29] The *Problemata Physica* appeared in 1662 and consists of seven dialogues on problems in natural philosophy along with two mathematical appendices. The first appendix contains sixteen propositions on the quadrature of the circle, while the second is an expanded version of the cube duplication from 1661, along with a revised version of his replies to the objections of Wallis and Rooke.[30] It also contains an un-

---

28. Bird makes a similar point when he remarks that the distinction between construction and argumentation "allowed Hobbes tacitly to accept and correct fallacies in his argument while sticking to his quadrature and duplication claims. These were founded on the unchanged constructions, which, of course, embodied the all important generation of the desired quantities. The constructions, independently of the proofs, were embodiments of Hobbes's generative principle" (Bird 1996, 230).

29. It is not certain that Sorbière was Hobbes's intermediary in this project, but the letters between the two in the early 1660s do show that Sorbière was aware of Hobbes's mathematical struggles and that he served as his contact for many exchanges with French mathematicians.

30. The construction in the *Problemata Physica* differs minimally from that in the original *Duplication du Cube* of 1661 and the *Dialogus Physicus* of the same year. There

usual appeal "to foreign geometers" placed immediately after the dedicatory epistle and before the first chapter. In it, Hobbes declares that

> to these physical problems, which aim at nothing higher than
> verisimilitude, I have added some geometrical propositions that
> the professors of geometry and many others of our mathemati-
> cians do not accept, but think they have confuted by their arith-
> metical calculations. I alone (at least so far) contend against them
> that this arithmetic they use is suitable neither for the confirma-
> tion nor for the confutation of geometrical claims. I now appeal
> to you (readers of mathematics who have not yet condemned my
> works by prejudicial decree) to turn your faculties of reason to
> this discrepancy between my geometrical calculations and their
> arithmetical ones and (in the interest of mathematics itself) most
> humbly beseech and implore you to approve whichever one ap-
> pears true. (Hobbes 1662a, sig. A8)[31]

This appeal was supplemented by a pronouncement at the end of the
mathematical appendices to *Problemata Physica* that promised to end
the controversy once and for all:

> If the geometers into whose hands these things come deem their
> opinions of these geometric computations worthy of sending to
> the bookseller whose name and address are to be found on the
> first page, then in the future either my antagonists will be silent,
> or I shall be silent. (Hobbes 1662a, 139)

In the course of events Hobbes saw fit to remove from subsequent versions of the *Problemata Physica* both his appeal to Continental mathematicians and his offer to remain silent. The reason for this apparent
change of heart is not difficult to discern: his mathematics found no

---

is consequently no need to investigate them in detail. The circle quadrature does differ
from the earlier efforts in *De Corpore,* although an account of these differences is not
of concern here. The reply to Rooke no longer uses an argument based on the computa-
tional error of taking $(\sqrt{2})^3 = 2$, but otherwise keeps the main line of analysis intact.

31. The original Latin has been rather freely translated. It reads: "Ad Problemata
haec Physica, quae altiuɑs non spectant quam ad Veri-similitudinem, Propositiones ad-
junxi Geometricas alíquas, quas Professores Geometriae, & praeterea alii complures
Mathematici nostri non recipiunt, sed Calculis suis Arithmeticis videntur sibi confu-
tasse. Ego contrà Arithmeticam illam qua usi sunt, nec confutandis nec confirmandis
Pronunciatis Geometricis idoneas esse solus hactenus contendo. Vos ergo nunc appello
(qui nondum praejudicio edito mea haec condemnastis Lectores Mathematici) ut dissen-
sionem hanc inter Calculum meum Geometricum, & illorum Arithmeticam (ipsius Mathematicae causa) rationibis vestris componere, & veritati undecunque apparenti suc-
currere velitis, humillimè oro obsecroque."

supporters, and by 1668, when he was preparing his *Opera Philosophica* for publication at Amsterdam, Hobbes was no longer prepared to submit his geometric work to the judgment of prevailing mathematical opinion.

In the defense of his cube duplication in the *Problemata Physica* against the refutations of the "algebraists" Hobbes returns to the theme of the difference between geometric multiplication (the drawing of lines into lines) and the algebraic or arithmetical multiplication of one number by another. He complains that Wallis's refutation commits an equivocation, since it assigns the quantity *one* indifferently to a line, a surface, and a solid: "In this calculation, if we follow the sound of the words, unity does not change, but (as is customary and necessary to do arithmetic) it is kept the same, namely the line *DV*. . . . But if unity is understood to be changed, as it is first one line, then one plane, and lastly one solid, . . . the calculation can come close to the truth but not attain it" (*PP; OL* 4:382–83). The "small error" that Hobbes claims to find arises, as before, from double-counting of the vertices of a cube: "But when *AS* is converted into numbers, these numbers will be parts of the cube thus made, and the same is counted three times, whence arises the small difference between 1682 and 1681" (*PP; OL* 4:383).

Hobbes's hopes that the *Problemata Physica* and its associated mathematics might receive a favorable hearing from Continental mathematicians were in vain. Huygens (evidently at the request of several members of the Royal Society) responded to the invitation at the end of *Problemata Physica* to submit his thoughts on Hobbesian geometry to Andrew Crooke (Hobbes's bookseller). The result was a letter in which Huygens refuted both the duplication of the cube and the quadrature of the circle. In objecting to the duplication, Huygens observed that Hobbes's construction fails to achieve the desired proportionality. He then dismissed Hobbes's entire solution, together with its defense against the criticisms of Wallis and Rooke with these words:

> But it is extraordinary that he did not notice that his demonstration was flawed, given that it made no mention of what had been assumed in the construction of the problem, namely that AS was taken to be equal to the semidiagonal AI. This assumption, if it were valid, ought to have been a necessary part of the demonstration.
>
> Moreover, he seems to reject without cause the use of arithmetical calculation in the investigation of geometrical construc-

tions. (Huygens to Andrew Crooke for Hobbes, August 1662; *CTH* 2:532)

Predictably enough, Hobbes responded to these objections. His reply is now lost, and it is unclear exactly what form it took, whether a private letter to Huygens, a separate publication, or a letter to be appended to the version of *Problemata Physica* in the collection of Hobbes's works then being prepared at Amsterdam.[32] We can, however, infer that Hobbes refused to give ground on any point, adding a new "demonstration" to his earlier construction and challenging the notion that all aspects of the construction must figure in the associated demonstration. In an enclosure to a letter to Sir Robert Moray, read to the Royal Society on 31 December 1662/10 January 1663, Huygens offers a second refutation of Hobbes's efforts. He repeats the charge that Hobbes's construction is inadequate to the case at hand, observing that "in order to let geometers understand just how flawed and absurd this whole attempt at a duplication of the cube is, it is sufficient to know that after he has posited that the line *AS* is equal to half *AC,* and *DV* equal to half *AD,* these things are never mentioned again in the demonstration, nor does he mention anything deduced from them. And since that never happens, anyone can see that the same demonstration by Hobbes would fit a construction in which *AS* and *DV* were taken to be of any length you liked" (Huygens to Moray for Hobbes, 10/20 December 1662; *CTH* 2:538–39).

The Belgian mathematician René-François de Sluse was also asked to comment on Hobbes's efforts in the *Problemata Physica* and rendered an opinion no more favorable than that of Huygens, although his language was less dismissive. Through Sorbière as intermediary, de Sluse sent a letter to Hobbes in November of 1663 containing an exhaustive refutation, showing (by exactly the same kind of calculation employed by Wallis and Rooke), that the Hobbesian duplication of the cube led to absurdity. He observes that "if one assumes that the extremes [in the proportionality] are 1 and 2, it will follow that the cube

---

32. Moray to Huygens, 7/17 November 1662, reports that "[Hobbes] has made a response [to your refutation], which he has had printed, a copy of which I have addressed to Mr. Bruce to give to you" (*HOC* 4:261). In a letter of 30 November/10 December 1663 to Sorbière, Hobbes mentions a "little treatise that I called 'Epistola anonymi,'" that he had addressed to Huygens, but that he no longer wants removed from the edition of his works being prepared by Johann Blaeu in Amsterdam (*CTH* 2:577). No trace of this tract survives, although the editors of Huygens's *Oeuvres Completes* refer to a work entitled *De duplicationi cubi ad defensionem problematum geometricorum Th. Hobbes contra C.H.,* with a date of 1662 (*HOC* 4:261).

of 3 − √2 equals 4" (de Sluse to Sorbière for Hobbes, 28 September/9 October 1663; *CTH* 2:566). To make his case even stronger, de Sluse proposes that "the disproof will perhaps be clearer if we use larger numbers." Assigning the numbers 29 and 58 to *DV* and *AD* (figure 6.1), he calculates that *AS* is √1682 and then shows that Hobbes procedure results in the claim that 97,336 = 97,556 (de Sluse to Sorbière for Hobbes, 29 September/9 October 1663; *CTH* 2:567).[33]

Hobbes's reaction to the steady stream of refutations of his duplication was to stand firm and admit no error. In forwarding de Sluse's objections, Sorbière had advised Hobbes "in the interests of your reputation that you should consider his objections calmly and attentively, so that you should not cling firmly to that paralogism which nearly all the mathematicians agree in claiming to find in your demonstration, and that you should honestly strike out your mistake, or demonstrate the mistake of the other mathematicians so clearly and evidently that there can be no further possibility of refutation" (Sorbière to Hobbes, 30 October/9 November 1663; *CTH* 2:565). Hobbes responded to this caution by insisting that "so far as the reputations of de Sluse and Huygens are concerned, and of all those who think my proofs are paralogisms, I shall show that the proofs which they published in criticism of me are paralogisms, and shall make this clear both to them and to all other men—except to those who think that ten times one stone equals ten square stones" (Hobbes to Sorbière, 30 November/10 December 1663; *CTH* 2:577). His reply to de Sluse on this particular point is one of his most extensive discussions of his doctrine of roots and the associated thesis that geometric constructions cannot be challenged on the basis of arithmetical or algebraic calculation. He writes of de Sluse's calculations:

> I think that because he pays too much reverence to the geometers of our age, he did not think it worth studying geometry, seeing that it has been spurned by them. For otherwise he could have learned from it that when one is applying arithmetic to geometry, it is necessary to refer to different units, especially when making calculations about surfaces and solids; so that in using the same "one" to refer to a solid or a surface or a line necessarily makes the arithmetical calculations faulty. But in case he says that when he refers to multiplying the number 46 by 46 he does not mean

33. De Sluse also argued (as had Huygens and Wallis) that Hobbes's argumentation is also vitiated because it does not make use of the stipulated length of the various lines in the construction. I leave these details aside.

simply multiplying the number, but rather applying the numbered things, namely 46 lines to 46 lines, I shall show that the calculation is false even on that basis. For in that case, the product of the first multiplication, namely 2,116, will be so many squares of one of the said lines; and therefore even if its side is 46 lines, its square root will still be 46 squares on the said lines. For in every extraction of a numerical square root the numbered thing is the same as in the number from which the root is extracted. For example, the mean proportional (which is the root) between 2 squares and 8 squares is 4 squares, not 4 lines. . . . By neither of his calculations, therefore, has he undermined my duplication of the cube. Nor could it be undermined by the use of numbers even if it were false—unless the squares are divided into an infinite number of parts, so that an infinitely small part (if one may use such a phrase) can be said to be equal both to its square and to its cube. (Hobbes to Sorbière, 19/29 December 1663; *CTH* 2:582–83)

The crux of the matter here is Hobbes's declaration that "in every extraction of a numerical square root the numbered thing is the same as in the number from which the root is extracted." To unravel the doctrine, it is necessary to recall Hobbes's definitions of the terms *point, line,* and *square.* A point is a body whose magnitude is not considered in a demonstration; a line is the trace of a moving point; and a square is a figure generated by drawing a right line through another right line. This means that the side of a square is "inconsiderable" with respect to the area of the square, i.e., there can be no ratio or comparison between them. Hobbes then assumes that the root of anything must be homogeneous with or comparable to the thing whose root it is. Thus, the root of a number is a number, the root of a surface is a surface, etc. This requirement of homogeneity does have some plausibility if we think of roots as *parts* of the quantities of which they are roots; but Hobbes needs some kind of motivation for this doctrine. Further, as attractive as this doctrine may sound when stated abstractly, it amounts to the declaration that arithmetic or algebra cannot be applied to the investigation of geometry. De Sluse was naturally not impressed by such a nonstandard usage of the term *root* and replied that he had calculated in the way that anyone who understands arithmetic, geometry, or algebra would calculate.[34]

34. De Sluse writes, "The only reply I think I should make here is that I calculated in the same way that is normally used by all the arithmeticians who exist, who have

Hobbes's remark that algebra could apply accurately to geometry only "if the squares are divided into an infinite number of parts" suggests that he thought of the size of the algebraic "error" as depending upon the number assigned to a line in a geometric calculation. Recalling his thesis that the error arising from the application of numbers to geometric magnitudes arises from a "double counting" of the vertices of a figure, it appears that Hobbes held that the larger the number assigned to the side of a square, for example, the larger the number of square units it contains, and thus the proportion of squares that overlap in the four vertices of the large square becomes less significant. The problem with such a view is obvious enough: it simply makes no sense. All that is necessary in order to assign numerical values to geometric objects is that they are the sort of thing that can be measured. In fact, nobody ever actually held that the root of a square is its side. Instead, the square root of the quantity that is the *area* of the square (measured in square units) is the number assigned to the *side* of the square (measured in linear units).

The uniform rejection of his mathematical efforts from every corner of the learned world eventually soured Hobbes on the whole of mathematics, at least as it was understood by his contemporaries. We can see a decisive move in this direction in a letter from March of 1664 to Sorbière, where Hobbes declares that he does not "think it worth replying to the critics of my demonstrations (which, after so many explanations, are now being published)." He dismisses all of his critics in language that is clearly that of a man with no interest in reconciling his differences with the tradition:

> I do not want to change, confirm, or argue any more about the demonstration which is in the press. It is correct; and if people burdened with prejudice fail to read it carefully enough, that is their fault, not mine. They are a boastful, backbiting sort of people; when they have built false constructions on other

existed, and (I confidently predict) who will exist in years to come. Nor is there any reason why he should say that I am constrained by reverence or prejudice, since in these sciences nothing can make me submit to authority, and no authority can prevent me from refusing to accept this duplication of the cube of his—even though I value his authority (which is valued by the learned) in literary matters" (de Sluse to Sorbière for Hobbes, 18/28 January 1664; *CTH* 2:614). Hobbes's reaction was to denounce de Sluse as one who "having been warned of his error," refused to embrace the truth (*PRG* 18; *OL* 4:440). Observing this apparently irreconcilable difference in opinion between the two, Sorbière thought he might enlist Fermat as a judge in the contest—a plan de Sluse insisted was not worth undertaking and was in any case prevented by Fermat's death.

people's principles (which are either false or misunderstood), their minds become filled with vanity and will not admit any new truth. Can a man who believes in the following prodigies really be a suitable judge of my proposition (which is a little more deeply thought out): ten times ten lines are 100 squares; the side of a square and the root of a square number are the same thing; a ratio is a quotient; the same point can be on a line, and outside it, and inside it? (Hobbes to Sorbière, 7/17 March 1664; *CTH* 2:603)

This very hard line against his mathematical opponents is followed by Hobbes's remarkable expression of "a doubt concerning Euclid, book 1, proposition 47" in the very same letter to Sorbière (Hobbes to Sorbière, 7/17 March 1664; *CTH* 2:608–10). This proposition is none other than the Pythagorean theorem, the very result whose demonstration supposedly was the source of Hobbes's infatuation with geometry several decades earlier.

The source of Hobbes's "doubts" is the same as his dismissal of his critics, namely his insistence that the doctrine of roots must be interpreted in accordance with his principles. By this time Hobbes had convinced himself that the entire tradition of geometry—including such classical figures as Euclid and Archimedes—had been misled by thinking that arithmetic or algebra could be applied to geometry.[35] The inevitable result was that Hobbes's program for the reform of mathematics collapsed into self-contradiction and incoherence.

6.2 HOBBES UNREPENTANT

The final stage in the demise of Hobbes's geometrical program is closely connected with the Royal Society and can best be interpreted against the background of his conflict with that institution. In point of fact, some of what we have already seen concerning the duplication of the cube is related to Hobbes's conflicts with the Royal Society. Huygens's refutations of Hobbes's cube duplication, for example, were both solicited by Sir Robert Moray on behalf of the society, with the second one being warmly received at its meeting of 31 December 1662/ 10 January 1663 (*HOC* 4:295; Birch [1756–57] 1968, 1:167). Hobbes's isolation from the mathematical and scientific community

---

35. In the letter replying to de Sluse's objections, Hobbes remarks that Archimedes himself "was mistaken in his application of numbers to geometry" (Hobbes to Sorbière, 19/29 December 1663; *CTH* 2:584).

became steadily more pronounced through the 1660s and 1670s, and his unhappy relationship with the Royal Society is emblematic of his increasing isolation. I propose first to address the more general issue of Hobbes's relationship to the Royal Society, then to examine the mathematical doctrines in his essay *De Principiis et Ratiocinatione Geometrarum,* and briefly consider some of his final mathematical publications addressed to the Royal Society.

6.2.1 Hobbes and the Royal Society

The Royal Society for Improving of Natural Knowledge grew out of a series of informal meetings of English scientists in Oxford in the 1640s—a group that included Wallis, Ward, and Wilkins, as well as other scientific luminaries of the age. Its actual founding took place in London late in 1660, when the group undertook to meet regularly at Gresham College for the discussion of scientific matters and the presentation of experiments in natural philosophy. The group obtained a royal charter in 1662. The society has long been the object of scholarly attention, and the history of its founding need not be repeated here in any detail.[36] For my purposes the most salient fact about the society is that Hobbes was not a member, notwithstanding the fact that he was on good terms with many of its fellows and his scientific achievements exceeded those of many who were granted membership. Indeed, Hobbes was one of very few who could claim personal acquaintance with Francis Bacon, the figure who has been aptly characterized as "the Society's patron saint and arch-bestower of intellectual respectability" (Malcolm 1988, 51).

Hobbes's exclusion from the Royal Society has been thought by some to be a matter that cries out for explanation, and various explanatory strategies have been pursued. Quentin Skinner reasoned that, in light of his scientific achievements, his acceptance of the seventeenth century's "mechanical philosophy," and his friendship with members of the society, Hobbes must have been excluded for personal reasons, i.e., for reasons having less to do with his scientific work than his personal qualities. Skinner concludes that Hobbes's reputation for dogmatism and an argumentative temperament would have made him unwelcome in the clubby atmosphere of the Royal Society; thus, his failure to be elected "is readily explained: nobody wants to encourage

---

36. See Birch [1756–57] 1968 for the classic account of the founding of the society, usefully supplemented by a modern editorial introduction. Further studies can be found in Hunter 1981 and Purver 1967.

a club bore" (Skinner 1969, 238). Shapin and Schaffer find this explanation superficial and demand ideological reasons for Hobbes's exclusion. They conclude that Hobbes's alleged "dogmatism" was a threat to the form of life exemplified by the society because it represented a rationalistic, absolutist program for the organization of society and the production of knowledge. They insist that his approach to questions of science and politics (which they regard as inseparable) was at odds with the experimentalism and moderate political program of the Royal Society. In consequence, the exclusion of Hobbes was really an attempt to protect the Royal Society's preferred form of politics from an outsider's challenge (Shapin and Schaffer 1985, 139). In other words, complaints about Hobbes's dogmatic and argumentative temperament are really a cover for deeper concerns having to do with his opponents' shared view of how best to maintain a stable social structure.

Noel Malcolm offers a much more nuanced interpretation of Hobbes's exclusion, one that fits together both personal and "ideological" factors. He observes that much of the opposition to Hobbes seems to have been rooted in the contingent accidents of history—Hobbes made enemies for reasons having little to do with "deeper" world-historical forces, and these enemies happened to be leading lights of the British scientific establishment.[37] On the other hand, Hobbes's theology and politics were highly controversial, and it was prudent for those seeking public support for scientific inquiry in the 1660s to distance themselves from one, like Hobbes, who had a reputation as a proponent of dangerous views. Malcolm argues that it is precisely the very significant points of agreement between Hobbes's scientific project and that of the society, combined with his reputation of atheistic materialism, that made it imperative for the society to exclude him for the sake of a proper public image. He writes: "Hobbes was becoming an increasingly disreputable figure, both politically and theologically; and the people who felt that it was most in their interests to blacken his reputation further were the ones who were vulnerable to embarrassing comparisons between his position and their own" (1988, 60). The result, quite naturally, is that there was little sentiment for including Hobbes in the Royal Society.

37. Malcolm writes, "Looking back on Hobbes's disputes with Ward, Wilkins, and Wallis in the 1650s, one is struck at first by the contingency of it all; if only the Universities had not felt politically threatened in 1653–54, one feels, Hobbes would never have become embroiled in these disputes, and would never have suffered the running sore of his mathematical controversy with Wallis—one which did in the end damage his reputation as a scientist" (Malcolm 1988, 57).

I think that Malcolm's analysis is essentially correct, but go further and argue that Hobbes's exclusion from the Royal Society was actually overdetermined. We should be wary of looking for one key factor that explains Hobbes's exclusion when it is quite plausible that there are a number of factors that, even taken separately, can adequately account for it. It is worth recalling that Hobbes's enemies Wallis, Ward, and Wilkins (the "worthy triumvirate" to which Kenelm Digby referred) were all instrumental in the founding of the Royal Society. It is particularly significant that Wilkins was one of its original two secretaries (the other, Henry Oldenburg, was also on very good terms with Hobbes's opponents). Wilkins used his position to propose others for membership in the society, and he was by far the most active fellow when it came to proposing new members.[38] Although the society had no provision for blackballing a prospective member, the proposing of new members was not a matter to be taken lightly, and these three could easily have exercised sufficient influence to bar Hobbes's membership.[39] This alone would account for Hobbes's exclusion.

Quite apart from Hobbes's misfortune in having such well-placed enemies in the fledgling society, his scientific reputation had been dealt a series of blows by the time the society was chartered in 1662. In the wake of the numerous refutations of his mathematical writings and the combined campaign of Ward and Wallis to discredit every aspect of his philosophy, Hobbes's standing as a man of science had been seriously damaged. The disaster of the circle quadrature in *De Corpore,* combined with Hobbes's continued public mathematical failures, had made him a laughingstock among the scientific establishment by the early 1660s.[40] Thus, Huygens could declare in December of 1662 that

38. Hunter (1982, 59) reports that among those who proposed new members "[i]n the early years of the Society three men stand out: John Wilkins, who proposed forty-one candidates between 1661 and his death in 1672, more than twice as many as anyone else in the whole period; Sir Robert Moray, who proposed nineteen, all but one before 1666; and Seth Ward, who proposed sixteen between 1666 and 1675." All three of these men are known to have been hostile to Hobbes.

39. This is exactly Aubrey's interpretation of the matter when he remarks that Hobbes "would long since have been ascribed a member there, but for the sake of one or two persons whom he took to be his enemies." (Aubrey 1898, 1:371). Ironically, Aubrey himself was proposed for membership at the society's 31 December 1662/10 January 1663 meeting, which saw his friend Hobbes humiliated by the reading of Huygens's refutation of the cube duplication from *Problemata Physica* (Birch [1756–57] 1968, 1:166).

40. Robertson ([1886] 1910, 180) sums up this problem when he notes that "[i]n the new combination of 1660 [i.e., the actual founding of the Royal Society], there could

"through his frequent mistakes, he has so diminished his credit with everyone, that almost as soon as they see a new problem propounded by Hobbes, they declare that a new ψευδογράϑημα has appeared" (Huygens to Moray for Hobbes, 10/20 December 1662; *CTH* 2:537). Wallis uses practically the same language when he gloats that "the world is not, for the future, likely to be imposed upon by his Paralogisms, and Ψευδογράϑημα (the *name* of *Hobbes* not bearing now any great authority with intelligent persons)" (*HHT* 2). Thus, even if his enemies had not been prominent and influential fellows of the Royal Society, Hobbes's scientific reputation in 1662 was not the sort on which one could plausibly pin hopes of election.

Ever solicitous of a positive public image, the society took pains to make clear that it posed no threat to (properly reformed) religion and stood stoutly against all dogmatic systems, most especially those associated with intolerance or impiety. Michael Hunter has observed, "[T]he alarm that scientists felt at the possibility of being tarred with the brush either of illuminism or atheism is shown by the eagerness with which they sought to dissociate themselves from both" (1979, 189). By the time the Royal Society was chartered, Hobbes had earned a reputation as a foe of all established religion. As a favorite polemical target of Restoration divines concerned to combat the atheistic and libertine tenets the Monster of Malmesbury was supposed to uphold, Hobbes was simply too controversial a figure to be welcomed into a society that sought to reconcile the promotion of empirical science with the demands of political and religious orthodoxy—all the more so when his once-great scientific reputation had been destroyed. In light of all these factors it seems that, rather than puzzle over why Hobbes was denied membership in the society, one might more sensibly ask what possible grounds the society could have had for even considering his election.

It is remarkable that even Hobbes's good relationship with Charles II, the royal patron of the society, was not enough to gain him entrance to the group. Hobbes was granted a royal pension of approximately one hundred pounds per annum and was allowed personal access to the king, which he evidently used in part to press his case against some

---

have been no thought among the chief movers of including an Ishmael like Hobbes, after the proof he had given of mathematical incompetence and of disinclination for the laborious experimental work that was meant (in the spirit of Bacon) to be pursued; but it was not unnatural that Hobbes should resent the exclusion, dictated, as he believed, by the spite of Wallis."

of his enemies.[41] One might expect that the society's eagerness to curry the favor of its patron would have given Hobbes the leverage to be admitted, and Hobbes in fact sought to exploit his royal connection for just this purpose by trying to impress the monarch with his cube duplication. In the summer of 1661 Hobbes presented a version of his cube duplication to Charles II, who forwarded it to the society to be examined. This action surely hints at a campaign by Hobbes to improve his scientific reputation in the king's eyes. Sadly for Hobbes, Viscount Brouncker produced a refutation of the duplication (which Hobbes in the meantime had disowned), and this refutation was presented to the society along with Hobbes's original effort in September of 1661.[42] The whole episode thus served only to harm Hobbes's prospects further and gave him still more reason to resent his treatment by the society.[43] Hobbes also dedicated his 1662 *Problemata Physica* to Charles II, taking the opportunity to apologize for any offense that *Leviathan* may have given and to defend himself against charges of atheism. Such actions certainly suggest that Hobbes hoped to impress

41. For Hobbes's relationship with Charles II, see Malcolm's biographical register to the correspondence (*CTH* 2:817–20). Malcolm there relates an example of Hobbes's use of his royal connection to assist in his campaigns against his enemies, namely his making a personal appeal to the king to authorize the printing of a letter of complaint against Dr. John Fell, who had made unflattering revisions to Anthony Wood's account of Hobbes's life in the *Athenae Oxoniensis*.

42. Birch reports that at the meeting of 4/14 September "A proposition of Mr. HOBBES *for finding two mean proportionals between two strait lines given,* was delivered to the society by Sir PAUL NEILE from the king, indorsed with his majesty's own hand, and was ordered to be registered; as was afterwards the answer to the problem, by lord viscount BROUNCKER (Birch [1756–57] 1968, 1:42).

43. In a letter from 13/23 September to Huygens, Robert Moray included a copy of Hobbes's duplication and Brouncker's refutation, observing that "I am also sending you what I had promised. It is a proposition of Mr. Hobbes. He thinks he has found a demonstration of two mean proportionals between two given lines. He had enough confidence to place it in the hands of the king. His majesty gave it to Sir Neile to be examined by our Society. But Mylord Brouncker, having seen it before it was produced in public, very quickly produced a refutation of it, a copy of which in his own hand is enclosed, which he sends with his most humble service" (*HOC* 3:336). The incident is confirmed in Birch [1756–57] 1968, 1:42. Hobbes complained of this treatment in an addition to the 1668 version of the *Dialogus Physicus:* "I thought I had found a method of interposing two mean proportionals between any two right lines. I wrote it down, having worked on it in the country, and sent it to a friend in London so that he could show it to our geometers *[copiam ejus faceret].* The next day it happened that I noticed it was wrong, and I wrote a recantation of it. It was one of [the Greshamites] who, seeing the same fault in the meantime (which was easy to do), refuted it. They recorded this refutation in the archives of the society, even knowing it to be condemned by the author himself. What a noble and generous deed!" (*OL* 4:237). Wallis reports essentially the same version of events (*HHT* 147).

the king with his scientific credentials, and one salient reason for such self-promotion is Hobbes's interest in making his way into the society. That Moray would seek Huygens's opinion on the matter, and that Brouncker would be so eager to refute Hobbes's efforts, also indicates that there were members of the society who were interested in making sure that his work was not taken seriously. Charles was evidently unimpressed by Hobbes's campaign, as we can glean from Sorbière's account of an audience with the monarch. He reports that they both agreed that "if he had been a little less dogmatic, he would have been very useful to the Royal Society. For there are few who can see farther into things than he" (Sorbière 1664, 97). Hobbes's "dogmatism" is a theme that recurs throughout the polemics of Wallis and Ward, and in light of Hobbes's stubborn refusal to give ground on the question of the cube duplication, it is clear that Hobbes deserved this much of his reputation.

The role of the cube duplication in Hobbes's battles with the Royal Society is illustrated in an interesting note from the Danish Scholar Ole Borch, who was a correspondent and acquaintance of Sorbière. His journal records a meeting with Sorbière in 1664 in which they discussed a letter in which "Hobbes tries to show that he has found a duplication of the cube; but Sorbière said there is someone else who thinks that he is playing with a paralogism. To which, however, Hobbes has already replied that that man is not a fellow of the Royal Society, and that it is against that society that Hobbes is arming himself." [44] It therefore appears that Hobbes saw his cube duplication as a means both to gain admission to the Royal Society and, by successful defense of his claims, to exact some revenge upon his enemies.

Whatever the full story of Hobbes's relationship with the society, he regarded the "Greshamites" with a mixture of longing for acceptance and resentment at his exclusion. Hobbes's public controversy with the Royal Society opened in 1661 when he attacked the experimental philosophy of Robert Boyle with his *Dialogus Physicus*—a work that, as we have seen, contained a revised version of his earlier cube duplication. This drew two printed responses: Boyle's *Examen of Mr. T. Hobbes, his Dialogus Physicus de Natura Aëris* (Boyle 1662) and Wallis's *Hobbius Heauton-timorumenos; or A Consideration of Mr. Hobbes his Dialogues in An Epistolary Discourse Addressed to the Honourable Robert Boyle, Esq.* (Wallis 1662). The latter is a full-scale

---

44. This report from *Olai Borrichii itenerarium* is taken from Malcolm's notes to Hobbes's correspondence (*CTH* 2:584, note to letter 161).

assault on Hobbes's scientific and mathematical pretensions reminiscent of the *Elenchus* from seven years earlier, although the great bulk of it is devoted to a point-by-point rebuttal of Hobbes's 1660 *Examinatio et Emendatio Mathematicae Hodiernae.*[45] There is no need to go through the many points and counter-points in this treatise, but it is important to stress both the cumulative character of its assaults and its obvious intention of destroying what little may have remained of Hobbes's intellectual reputation. The theme of Hobbes's dogmatism, vanity, and imperviousness to criticism (always a staple of Wallis's anti-Hobbesian publications) takes center stage in *Hobbius Heautontimorumenos,* and Wallis spares no opportunity to stress how poorly Hobbes's reputation has fared. He notes that, had Hobbes died before the publication of *De Corpore,* the learned world might have regarded him as a great philosopher, but "he might have taken time sooner to go off the Stage, with more Advantage, than now he is like to do. And tis (you know) no small Mis-fortune, for a man to *Out-live his Reputation*" (*HHT* 9–10). The effect of such attacks was to identify Hobbes publicly as an enemy of the Royal Society and to heighten his sense of intellectual isolation.

Hobbes's resentment against the society was greatly intensified by the fact that his opponents (particularly Wallis) could use the *Philosophical Transactions* to publish critical comments on his mathematical writings while depriving him the use of that forum to respond. In a 1672 letter to Oldenburg, Hobbes complains about Wallis's review of the *Rosetum Geometricum* published in the previous *Philosophical Transactions,* and then asks whether "if hereafter I shall send you any paper tending to the aduancement of Physiques or Mathematiques, and not too long, you will cause it to be printed by him that is Printer to the Society, as you haue done often for D.ʳ Wallis" (Hobbes to Oldenburg, 26 November/6 December 1672; *CTH* 2:726). Oldenburg's response was to remark that he could not present Hobbes's anti-Wallis invective to the society, but that he was willing to consider any of Hobbes's papers that were "not too long, nor interwoven with per-

---

45. More than 120 of the 160 pages of *HHT* are concerned with the dialogues in Hobbes's *Examinatio,* and it is likely that Wallis had prepared them before the *Dialogus Physicus* appeared in 1661. In any event, Wallis portrays himself as having been reluctant to pursue his quarrel with Hobbes: "there is not *Now,* the like need of a serious Reply, to what he Writes, as when he first *Began.* . . . And upon this Account it was, that, when he published his first Six Dialogues the last year; though fronted against my self; I did not think my self obliged to make any Reply, because 'twas known sufficiently, by what Person, and how Affected, the Dialogues were so written; Besides, that the Contents thereof were not worth a *Book,* much lesse *Two*" (*HHT* 14).

sonal reflexions" (Oldenburg to Hobbes, 30 December 1672/9 January 1673; *CTH* 2:727). Some two years later, responding to some inquiries by Hooke, Hobbes wrote to Aubrey that

> [i]f I had any thing now in my hands towards the aduancement of that Learning which the Society pretendeth to, I could be content it should be published by the Society much rather then any other, prouided that they continually attend the businesse, and are of the Society vpon no other account then of their Learning, either had forborn to do me iniury or made me reparation afterward. Do they thinke, that no body takes his Learning to be an honour to him, but they? But what reparation can they make? As for the Members, I haue amongst them for the most part a suficient reputation, and I hope I haue so of Mr Hooke; and amongst the Learned beyond the Seas a greater estimation then the Society can suppresse; but that is nothing to the Body of the Society, by whose authority the euil words and disgraces put vpon me by Dr Wallis are still countenanced, without any publique Act of the Society to do me right. (Hobbes to Aubrey, 24 February/6 March 1675; *CTH* 2:751–52)

Hobbes's mathematical battles with the Royal Society included several papers that Hobbes's addressed to the society in 1671 (*EW* 7:429–48), evidently with the faint hope that he could persuade the fellows that the mathematical principles endorsed by Wallis could not stand up to scrutiny. This episode, however, can be left aside here, as it involved nothing beyond the issues we have already covered.

6.2.2 *Hobbes's* De Principiis et Ratiocinatione Geometrarum and the Reply to Huygens

By 1666 Hobbes realized that he had become the ultimate mathematical outsider. He summarized his disagreement with the mathematical tradition in his treatise *De Principiis et Ratiocinatione Geometrarum*, whose avowed aim is to show that there is as much disagreement, conflict, and uncertainty among geometers as among writers on morals or politics. His feelings of isolation and vulnerability were by no means confined to the fields of natural philosophy or mathematics. The Monster of Malmesbury faced powerful and determined opposition from other quarters as well, and they could do him considerably more harm. Much of London had been destroyed in the great fire of September 1666 and in October the House of Commons concluded that the fire was probably divine punishment for the country's toleration of the (alleged) atheism of Hobbes and Thomas White. Parliament empowered a committee to examine their writings and determine whether they should be tried for heresy. Hobbes was spared further investigation or harm by the action of his patron Henry Bennet, baron of Arlington, who intervened on his behalf to stop the proceedings. Still, Hobbes was sufficiently concerned by the episode that he destroyed a large number of his papers.

The dedicatory epistle to *De Principiis et Ratiocinatione Geometrarum* is addressed to Arlington and vividly portrays Hobbes's sense of isolation. He describes this book as attacking "the whole nation of geometers," and admits that "my reputation faces a great danger, since I dissent from the opinions of nearly every geometer." The inevitable conclusion is that "of those who have held forth on the same things as I, either I alone am insane, or I alone am sane. There is no third alternative, unless, as someone might say, we are all insane" (*PRG* epistle; *OL* 4:387). In the introduction to the treatise Hobbes explains that he writes "against geometers, not against geometry," but he complains that previous authors (such as Clavius) who have lavished praise on geometry have in fact praised "the masters and not the art itself." Geometry has no privileged place in the Hobbesian scheme of the sciences because any body of knowledge that could be properly demonstrated would be both true and certain. "The certitude of all sciences is equal," he declares, "and otherwise they are not sciences, since *to know* does not admit of more and less." The result is that "physics, ethics, and politics, if they are well demonstrated, are no less certain than the pronouncements of mathematics, just as mathematics is no more certain than other sciences unless its pronouncements are properly demonstrated" (*PRG* introduction; *OL* 4:390).

The bulk of *De Principiis et Ratiocinatione Geometrarum* is devoted to an exposition of Hobbes's materialistic philosophy of mathematics and his criticism of the traditional conception of geometry. A recurring theme in this treatise is the incoherence of the doctrine of dimensionless points, breadthless lines, and other notions that do not treat geometric objects as material bodies. I have already considered such criticisms in my exposition of Hobbes's philosophy of mathematics and its intended applications and will therefore not be greatly concerned with them here. However, there is one important feature of this work to which we must turn our attention: Hobbes's reaction to Huygens's work on circle quadrature. The *Problemata Physica* of 1662 contained a circle quadrature that concluded that the arc of a quadrant of the circle was equal to the radius plus the tangent of 30°. This implies that π is greater than 3.154, and Huygens had refuted it in his letter to Sir Robert Moray, which was read before the Royal Society in December of 1662. He there challenged Hobbes to "look, then, at my theorem on that subject in my little book *De Circuli Magnitudine Inventa,* where he will indeed find it done without numbers or any numerical calculation whatsoever" (Huygens to Moray for Hobbes, 10/ 20 December 1662; *CTH* 2:538). Hobbes's response was the twenty-first chapter of *De Principiis et Ratiocinatione Geometrarum,* but before we turn to a consideration of it, it is necessary to have some acquaintance with the basic ideas behind Huygens's work.

Huygens's 1654 treatise *De Circuli Magnitudine Inventa* was a decisive advance in the attempt to determine the value of π, or, equivalently, to determine the area of the circle exactly.[46] Archimedes had given a method of calculating π to within any desired degree of accuracy, but the method was as cumbersome as it was rigorous. The Archimedean procedure (which I outlined briefly in chapter 1) is based on the principle that the perimeter of a regular polygon inscribed in a circle is less than the circumference of the circle, while the perimeter of a similar polygon circumscribed about the circle exceeds the circumference. By increasing the number of sides in the inscribed and circumscribed regular polygons, these upper and lower bounds can be systematically improved and made to converge upon the value of π. Thus, using Archimedean methods, we can construct a sequence of inscribed regular polygons with perimeters $p_1, p_2, \ldots, p_n \ldots$ and a sequence of circumscribed polygons with perimeters $P_1, P_2, \ldots, P_n \ldots$ such that (assigning the diameter of the circle a unit value) $p_1 < p_2 < \ldots < p_n \ldots < \pi < \ldots P_n < \ldots P_2 < P_1$. Archimedes himself inscribed and circumscribed hexagons about a circle, then successively doubled their sides to produce two ninety-six-sided figures, which yields the inequality $3.14084 < \pi < 3.142858$.

Although the Archimedean procedure can give arbitrarily precise approximations to π, it is quite cumbersome and involves horrendously difficult calculations because the approximating sequences converge very slowly to the desired value. In a truly remarkable exhibition of perseverance through sheer tedium, the Dutch mathematician Ludolph van Ceulen calculated π to thirty-five digits, using Archimedes' procedure. Willebrord Snell, another Dutchman, sought a faster means to determine the value of π. In his *Cyclometricus* Snell used, although he

46. See Hoffman 1966 for a detailed study of this work, its background, and its influence.

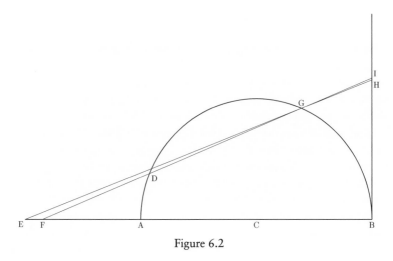

Figure 6.2

did not rigorously prove, geometric principles that enabled the value to be calculated more efficiently.[47] Huygens found rigorous geometric proofs of Snell's principal results, which can be stated as follows. If the line *EA* is chosen so that $EA = AB/2$ (as in figure 6.2), then $\overset{\frown}{BG} >$ the tangent *BH;* similarly if *DF* is chosen so that $FD = AB/2$ (again as in figure 6.2), then $\overset{\frown}{BG} <$ tangent *BI.* The effect of instituting such bounds is to accelerate the convergence of the sequences $\{p_n\}$ and $\{P_n\}$ toward the common limit of π and thus to give much more precise approximations toward the sought value.

In addition to giving a rigorous derivation of Snell's methods, Huygens also established a number of other results concerning the area of a segment of the circle and that of inscribed or circumscribed triangles. These can be stated with reference to a segment *ABC* less than a semicircle (whose area we can designate as *a*), triangle *ADC* (formed by the base and tangent lines *AD, CD*) and triangle *ABC* (the largest triangle that can be inscribed in the segment), as in figure 6. 3. Huygens showed that 2/3 triangle $ADC > a > 4/3$ triangle *ABC.* With the aid of these and other theorems, Huygens showed that the Archimedean approximation procedure can be accelerated. In particular, under the appropriate choice of inscribed and circumscribed regular polygons, if $p_n < \pi < P_n$, we can improve the approximation by exploiting the fact

47. These two results appear as propositions 28 and 29 in Snell 1621, 42–45. They are demonstrated in theorem 12 of Huygens's *De Circuli Magnitudine Inventa* (*HOC* 12:156–59). Cantor [1900–1901] 1965, 2:704–17) contains an overview of this and related material.

that π lies in the first third of the interval; i.e., $p_n < \pi < (p_n + P_n/3)$. This fact enabled Huygens to derive the inequality $3.1415926533 < \pi < 3.1415926538$. The equivalent degree of precision using Archimedes' procedure would have required the construction of polygons of 240,000 sides. Huygens can achieve the result with a regular polygon of 60 sides.[48] Perhaps the most striking feature of Huygens's treatise is the fact that its results are obtained without the use of infinitesimal methods or analytic geometry, but rely exclusively on the techniques of classical geometry.

Hobbes had little understanding of Huygens's results, but he nevertheless attempted a critique of his procedure and a defense of his own alleged quadrature.[49] Hobbes's negative evaluation was no doubt intensified by Huygens's refutation of his efforts to double the cube and square the circle. Indeed, Hobbes direct acquaintance with Huygens's treatise *De Circuli Magnitudine Inventa* was probably due to the fact that the Dutch mathematician sent him a copy of it in order to persuade him of the error in his circle quadrature in the *Problemata Physica*. Hobbes's failure to grasp the import of Huygens's work is evident

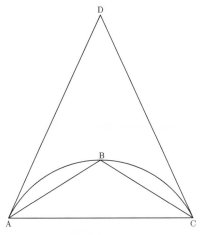

Figure 6.3

48. The main result is contained in *HOC* 12:179. For more on Huygens's approach to the calculation of π see Vavilov 1992.

49. Hobbes's opinion of Huygens was not always negative, however. Loedwijk Huygens, the mathematician's brother, reported a meeting with Hobbes in February of 1652, where Hobbes "at once began to speak about my brother Christiaan's *De Quadratura Parabolis et Hyperbolis,* etc. which Mr. Brereton had given him a few days earlier. He praised it abundantly and said that in all probability he would be among the greatest mathematicians of the century if he continued in this field" (Huygens 1982, 74–75).

from the summary he gives of the principal result. In Hobbes's recapitulation, Huygens is supposed to have argued as follows:

> Let a circle segment ABC [figure 6.4] less than a semicircle be described, and let it be divided in two equal parts by the perpendicular FB. And further having bisected the arcs AB, BC at D and E, and having drawn their chords CE, EB, BD, and DA, and the tangents CH, BH, BI, IA, and CK, KE, EL, LB, and BM, MD, DO, OA; and letting these be successively bisected as far as can be understood *[quantum intelligi potest]*, he demonstrates (and, as far as I can see, correctly) that the circle segments CEC, EBE, BDB, DAD are less than the triangles CKE, ELB, BMD, DOA. (PRG 21; OL 4:451)

This corresponds to no particular theorem in Huygens's treatise, although it may be Hobbes's attempt to reproduce the exhaustion proofs that establish some of the ratios between circular segments and inscribed and circumscribed triangles.[50] Hobbes's defense of his results against Huygens's reasoning is truly remarkable. He insists:

> But when he likewise infers that, if by perpetual bisection an infinite number of segments are made, these also taken together will be less than all of the triangles, corresponding to the segments, taken together, he infers badly, unless the right line IBH is outside the circle, so that the point B is not common to both the right and curved lines but between them. For if B is common to both lines, then all the infinite number of tangents constitute the arc ABC. But if B is outside the circle, however much it may touch the circle, the chords AB, BC will not cut the circle in the same point in which they cut the right line FB, but will be short of it on both sides. But Euclid teaches (*Elements* 3, prop. 2) that the right line CB is entirely inside the circle, and (*Elements* 3, prop. 16) that the tangent is entirely outside the circle. And so no right line except FB can go through the arc and tangent at the same point B, namely at the point that is called that of *contact*, unless both lines are attributed some latitude. (PRG 21; OL 4:452)

50. Particularly that in theorems 3 and 4 of *De Circuli Magnitudine Inventa*. Hobbes himself admits the likely inadequacy of his presentation: "Consult his book itself, my copy of which is not with me as I write; nor if it were here, would his demonstrations easily be transcribed" (PRG 21; OL 4:452).

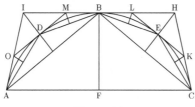

Figure 6.4

Hobbes then declares that from the use of the false conception of points as "nothing,"

> after many demonstrations, [Huygens] inferred that "the differ-
> ence between a third part of the arc of a quadrant and the chord,
> is to the difference between the chord of the same third part of
> the arc and its sine . . . in a ratio greater than 4 to 3," which is
> insufficient to confirm what he wanted to refute. But when he
> inferred that the right line composed of a radius and the tangent
> of thirty degrees is greater than the arc of the circle, he fell into
> error, because he thought that the radius of the circle is not less
> than that which is drawn from the center to a tangent, which is
> outside the circle. (*PRG* 21; *OL* 4:452)

This response can only be interpreted as an act of desperation. In effect, Hobbes seeks to protect the validity of his quadrature by attributing any difference between his and Huygens's results to the latter's neglect of point extensions. In this context it is important to recall that
Hobbes had voiced no qualms about Huygens's central theorems in
1656, when Mylon had used them as the basis for objections to the
treatment of parabolic arcs in chapter 18 of *De Corpore*. In the in-
tervening ten years Hobbes had obviously changed his estimation of
such methods.

The price to be paid for such an insistence on the correctness of his
quadratures is Hobbes's ultimate rejection of such geometric basics as
the Pythagorean theorem. We have already seen how, in a 1664 letter
to Sorbière, Hobbes expressed "doubts" about the Pythagorean theorem. By 1666, he was more emphatic. He remarks in Chapter 23 of
De Principiis et Ratiocinatione Geometrarum that "if these things of
mine are correctly demonstrated, there are some further things that
you should consider. First, the greater part of the propositions that
depend on proposition 47 of book 1 of Euclid (and there are many)

are not yet demonstrated. Second, the tables of sines, tangents, and secants are wholly false" (*PRG* 23; *OL* 4:462–63).

The publication of *De Principiis et Ratiocinatione Geometrarum* did nothing to enhance Hobbes's reputation. In a brief and dismissive review published in the *Philosophical Transactions,* Wallis highlighted the essentially hopeless condition of Hobbes's geometric program. He recounts the difficulties facing the alleged quadratures:

> For finding himself reduced to these inconveniences; 1. that his *Geometrical Constructions,* would not consist with *Arithmetical calculations,* nor with what *Archimedes* and others have long since demonstrated: 2. That the *Arch* of a Circle must be allowed to be sometimes *Shorter* than its *Chord,* and sometimes *longer* than its *Tangent:* 3. That the Same Straight Line must be allowed, at one place onely to *Touch,* and at another place to *Cut* the same Circle: (with others of a like nature;) He findes it necessary, that these things may not seem Absurd, to allow his *Lines* some *Breadth,* (that so, as he speaks, *While a Straight Line with its Out-side doth at one place Touch the Circle, it may with its In-side at another place cut it,* &c) But I should sooner take this to be a *Confutation* of his *Quadratures,* than a *Demonstration of the Breadth of a* (Mathematical) *Line.* (Wallis 1666, 290–91)

It is hardly necessary to stress that Hobbes was not taken seriously by the mathematical world after the publication of *De Principiis et Ratiocinatione Geometrarum.* Where he had been both able and willing to gain a hearing for his mathematical work before the mid-1660s, he no longer corresponded with mathematicians in England or on the Continent, and his mathematical works were generally ignored.

6.3 ENDGAME: HOBBES'S LAST PUBLICATIONS ON MATHEMATICS

From 1668 until his death in 1679, Hobbes continued to publish an astounding amount of mathematical material, but he had set himself so firmly against prevailing mathematical opinion that he was entirely impervious to any criticism. These publications include two compendia of Hobbesian mathematical writing, *Rosetum Geometricum* and *Principia et Problemata,* as well as four papers addressed to the Royal Society, the *Lux Mathematica,* two different editions of a 1669 pamphlet claiming to square the circle, and a mathematical appendix to the *Decameron Physiologicum.* There is nothing to be gained by a detailed examination of all these works, but some key points can be clarified by looking into some of them briefly.

In 1668 Johannes Blaeu published a collection of Hobbes's works in Amsterdam under the title *Opera Philosophica*. This gave Hobbes the opportunity to revise his earlier publications, and it was also the occasion for the Latin translation of *Leviathan*. The 1668 revisions to the mathematical sections of *De Corpore*—especially the circle quadrature in chapter 20—are especially relevant because they document Hobbes's retreat from the principles and practices of European mathematics. In the original (1655) version of chapter 20 (and in the English version from 1656) Hobbes began his attempted quadrature with a brief resume of earlier approaches to the problem, whose validity he clearly accepted. Hobbes there informs us that "[a]mongst those Ancient Writers whose Works are come to our hands, *Archimedes* was the first that brought the Length of the Perimeter of a Circle within the limits of Numbers very little differing from the truth; demonstrating the same to be less than three Diameters and a seventh part, but greater than three Diameters and ten seventy one parts of the Diameter" (*DCo* 3.20.1; *EW* 1:287). Such a positive evaluation of previous efforts is not confined to Archimedes, for Hobbes mentions "*Ludovicus Van Cullen* & *Willebrordus Snellius*" as having "come yet neerer to the truth; and pronounced from true Principles, that the Arch of a Quadrant (putting, as before 10000000 for Radius) differs not one whole Unity from the number 15707963" (*DCo* 3.20.1; *EW* 1:287–88).

The 1668 version of chapter 20 of *De Corpore* opens, not with a review of previous approximations to π, but with an argument designed to show that the sine of 30 degrees (taken to seven decimal places) is .5811704. This value differs significantly from the accepted value of .5773503, and Hobbes takes it upon himself to explain why his result should differ from the commonly accepted one. His opponents are in error, Hobbes explains, and their error has its origin in the view that the multiplication of numbers is the same thing as the multiplication of lines:

The cause of the error is not that the numbers were falsely counted up, but that a number multiplied into a line never produces the same thing as a line drawn into just as many lines. And thus the root of a square number is never the side of a square figure, whatever the number of its parts may be. If you divide a right line, for example, into three equal parts, and you multiply the same line by the number three, there will not be produced a square but nine lines. Or if you divide this line into 10,000,000 parts, the product of these parts and [the number] 10,000,000 will be so many lines, not so many squares on these parts. For in order to make a rectangle from a line and a number it is necessary that the number be infinite. (*DCo* 3.20.1; *OL* 1:243–44)

This by now familiar doctrine is underwritten by Hobbes's insistence that "in every extraction of roots from a number, the root will be an aliquot part of its square"; similarly, the traditional method of computing sines is faulted for its use of arithmetical calculation in a geometric setting. Archimedes' result, which Hobbes had earlier accepted, is now rejected because "even Archimedes' measure of the circle, which depends from the beginning upon this calculation, is less than the true value; nor can it succeed in refuting a geometric calculation in which the perimeter of the circle happens to be found greater than that found by him" (*DCo* 3.20.1; *OL* 1:245).

Hobbes's 1669 pamphlet *Quadratura Circuli, Cubatio Sphaerae, Duplicatio Cubi, Breviter demonstrata* deserves a brief mention in this context because it contains a new assault on the quadrature of the circle that (perhaps not surprisingly) yields a result inconsistent with the quadratures Hobbes had defended just three years earlier.[51] Take the circle *BCDE* (figure 6.5) with center *A*, divided into four equal sectors by the diameters *BD* and *CE*. Construct the square *GHIF* to circumscribe the circle and touch it at the points *B*, *C*, *D*, and *E*. Draw the diagonals *GI* and *FH*, cutting the circle at points *K*, *L*, *M*, *N*. Bisect the line *GC* at *O*, and draw the line *AO*, cutting the circle at *P*. Through *P* draw a line parallel to *GC*, to intersect *AG* in *Q* and *AH* in *R*. With *QR* as a side, construct the square *QRST*. Hobbes asserts that the *QRST* is equal in area to the circle *BCDE*. The argument for this remarkable result is not sufficiently interesting to detain us; the salient fact is that Hobbes derives the conclusion by introducing a key assumption in the form of a disjunction, viz.: "Either there can be no right triangle in *ACG* with vertex *A* equal to the sector *ACL*, or the areas *PQL*, *CYP* are equal" (Hobbes 1669a, 2). Because the first disjunct is obviously false (i.e., there must be *some* rectilinear triangle similar to *ACG* with vertex *A* and equal in area to *ACL*), Hobbes concludes that the areas *PQL*, *CYP* are equal, and then quickly concludes that the square *QRST* is equal to the circle *BCDE*.

However, Hobbes conveniently overlooks the possibility that both

---

51. The full argument for this quadrature is contained in section 6 of the appendix. My treatment of it here is consequently quite brief.

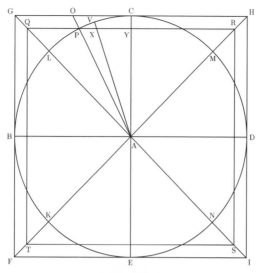

Figure 6.5

disjuncts of his assumption are false, and indeed, his argument does little more than beg the question at issue. Furthermore, his argument has the unfortunate consequence of setting the value of π at 3.2—a result that Wallis happily demonstrated in his rebuttal *Thomae Hobbes Quadratura Circuli, Cubatio Sphaerae, Duplicatio Cubi Confutata* (Wallis 1669a, 3). Hobbes was evidently prepared for such a reaction and in the second edition of the *Quadratura* he replies to Wallis's objection. This work was dedicated to Prince Cosimo de'Medici, and in a letter to the prince Hobbes explains that "I did not want to abuse the patronage of such a great prince by invoking it on behalf of false or uncertain doctrine. So, in order that the earlier copies of this pamphlet should be subjected to the most minute examination, I submitted them not to friends but, more cautiously, to torturers—putting them, as it were, in a purifying fire" (Hobbes to Cosimo de'Medici, 6/16 August 1669; *CTH* 2:711). Hobbes's reply to the central objection is to complain that when Wallis

> proves that the triangle *AYQ* and the sector *ACL* are not equal, he assumes as demonstrated by Archimedes that the perimeter of the circle to the diameter is in a ratio less than that of 22 to 7. But this is neither demonstrated by Archimedes, nor can it be demonstrated by his method. For it proceeds from the extraction of roots, and assumes that the root of a number of squares con-

tained in a greater square is the side of the whole square, which is manifestly false. For the root of a number of squares is some other number of squares, just as *[non aliter quam]* the root of 100 stones is 10 stones. (Hobbes, 1669b, 2; OL 4:508)

The utter collapse of Hobbes's geometric ambitions was complete by 1670. He continued to publish mathematical tracts and defenses of his various alleged results, but seems to have had no readership. Hobbes undertook the hopeless task of gaining a hearing for his views from the Royal Society, to whom he addressed a collection of papers and his *Lux Mathematica,* but the society's only response was Wallis's scathing reviews of his work in the *Philosophical Transactions.* He also printed three separate papers addressed to the Royal Society, specifically taking issue with Wallis's mathematics and defending his own thesis that algebraic calculation can never challenge a geometric result. These also gained no hearing for his views other than giving Wallis a further opportunity to disparage his opponent in the pages of the *Philosophical Transactions.* Some ten years before he actually died, Hobbes was a mathematical casualty and his program for geometry did not survive the death of its founder and lone adherent.

CHAPTER SEVEN

The Religion, Rhetoric, and Politics of Mr. Hobbes and Dr. Wallis

I can, after all, write the remaining parts of my Philosophy, and at the same time write letters to you, and moreover give a thrashing in passing to some worthless fellow who treats me impertinently. My quarrel with him is not like the quarrel between Gassendi and Morin or Descartes. I was dealing at the same time with all the ecclesiastics of England, on whose behalf Wallis wrote against me. Otherwise I would not consider him the least bit worthy of a reply, whatever is thought of his books by certain rather famous geometers—who are also rather ignorant of the art they teach, as will perhaps become more obvious shortly.

—Hobbes to Sorbière, 29 December/8 January 1657

The preceding six chapters have been primarily concerned with the mathematical aspects of the war between Hobbes and Wallis. There is good reason for this, since their dispute was first and foremost over mathematical matters. Nevertheless, it would be misleading to characterize the exchange of polemics as concerned exclusively with mathematics. As I showed in chapter 2, nonmathematical issues involving religion and politics were crucial to the early stages of the controversy. In fact, it is unlikely that Wallis or Ward would have bothered refuting Hobbes's mathematical claims so publicly and at such length had they had not seen him as a danger to the universities and religion. Indeed, we saw that Hobbes's decision to publish his circle quadratures in *De Corpore* was in all likelihood prompted by Ward's baiting in the *Vindicae Academiarum*. But political and religious issues were not confined to the opening stages of the controversy. Extramathematical material appears alongside debates over purely mathematical issues in many of the exchanges between the two combatants. Thus, to take one example, the question of how best to interpret the doctrine of proportions is taken up in the first part of Hobbes's Στίγμαι, to be followed several pages later by his attack upon Wallis's conception of Christian ministry, which itself is followed by a seemingly unrelated discussion of subtleties of Latin grammar. My task in this chapter is to cover those aspects of the Hobbes-Wallis controversy that do not concern mathematics but are nevertheless important components of the war between these two.

This chapter is organized into three sections that cover the most salient extramathematical aspects of the dispute. The first covers the thorny issues of religion and church government. The second is a summary of the various charges of grammatical and rhetorical infelicity traded by the disputants. The third deals with political questions, and particularly the issue of loyalty to the restored monarchy. I make no claim to comprehensive coverage of these aspects of the dispute, but I offer an overview complete enough to acquaint the reader with the relevant contested issues.

7.1 HOBBES AND WALLIS ON RELIGION

Religious conflicts were practically omnipresent in seventeenth-century Europe, and they provide something like a fixed frame of reference for the intellectual historian. One way of locating a philosophical figure from this period on the conceptual map is to investigate his stand on the contested religious questions of the day. Indeed, it is arguably the case that almost any significant seventeenth-century controversy contains a religious element. Hobbes and Wallis disagreed about nearly everything, so it is no surprise that their views should diverge on questions of religion. Hobbes's religious opinions stood far outside the mainstream of seventeenth-century thought, although the extent of his heterodoxy remains a matter of debate and will be of concern to us later. Wallis, in contrast, fits fairly neatly into the reformed tradition, and particularly in the Presbyterian strain of English Puritanism. Here again, as in many mathematical aspects of their dispute, their conflict pits the "outsider" Hobbes against the traditionalist Wallis. This is not to say that the two men's doctrinal differences made them incapable of understanding each other when issues of religion arose, but it does reinforce the idea that Hobbes's philosophical, mathematical, and religious doctrines placed him in opposition to received views.

Two important issues are at the heart of the religious dimension of the Hobbes-Wallis dispute. The first is that of church government, i.e., how the church ought to be organized and the extent of the powers granted to the church hierarchy to enact and enforce laws binding on the populace. The second is the charge (raised by others besides Wallis) that Hobbes is an atheist, notwithstanding his numerous professions of religious belief. These two topics merit separate treatment and will be covered in two subsections.

7.1.1 The "Scottish Church-Politics" of Dr. Wallis and the Limits of Sovereign Power

The issue of church government was of paramount importance in seventeenth-century England. The Civil War was, among other things, fought over questions of how the church should be organized and who should hold the power of determining its doctrine, liturgy, and forms of worship. I cannot give an exhaustive account of the issues in play in this disputed area, but the main lines of the conflict can be discerned quite readily. There were three principal forms of Christianity in mid-seventeenth-century England: Catholicism, Anglicanism, and Puritanism. These obviously differ in many respects, but one of the most important points of difference is on the question of church government, and it is against the background of such differences that Hobbes and Wallis traded polemics on the question of church authority. Although officially suppressed in England after the reign of Mary (1553–58), Catholicism survived surreptitiously into the seventeenth century, and with it the doctrine of the pope's supreme authority in matters of religion. Catholicism also had a distinctive episcopal structure in those countries where it was officially established. Catholic bishops and canon courts claimed authority in church matters and constituted essentially a parallel governmental structure alongside that of the civil sovereign.[1]

---

1. Hobbes is an excellent source to outline this division between civil and religious powers under Catholicism. In *Behemoth*, he succinctly catalogs the claims made by the pope against the rights of civil sovereigns. These include: "[f]irst, an exemption of all priests, friars, and monks, in criminal causes, from the cognizance of civil judges. Secondly, collation of benefices on whom he pleased, native or stranger, and exaction of tenths, first fruits, and other payments. Thirdly, appeals to Rome in all causes where the Church could pretend to be concerned. Fourthly, to be the supreme judge concerning lawfulness of marriage, that is concerning the hereditary succession of Kings, and to have the cognizance of all causes concerning adultery and fornication. . . . Fifthly, a power of absolving subjects of their duties, and of their oaths of fidelity to their lawful sovereigns, when the Pope should think it fit for the extirpation of heresy." The result is that "[t]his power of absolving subjects of their obedience, as also that other of being

Anglicanism retained the episcopacy but placed this structure under the control of the monarch, who was the supreme authority in civil and religious questions. The 1534 Act of Supremacy established Henry VIII as the head of the English church and seized papal lands for the crown. After a period of reestablished Catholicism under Mary, the Elizabethan Settlement of 1559 issued in Acts of Uniformity and Supremacy that made the monarch the supreme governor of the Anglican Church and imposed uniformity of observance throughout the nation. The result was a national church decidedly Protestant in its theology, with an episcopal structure subordinated to the crown but with ceremonies and liturgy that retained many Catholic elements.

By the seventeenth century Anglicanism was under attack from Puritans who regarded it as insufficiently reformed or fatally tainted by "popish" elements in doctrine, liturgy, governmental structure, and ceremony. Resentment against the episcopacy is a common thread in Puritan writings of the period, and calls for the abolition of the episcopal structure were a commonplace among the reformers of the 1630s. There was considerably less uniformity among Puritans about what form of church polity should replace episcopacy, because the term *Puritan* means little more than "non-Anglican Protestant" and is therefore compatible with a variety of models of church government. For my purposes the principal division in Puritanism is that between Presbyterianism and Independency. The differences between these two models of church government are sufficiently important to be set out briefly, and the best way to do this is to see their interaction in the conflict between the monarchy and Parliament that turned into the English Civil War.

Presbyterianism in England took its lead from the Church of Scotland, which had been reformed along strict Calvinist lines in the sixteenth century. The preferred form of church government among Calvinists was a ruling body of ministers and elders—the presbytery—who answered not to the civil authority, but directly to God. The Presbyterian Church claimed authority over the populace as a whole, and it assumed the power to exact tithes and enforce discipline as it saw fit, without reference to the claims of the civil sovereign, whom Presby-

---

judge of manners and doctrine, is as absolute a sovereignty as is possible to be; and consequently there must be two kingdoms in one and the same nation, and no man be able to know which of his masters he must obey" (*Behemoth* 1; *EW* 6:172–73). A lucid study of church-state issues in the context of Hobbes's political theory is Sommerville 1992, chap. 5.

terian doctrine takes to be simply another parishioner. In its official doctrine Presbyterianism also has no place for episcopacy, but the reform of Scotland was not so thoroughgoing as to effect the absolute abolition of bishops. Indeed, James VI of Scotland (the future James I of England) managed through a series of measures to retain some portion of episcopal structure in the church and thereby keep it from the most rigidly Presbyterian model (Mullan 1986). No monarch with aspirations to supreme authority could welcome the claims of Presbyterianism, and when James I ascended to the English throne in 1603 he took steps to dilute the power of the Presbyterians and bring Scotland into still closer conformity with the Anglican model. The result was naturally a contentious relationship between the crown and the Scots church, which only intensified with the ascent of Charles I to the throne in 1625. Charles supported a variety of policies that aimed at increasing the power of the crown over the church while seeking a rapprochement with the church of Rome. William Laud, whom Charles made archbishop of Canterbury in 1633, was closely identified with a campaign against Puritan influences and the reintroduction of "Romish" practices in Anglican liturgy and ceremony. With the attempt to impose the English Book of Common Prayer in Scotland in 1638, Scottish Presbyterians rose in open rebellion.

Needing money to prosecute the first of two Bishops' Wars against the recalcitrant Scots, Charles I (who had ruled without a Parliament since 1629) called Parliament into session in the spring of 1640, but adjourned it almost immediately when it began to list grievances against his "personal" (i.e., nonparliamentary) rule. Renewed fighting with the Scots in the second Bishops' War led to the occupation of northern England by Scottish forces, and Charles was forced to call Parliament into session in November of 1640. This Long Parliament would remain in session throughout the Civil War in the 1640s until it was purged by the army in 1648, leaving the radical Rump Parliament, which presided over the execution of Charles I and the abolition of the monarchy. One of the Long Parliament's more important actions was the presentation of a Grand Remonstrance to Charles in December of 1641 that listed a comprehensive catalog of grievances against the crown. On the matter of church government, the members of Parliament insisted that they had no intention "to let loose the golden reins of discipline and government in the Church," because "we hold it requisite that there should be throughout the whole realm a conformity to that order which the laws enjoin according to the Word of God"

(Prall 1968, 72). Nevertheless, the authors of the Grand Remonstrance objected to the episcopalian model as implemented by Laud, and demanded that

> the better to effect the intended reformation, we desire that there may be a general synod of the most grave, pious, learned and judicious divines of this island; assisted with some from foreign parts, professing the same religion with us, who may consider of all things necessary for the peace and good government of the Church, and represent the results of their consultations unto the Parliament, to be there allowed of and confirmed, and receive the stamp of authority, thereby to find passage and obedience throughout the kingdom. (Prall 1968, 72)

This demand for a general synod eventually resulted in the creation of the Westminster Assembly of Divines, which was first convened on 1 July 1643. Charles had opposed the institution of the assembly as an obvious imposition upon his right, as head of the Church of England, to decide matters of church governance. In the course of events, however, Charles's opinion had become largely irrelevant. After the outbreak of general hostilities between Parliamentary and Royalist forces in September of 1642, there was a state of open war between the two sides and the Long Parliament no longer answered to the king.

The Westminster Assembly of Divines was charged by Parliament with advising it on how to institute such a form of church government "as may be most agreeable to God's holy word, and most apt to procure and preserve the peace of the Church at home, and nearer agreement with the Church of Scotland and other Reformed Churches abroad" (Mitchell and Struthers 1874, ix). By the late summer of 1643 the Parliamentary cause was faring badly. Seeking assistance from Scotland, Parliament agreed to a "Solemn League and Covenant" with the Scots, the object of which was "the preservation of the reformed religion in the Church of Scotland, in doctrine, worship, discipline, and government, against our common enemies; the reformation of religion in the kingdoms of England and Ireland, in doctrine, worship, discipline and government, according to the Word of God, and the example of the best reformed Churches" (Prall 1968, 104). One of the key elements of the Solemn League and Covenant was its commitment to "the extirpation of Popery [and] prelacy (that is, Church government by Archbishops, Bishops, their Chancellors and Commissaries, Deans, Deans and Chapters, Archdeacons, and other ecclesiastical officers depending on that hierarchy)" (Prall 1968, 105). The Solemn

League and Covenant was generally understood to oblige Parliament to introduce a Presbyterian form of church governance, but the deliberations of the assembly showed that there was hardly unanimity of opinion on this point.

The label *Independent* was first used to describe those members of the assembly who resisted the institution of Presbyterianism. The clerical leadership of the Independents has been described as "a small group of members of the Westminster Assembly fighting a rearguard action against the apparently inexorable implementation of Presbyterianism between 1643 and 1647" (Mullett 1994, 196). Although Independency had a small following in the early stages of the Civil War, the Independents and their allies ultimately held positions of great power in the army and under the Protectorate. No less a figure than Cromwell was closely identified with the Independent faction, and by the time he assumed the title of Lord Protector in 1653 the Independents were the dominant party while the Presbyterians had assumed the uncomfortable role of an embattled minority. The fundamental point of difference between Independency and Presbyterianism is the autonomy of the congregation. The Independents generally held that the freely gathered congregation was the only true model of the New Testament Church and they resisted attempts to subordinate the congregation to an ecclesiastical hierarchy. Presbyterians regarded such autonomy as intolerable laxity no better than the outright abolition of church government. As ever-more radical sects proliferated in the late 1640s and 1650s, the Presbyterians accused the Independents of fomenting schism and offering no means of controlling the dissolution of religious order.[2]

Wallis was appointed to the Westminster Assembly in 1644, serving as a secretary until the final plenary session in February of 1649. He reports in his autobiography that "I was one of the Secretaries to the *Assembly of Divines* at *Westminster*. Not from the first sitting of that *Assembly;* but some time after, and thenceforth during their Sitting" (Wallis 1970, 31). His sympathies were strongly with the Presbyter-

---

2. For example, in attacking Independency as a form of schism, Daniel Cawdrey complained that "Toleration (which is our *present condition*) hath done much more towards the *rooting of Religion,* out of the hearts of many men in 7 yeares, than the enforcing of *uniformity* did in 70 yeares." He concluded that the toleration sought by John Owen and the Independents was the source of complete religious disorder: "Let experience speake; If since the men of *his* [i.e., Owen's] *way* have gotten a *Toleration* for themselves, they have not opened a *doore* for all *errours, heresies,* and horrid blasphemies, or *profanenesse*" (Cawdrey 1657, 14, 16).

ian faction, although after the Restoration he preferred to take the label "Presbyterian" to signify "anti-Independent" rather than "anti-Episcopal" (Wallis 1970, 35). Wallis had attended Emmanuel College, Cambridge (taking the degree B. A. 1637 and M. A. in 1640), and was probably exposed to important Puritan influences during his education there, although the autobiographical report of his religious upbringing makes him seem to have been a Puritan from the beginning.[3] With evident distaste, Wallis describes the Independents as "against all *united Church Government* of more than one single Congregation; holding that each single Congregation, voluntarily agreeing to make themselves a *Church,* and chuse their own Officers, were of themselves *Independant,* and not accountable to any other *Ecclesiastical Government;* but only to the Civil Magistrate, as to the Publick Peace" (Wallis 1970, 34).

Wallis's dislike for Independency did not disappear with his intrusion into the Savilian Chair of Geometry in 1649. The contest between Presbyterians and Independents at the national level was mirrored in the local struggles at Oxford over control of the university (Green 1964, 137–40). With Owen's appointment as vice-chancellor in 1652, Oxford witnessed many skirmishes between Presbyterian and Independent factions (Shapiro 1969, chap. 4). Wallis's firm commitment to Presbyterianism led inevitably to conflict with Owen and his allies. One such episode involved a public disputation in July of 1654, which took place as part of Wallis's being granted the degree of doctor of divinity. In this disputation Wallis argued for the negative answer to the question "whether the powers of the ministers of the gospel extend only to the members of one particular church?" The topic may strike the contemporary reader as arcane and unlikely to arouse much controversy, but it poses the very question of church governance at the heart of the contest between the Independents and the Presbyterians. On the Independent model of the church, ministers of the Gospel are called to serve a local congregation, and their ministerial authority does not extend beyond its confines. On the other hand, Presbyterian

3. Wallis recalls, "As to Divinity, (on which I had an eye from the first,) I had the happiness of a strict and Religious Education, all along from a Child. Whereby I was not only preserved from Vicious Courses, and acquainted with Religious Exercises; but was early instructed in the Principles of Religion, and *Catachetical* Divinity, and the frequent Reading of *Scripture,* and other good books, and diligent attendance on Sermons (And whatever other Studies I followed, I was careful not to neglect this.) And became timely acquainted with *Systematick* and *Polemick* Theology. And had the repute of a good Proficient therein" (Wallis 1970, 30).

theory regards the ministerial office as part of a system whose authority is guaranteed by the word of God and is not restricted to the members of a single congregation. A proper Presbyterian minister is not
simply someone responsible for a specific congregation, but one who
has been granted his authority by Christ's universal church and can, at
least in principle, authoritatively exercise his ministerial function anywhere.

In arguing for his negative response to the proposed question Wallis
first explains that by the term "ministers of the Gospel" he understands "those of whom the preaching of the Gospel is demanded, ex
officio, by Christ" (*Mens Sobria* 136). Although the primitive church
of the New Testament included apostles and prophets among its ministers, Wallis assumes that in the contemporary context the class of ministers would include only "pastors and doctors" of the church. The
powers of interest here are those "which belong to the Gospel minister,
by virtue of the office he exercises, in so far as they uphold his public
function and distinguish him from a private Christian" (*Mens Sobria*
136). Such powers include "the preaching of the Gospel, the adminis-
tration of sacraments, the exercise of ecclesiastical censures, the ordination of ministers, and such things of this sort" (*Mens Sobria* 136–
37). Wallis remarks that some (i.e., the Independents) conceive of the
relationship between the minister and congregation as closely analogous to that between a shepherd and his flock—just as a shepherd is a
shepherd only in relation to a specific flock and cannot properly exercise power over the flock of another, so the ministerial powers must be
restricted to a specific congregation. He objects that, on the contrary,
"the relation of a minister of the Gospel to a particular church is neither adequate nor his primary relation, which is rather that which
he bears to Christ, whose minister he is" (*Mens Sobria* 140). As is
customary in such disputations, Wallis builds his case on an analysis
and interpretation of scriptural passages; he reads references to "the
church" or to "Christ's flock" as referring to a universal church that
cannot be identified with a specific congregation. Even in those cases
where relevant biblical passages clearly seem to intend a specific congregation, Wallis contends that such references

are made synecdochically, that is by the application of the name
of a whole to some of its parts. Just as Christ himself is said to
have preached unto the world, although he never, as far as we
know, preached outside of the land of the Israelites [*terra Israe-
litica*]. . . . And so, if some particular church is on occasion syn-

ecdochically called Christ's family, or the flock of Christ, or even the body of Christ, nevertheless when several are spoken of they are not called the flocks of Christ, or his families, or his bodies, or even his brides. Instead, when these terms are used for one or more or even all, they are understood to be used only to indicate one family, one body, one flock, one bride, and one edifice. (*Mens Sobria* 142–43)

The consequence of this interpretation is that "those titles by which ministers [of the Gospel] are everywhere distinguished do not signal a relationship to this particular church, but instead either to Christ himself, or also to the universal Church, or to the absolute authority of preaching the Gospel" (*Mens Sobria* 146). Wallis concludes his account of the ministry with the declaration that "it is manifest that the ministry of the Gospel is not only of one particular church, but rather the ministry of the whole universal Church; or rather, speaking absolutely, the minister of Christ is not ordained for the good of only one particular church, but for the good of the whole universal Church" (*Mens Sobria* 152).

Such a forthright attack on the fundamental tenets of Independency found no favor with Owen, and Oxford's vice-chancellor subsequently became embroiled in controversy with the Savilian professor of geometry. Daniel Cawdrey reports that "when the learned DOCTOR WALLIS, had brought to him as *Vicechancellor,* that Question to be defended *negativè* in the *vespers* of the publick Act at *Oxford,* 1654. '*An potestas Ministri evangelici, ad unius tantum Ecclesiae particularis membra extendatur,*' this Reverend Doctor [Owen] said thereupon that *Doctor Wallis* had brought him a challenge, adding, that if he did *dispute* upon that Question, he must dispute *ex animo*" (Cawdrey 1657, 129). It is interesting to observe that a year later (1655) Wallis dedicated his *Elenchus* to Owen, and it is natural to see this as an attempt to improve his standing with the vice-chancellor by making common cause against the notorious Hobbes.[4] Nevertheless, Wallis aroused the ire of Owen and his allies when he subsequently published his critique of Independency as part of the *Mens Sobria* in 1656, an action that Owen took to be a direct challenge to his own authority and to Independency in general.

---

4. Hobbes noticed this dedication and took the chance to denounce Wallis for attempting to "curry favor" with Cromwell's ministers, "as you did by Dedicating a Book to his Vice-Chancellor, *Owen*" (*MHC; EW* 4:416).

Henry Stubbe wrote to Hobbes in 1656 and mentioned the ongoing controversy between Wallis and Owen. According to his report,

[t]here is now a matter of greater diuertisement happened; wch is this: D:^{r} Wallis who is now putting out a most childish answer, (as they say) of 16 sheets in a small letter, to yor *[Six Lessons]*, hath put out some theses ag^t a branch of independency: D:^{r} Owen hath wrote a booke of late in behalfe of Independency, & our du Moulin another, & it is intended that ye Presbyterians shall be mated here in Oxon: Wallis's theses are triuiall, & so D:^{r} Owen hath desired mee to fall upon him; & to seize him, as I thinke fit, because hee did under hand abuse d^r Owen lately. D:^{r} Wallis hath mistooke ye question, & so I haue order to tell him of it, & to reflect upon ye synod at westminster, whereof he was Scribe. this will take mee up for some weekes, for I intend to spend as good Latine as I can upon him, & I shall allow him ye ouerplus of one page or 2, in reuenge of yo^u, but not reflecting upon yo^u nor his bookes agt you. I know not well what wee shall driue at, but I haue receiued orders to study church-gouernem^{t,} & a toleration & so to oppose Presbytery. (Stubbe to Hobbes, 7/17 October 1656; *CTH* 1:311–12)[5]

Stubbe evidently produced a draft of a book opposing Wallis's published thesis in which he attacked the Savilian professor for having "built ye whole fabrique upon Metaphors, of a flocke, family, body, &c. & alledged impertinent proofes of ye Apostles &c. wch are no more pertinent to us y^n all ye other promises. & unlese hee proue y^e [pre]sent ministry to bee Apostles, ye question about their commi[ssion] hath no more influence upon us y^n ye debate betwixt Sigonius and Gruchius about ye power of [y^e] Roman tribunes" (Stubbe to Hobbes, 25 October/4 November 1656; *CTH* 1:333).[6] The tract seems never to have

5. The reference to "du Moulin" is to Lewis du Moulin, son of the prominent French Huguenot Pierre du Moulin; he was appointed Camden professor of history in 1648 and supported the Independent cause with his *Paranesis ad aedificatores imperii in imperio*. Owen's defense of Independency is his treatise *Of Schisme,* which attempts to argue that the Independent model of church polity does not invite schism in the church. It is unclear what "under hand abuse" Owen may have suffered at Wallis's hands, but Stubbe reports in a letter to Hobbes from 25 October/4 November 1656 that "in the 42^d page D:^{r} W hath touched him to ye quicke" (*CTH* 1:334). This refers to Wallis's denunciation of those who intolerantly insist upon formalities, especially at a university.

6. Noel Malcolm reports that Stubbe's reference to "Sigonius and Gruchius" is to a long-running dispute between the scholars Nicolas de Grouchy and Carlo Sigonio over the role of Roman tribunes (*CTH* 1:336 n. 10).

been published. In fact, it was apparently too harsh for Owen to permit its printing, although Stubbe reports to Hobbes that "D:ʳ Owen hath approued of my piece agᵗ [Wallis's] *thesis,* with all yᵉ harsh language agt yᵉ Synod, I know not when it will bee printed" (Stubbe to Hobbes, 17/27 March 1657; *CTH* 1:456). Cawdrey, however, claims that "when Doctor *Wallis's Thesis* . . . was since printed, this Reverend Doctor did imploy, or at least encourage (an *Amanuensis* of his) Mr. Stubbs of Christ-Church (now advocate for Mr. Hobs) to write against it: Though indeed, when that work written, was found a *Scurrilous ridiculous piece* (for so I heare, he is since pleased to style it) he did not thinke it fit to be made publick, because (they were his own words) *'he would not have that cause suffer so much as to be defended by such a Penne'"* (Cawdrey 1658, 129–30).

Hobbes accepted Stubbe's invitation to attack Wallis by "instanceing at yᵉ absurdity of one argument, or yᵉ tendency yᵗ it hath to confusion, or yᵉ erecting a power beyond yᵉ Papall in jurisdiction," although he did not, as far as currently available evidence attests, cooperate with his suggestion to "penne a short letter censuring yᵉ tract," which would have appeared as part of Stubbe's published refutation of Wallis's *Mens Sobria* (Stubbe to Hobbes, 25 October/4 November 1656; *CTH* 1:337). Rather than attack Wallis in an anonymous letter appended to Stubbe's polemic, Hobbes incorporated a critique of the "Scottish Church-Politicks of Dr. Wallis" into the Στίγμαι of 1657.

The brief but substantive critique of Wallis's Presbyterianism in Στίγμαι is based on key elements in Hobbes's political philosophy, specifically his views on what constitutes a church and the relationship between ministerial and civil power. It should be clear that Presbyterianism is antithetical to Hobbes's principles because it asserts the autonomy of the church and its exemption from the authority of the civil sovereign. Independency, in contrast, has a Hobbesian warrant, since it holds (to use Wallis's words) "that each single Congregation, voluntarily agreeing to make themselves a *Church,* and chuse their own Officers, [are] of themselves *Independant,* and not accountable to any other *Ecclesiastical Government;* but only to the Civil Magistrate, as to the Publick Peace" (Wallis 1970, 34). Because maintenance of the public peace is the sine qua non of Hobbesian political theory, it can accommodate Independency fairly easily. Moreover, because the Independents made no claim to exercise quasi-judicial powers over the populace, they did not attempt to lessen the authority of the civil sovereign by claiming the right to regulate the spiritual affairs of the commonwealth.

Hobbes's account of the relationship between church and state is a product of his definition of a church, and his understanding of the concept of a church makes room for Independency within his system. In chapter 39 of *Leviathan* Hobbes considers various senses of the term *church*; it can be taken to apply to a building, an actual assembly of believers, all believers whether or not assembled together, and in other ways. But there is only one sense of the term *church* in which the church can be said to be a person and have actions attributed to it, or as Hobbes puts it "to pronounce, to command, to be obeyed, to make laws, or to doe any other action whatsoever" (*L* 3.39, 247; *EW* 3:459). This is the sense in which the church is an assembled congregation of believers called together by some lawful authority. Hobbes thus defines the church as *"A company of men professing Christian Religion, united in the person of one Soveraign; at whose command they ought to assemble, and without whose authority they ought not to assemble"* (*L* 3.39, 248; *EW* 3:459). Because an assembly not expressly permitted by the civil sovereign must count as unlawful, Hobbes concludes that

> there is on Earth, no such universall Church, as all Christians are bound to obey; because there is no power on Earth, to which all other Common-wealths are subject: There are Christians, in the Dominions of severall Princes and States; but every one of them is subject to that Common-wealth, whereof he is himself a member; and consequently, cannot be subject to the commands of any other Person. And therefore a Church, such as one as is capable to Command, to Judge, Absolve, Condemn, or do any other act, is the same thing with a Civil Common-wealth, consisting of Christian men; and is called a *Civill State,* for that the subjects of it are *Men;* and a *Church,* for that the subjects thereof are *Christians. Temporall* and *Spirituall* Government, are but two words brought into the world, to make men see double, and mistake their *Lawfull Soveraign.* (*L* 3.39, 248; *EW* 3:460)

Given this account of what constitutes a church, Hobbes has no patience for the claims of Presbyterianism. Indeed, he made no secret of his opinion that the Civil War was a Presbyterian-inspired conflict that had brought ruin to England because its people had been misled by ambitious doctors of divinity.[7]

7. Hobbes's analysis of the causes of the Civil War is set out at greatest length in *Behemoth* (*EW* 6:161–418). In particular, he lays the greatest blame for the "corruption" of the people on a specific group who acted against the king's interest, namely "ministers, as they called themselves, of Christ; and sometimes, in their sermons to the people,

Hobbes's Στίγμαι levels several specific charges against the doctrines in Wallis's thesis of 1654. The first of these is that Wallis's definition of the Christian ministry illegitimately separates the church from the state. We saw that Wallis defines ministers of the Gospel as "those of whom the preaching of the Gospel is demanded, ex officio, by Christ" (*Mens Sobria* 136). Hobbes responds to this definition by asking "what do you mean by saying preaching *ex Officio* is *enjoyned by Christ?* Are they Preachers *ex Officio,* and afterwards enjoyned to Preach? *Ex Officio* adds nothing to the definition; but a man may easily see your purpose to disjoyn your self from the State by inserting it" (Στίγμαι 3; *EW* 7:395). The root of Hobbes's objection here is, of course, that such a definition of a minister is founded upon the error of taking the present (Presbyterian) church for the Kingdom of God, from which it follows that "there ought to be some one Man, or Assembly, by whose mouth our Savior (now in heaven) speaketh, giveth law, and which representeth his Person to all Christians" (*L* 4.44, 335; *EW* 3:606). Since Wallis and other Presbyterians deny that the ordination or recognition of ministers of the Gospel is a matter subject to the jurisdiction of the civil sovereign, they in effect arrogate this part of sovereign authority to themselves.

A second, and more interesting, objection is the charge that Wallis's principles provide no way to determine who is a minister of the Gospel. Hobbes takes it for granted that Christ himself does not immediately speak to his ministers and invest them in their offices. The bestowing of ministerial office must therefore come through the authority of some human intermediary such as a bishop or another minister. But in such a case there is no way to verify that this person's authority is legitimate, since it is impossible to tell whether a person who claims authority to invest ministers in their offices was indeed granted such authority immediately from Christ. Furthermore, the attempt to trace the line of transmission of Christ's authority back to the time of the Apostles immediately runs into difficulty because going back "but a mater of sixscore years, you will find your Authority derived from the *Pope;* which words have a sound very unlike to the voices of the Laws of England" (Στίγμαι 3; *EW* 7:395). The tradition holds that the author-

---

God's ambassadors; pretending to have a right from God to govern every one his parish, and their assembly the whole nation" (*Behemoth* 1; *EW* 6:167). Later in the same work Hobbes complains that "Presbyterians are everywhere the same: they would fain be absolute governors of all they converse with, having nothing to plead for it, but that where they reign, it is God that reigns, and nowhere else" (*Behemoth* 4; *EW* 6:372–73).

ity to invest ministers of the Gospel in their offices arises from the successive "imposition of hands" back to the age of the Apostles, but Hobbes takes such an opinion as "too rude to be endured in a state that would live in peace" inasmuch as it separates ecclesiastical authority from that of the civil sovereign. Hobbes concludes "you can never prove you are a Minister, but by the Supream Authority of the Common-wealth" (Στίγμαι 3; EW 7:395).

Because the office of the ministry depends upon the authority of the civil sovereign, Hobbes reasons that the powers pertaining to that office likewise depend upon the will of the sovereign. In a passage guaranteed to anger Wallis, Hobbes remarks:

> You can have no other power then that which is limited in your Orders, nor that neither longer than [the sovereign] thinks fit. For if he give it you for the instruction of his subjects in their duty, he may take it from you again whensoever he shall see you instruct them with undutiful and seditious principles. And if the Soveraign power give me command (though without the ceremony of imposition of hands) to teach the Doctrine of my *Leviathan* in the Pulpit, why am I not if my doctrine and life be as good as yours, a Minister as well as you, and as publick a person as you are? (Στίγμαι 3; EW 7:396–97)

Hobbes thereby reaches a conclusion familiar from his political theory: the civil sovereign is God's representative on earth, and that those who claim an authority directly from God and contrary to that of the civil power are agents of the "Kingdome of Darknesse" who conspire to make citizens ignore their principal civic duty.

To Wallis's suggestion that the ministers of the Gospel should, on their own authority, institute public preaching of the Gospel at public gatherings such as market days, Hobbes replies that such a practice can only be part of a plan for lessening public support for the civil sovereign. Proponents of such a scheme, he writes, do not know "that many teachers unlesse they can agree better, do any thing else but prepare men for faction, nay, rather you know it well enough; but it conduces to your end upon the Market-dayes to dispose at once both Town and Country, under a false pretense of obedience to God, to a Neglecting of the Commandments of the Civil Soveraign, and make the Subject to be wholly ruled by your selves" (Στίγμαι 3; EW 7:397).

Wallis's response to these criticisms of his Presbyterian church politics was brief and abusive. He regarded Hobbes's insistence on the supreme authority of the civil sovereign as, in effect, a requirement that the doctrines of *Leviathan* be established as the official state religion. He writes:

> You aske now whether *Christ do injoin those to preach, whose office is so to do?* Yes, he doth. You take notice further, That I do not there prove that *myself am a Minister;* nor tell you *who ordained mee,* nor *by what authority they acted;* that *I say nothing of the Civill Magistrate:* of the *Right of Ordination;* or *Excommunication;* nor *whether I be fit to govern Nations, &c,* (because all this was nothing to the purpose:) In summe, That *I do not teach the Doctrine of your Leviathan,* (for if I did I were *a Minister ipso facto:*) And then tell me what *you have heard* and what *you hope,* and what *you think,* which I think is nothing to the purpose; and so my *Thesis* stands as it was. (*Dispunctio* 42–43)

The main charge here against Hobbes is that he places the doctrine of *Leviathan* in place of the Gospel. This accusation is closely related to the allegation that Hobbes is an enemy of all established religion and that his philosophy is fundamentally atheistic. Indeed, after the Restoration of the Anglican Church in 1660, Wallis sought to downplay the extent of his Presbyterian sympathies and tried to deflect Hobbes's criticisms by suggesting that his opponent had no religion at all. Thus, in 1662 he could write "I am to be told ever and anon, That I am a *Presbyterian:* (Not because, he Knows 'tis *True;* but, because he thinks 'tis a *Reproach.*) But I shall be so far from Reproaching him for his Religion, whatever it be, that I shall not so much as *Charge him to be of any*" (HHT 18).

By placing the civil sovereign in a position of supreme power, Hobbes in effect denies the claims of those who see the true religion as something independent of the judgment of the civil magistrate. In Hobbes's political theory, only the sovereign has the authority to declare texts canonical scriptures, to decide how they are to be interpreted, and to determine what forms of worship are permitted. To many, including Wallis, these doctrines amount to an outright denial of all religion because (they argued) Hobbes thereby seeks to replace the worship of God with the worship of the sovereign. The result was the charge that Hobbes was really an atheist, which is an accusation we can now proceed to evaluate.

7.1.2 The Allegation of Atheism

Both Wallis and Ward publicly accused Hobbes of being an atheist, notwithstanding the fact that in *Leviathan* and other works he offers proofs of the existence of God and makes no explicit avowal of skepticism about the existence of a deity. There is nothing novel in this accusation. As I mentioned in chapter 2, the publication of *Leviathan* unleashed a torrent of anti-Hobbes literature, and a standard theme in these polemics was that of Hobbes's alleged atheism. Mintz has described atheism as the central issue around which Hobbes's otherwise diverse opponents could unite: "It was atheism then which was at the heart of the controversy about Hobbes, the source of all the fears and seething indignation which Hobbes's thought inspired, the single charge which is most persistently made, and to which all other differences between Hobbes and his contemporaries can be reduced" (Mintz 1962, 45). Although it is going too far to claim that all of Hobbes's differences with his contemporaries can be "reduced" to the question of atheism, it is certainly true that the allegation of atheism plays a leading role in many attacks on Hobbes. I cannot treat all of the different authors who detected atheistic tendencies in Hobbes's work and will confine my attention to the writings of Wallis and Ward. As it happens, Wallis's charge of atheism is closely connected with Ward's, and these two together offer something of a joint critique of Hobbes's religious opinions. After examining the arguments used by Ward and Wallis to support the charge of atheism, and considering Hobbes's defense against the charge, I will argue that, on balance, the preponderance of the evidence shows that Hobbes probably was an atheist.

Before we begin this investigation, it is worthwhile remarking that *atheist* was a general term of abuse in the seventeenth century. Consequently, those who accused others of atheism did not always literally think that their opponents denied the existence of God. Rather, the epithet was often applied quite indiscriminately, so that its use may indicate only that the alleged atheist's religious views are in some way objectionable. Because the charge of atheism was made so frequently in seventeenth-century England, even though open profession of atheism was outlawed, it is difficult to tell how widespread atheistic views may have been in Hobbes's day.[8] The sense of the term *atheism* that I

8. See Aylmer 1978 for an attempt to gauge the extent of unbelief in seventeenth-century England.

take to be central to this question is the root sense in which atheism is the outright disbelief in God or gods. A. P. Martinich has drawn attention to free use of the term *atheism* in the seventeenth century and has reminded us that Hobbes's opponents' labeling him an atheist does not, by itself, go very far toward showing that he is an atheist in the strict or literal sense (Martinich 1992, 19–22). Nevertheless, Hobbes publicly defended a number of views that were widely regarded as underwriting, if not actually implying, atheism in the strict sense. Michael Hunter has remarked that in the context of seventeenth-century thought

> atheists were seen as people who denied the existence of God, either directly or by implication. It was axiomatic that unbelief would be sustained by views—usually materialistic—of a natural world that had originated without a beneficent creator and in which God's activity was limited or completely absent. But in addition, numerous other arguments were seen as part of the atheists' armory: a denial of the immortality of the soul and of any absolute morality; a skepticism about the text of the Bible, based either on its internal inconsistencies or on the supposed irreconcilability of Holy Writ with pagan history; or the opinion—which orthodox polemicists repeatedly tried to turn back on itself— that religion had first been introduced as "a *meer politick Contrivance*." (Hunter 1990, 441)

Hobbes, it should be noted, subscribed to all of these "atheistic" views. He championed materialism, rejected providential history, denied the immortality of the soul, described a state of nature devoid of the usual moral absolutes, challenged the authenticity of much of Scripture, and defined religion as "*Feare* of power invisible, feigned by the mind, or imagined from tales publiquely allowed" (*L* 1.6, 26; *EW* 3:45). As we will see, a great deal depends on whether Hobbes can maintain such views without being an "atheist by consequence," i.e., one who endorses principles that imply atheism even if he never explicitly declares disbelief in God. An atheist by consequence must contradict himself whenever he avows belief in God, and there are many apparent contradictions between Hobbes's principles and his professions of Christian belief. This makes it at least minimally plausible that Hobbes was sincerely committed to an atheistic worldview and intended his expressions of religious belief ironically or disingenuously. Whether we can ultimately maintain such an "ironist" reading of Hobbes's religious pronouncements will be taken up shortly, but we must first investigate the charge of atheism as presented by Ward and Wallis.

Ward was the first of the Savilian professors to accuse Hobbes of atheism, but he did so rather obliquely in the *Vindicae Academiarum* while engaged in the task of defending the universities against Hobbes's charge that in them "the study of Philosophy . . . hath otherwise no place, then as a handmaid to the Romane Religion" (*L* 4.46, 370; *EW* 3:670). Ward took umbrage at this suggestion, remarking that the universities had shown themselves to be staunch opponents of Catholicism and adding that "it is said that Mr *Hobbs* is no otherwise an enemy to it, save only, as it hath the name of a Religion" ([Ward] 1654, 58). This portrayal of Hobbes as an enemy to all religion is an evident sharpening of Ward's earlier negative comments on Hobbes's philosophical enterprise. In his *Philosophicall Essay* of 1652 Ward had attacked Hobbes's materialism, contending that it undercut the principal grounds for proving the immortality of the soul and complaining that Hobbes gives no adequate reasons to rule out incorporeal substance. But in the *Vindicae Academiarum,* he has obviously raised the stakes in the dispute by suggesting that the Hobbesian system is fundamentally hostile to religion.

There is no doubt that the doctrines of *Leviathan* were the basis for Hobbes's reputation as an atheist, but some of the pronouncements in *De Corpore* also gave Wallis and Ward occasion to rebuke Hobbes for his religious opinions. In the dedicatory epistle to *De Corpore* Hobbes contrasts (his own) true philosophy with the absurdities put forward under the name *philosophy* in ancient Greece. He tells us that the mingling of Christian scripture with ancient philosophy led to disaster when, in addition to some "harmless" opinions of Plato, various "foolish and false" Aristotelian principles were endorsed by defenders of the Christian faith. This was the grave error of the early Church fathers, who by "bringing in the enemies, betrayed unto them the citadel of the Christian faith" (*DCo* epistle; *OL* 1:sig. h5v). The result is that "from that time, instead of the worship of God [ϑεοσεβεία], we have had this thing called school divinity *[scholasticam dictam ϑεολογίαν]*, walking on one firm foot, which is the Sacred Scripture, but with another rotten foot, which the Apostle Paul called vain, and might have called pernicious philosophy; for it has provoked innumerable controversies in the Christian world concerning religion, and from these controversies, wars" (*DCo* epistle; *OL* 1:sig. h5v).

Hobbes compares this hybrid school divinity to Empusa, a female specter in ancient Greek mythology who was supposed to have been sent by the goddess Hecate to frighten travelers or warn of impending evil. Empusa had the power to change forms but most frequently appeared with one leg of brass and the other of an ass—which Hobbes's compares with the "firm foot" of Scripture and the "rotten foot" of school divinity. Hobbes holds that no better exorcism can be found for the Empusa of school divinity than if "the rules of religion, that is, the rules of honoring and worshiping God, which are to be taken from the laws, are distinguished from the rules of philosophy, that is, the opinions of private men; and those things due to religion are granted to the Sacred Scripture, and those due to philosophy are granted to natural reason" (*DCo* epistle; *OL* 1:sig. h5v–h6r). Given Hobbes's absolutist theory of sovereignty, the task of honoring God therefore becomes no more than that of obeying the sovereign's decrees regarding worship. Hobbes further holds that once he has established the principles of true philosophy, the Empusa of school divinity will be dispatched, not by some sort of battle, but rather by letting in the light and frightening her away. These declarations against school divinity naturally annoyed both Ward and Wallis, who (as doctors of divinity) took them as attacks on Christianity itself.

In the *Elenchus* Wallis followed Ward's lead and imputed a contempt for all things religious and scriptural to Hobbes. In the epistle addressed to Owen he remarks that "with what arrogance *[ὑπερηφα-νεία]* and imperiousness *[pro imperio satis]* he tramples on all things both human and divine, writing terrible and horrible things of God, of sin, of the Holy Scripture, of all incorporeal substances in general, of the immortal soul of man, and of other weighty points of religion . . . is more to be lamented than doubted" (*Elenchus* epistle sig. A2). Wallis later elaborates on Ward's accusations in the context of commenting on a curious declaration in chapter 18 of *De Corpore,* where Hobbes remarks on the supposed success of his rectification of parabolic arcs. Hobbes there claims that "there are those who think that there is an equality between a right line and curve, but that it cannot now be found; now, they say, after the fall of Adam, without the special aid of divine grace" (Hobbes 1655, 161). Wallis takes these references to divine grace and Adam's fall as expressing Hobbes's disbelief in Scripture, and accuses him of making light of serious matters: "I realize how sceptically and sarcastically *[σαρκαστικῶς]* these things are said of divine grace. Indeed you boast that by your efforts you have found what others were expecting could not be found without 'the special aid of divine grace'" (*Elenchus* 89). Wallis adds that no right-thinking person could seriously doubt that the intellectual powers of man were diminished after Adam's fall, and that there is all the more reason to think that divine grace is necessary for any successful inquiry into nature. He concludes that Hobbes is implicitly denying the role of divine grace in the acquisition of knowledge, so that Hobbes's boasts of his abilities amount to the atheistic tenet that knowledge of the world can be obtained without God.[9]

Not content with objecting to Hobbes's comments on the role of divine grace in mathematics, Wallis launched an attack on the atheistic tendencies implicit in the political theory and metaphysics of the Malmesburian philosopher. The essentials of this charge are contained in the following long passage:

> But perhaps you take the whole story of Adam's fall for a fable, and smile ironically at its introduction. Which indeed need not to be marveled at in you, since you want that "Religion, that is, the rules of honoring and worshiping God are to be taken from the laws," and you openly hope that "by exorcism this Empusa" (the Christian religion) "is frightened off and driven away." Nor is this discordant with what you have elsewhere concerning the origin of the world, namely that it is not by the force of argu-

---

9. Hobbes's rebuttal is to emphasize that it was allegedly *special* grace that was necessary to find the equality between a line and curve, whereas he claims only to have discovered it by ordinary means. This, he insists, is consistent with holding that there is some need of divine assistance in developing any science, so that Wallis, "taking no notice of the word *Speciall,* would have men think I held, that humane Sciences might be acquired without any help of God" (*SL* 5; *EW* 7:320). Hobbes cites the Jesuit Antonius Lalovera as holding that special grace is needed for this part of mathematics, because in the prolegomena to his treatise *Quadratura circuli et hyperbolae* Lalovera declares that "although the quadrature of the circle is possible in its nature, theologians have nevertheless thought fit to inquire whether in these days, that is after the fall of Adam, man can attain knowledge of this matter without the aid of special divine grace. And they have declared that this truth is so shrouded in darkness that none can see it unless the clouds of ignorance flowing from the prevarication of our first parents be dissipated by a ray of divine light, which opinion I judge to be most true" (Lalovera 1651, 12–13). Hobbes finds that "he (supposing he had found that Quadrature) would have us believe it was not by the ordinary and Naturall help of God (whereby one man reasoneth, judgeth and remembereth better then another) but by a Special (which must be Super[na]turall) help of God, that he hath given to him of the order of *Jesus* above others that have attempted the same in vain. Insinuating thereby, as handsomely as he could, a Speciall love of God towards the *Jesuites*" (*SL* 5; *EW* 7:320). This reply may permit Hobbes to avoid Wallis's specific allegation, but it does little to diminish the appearance of arrogance in his pronouncements on his quadratures, nor does it unequivocally commit Hobbes to the view that divine assistance (albeit not special grace) is necessary for science. Hobbes removed the offending passage in the 1656 English version of *De Corpore* and it did not appear in the 1668 Amsterdam *Opera*.

ments or reasons that this matter is to be resolved, but that it is to be determined by the sovereign *[magistratus]:* as if this were not sufficiently agreed in the Holy Scripture *[codex],* but should depend entirely on the suffrage of sovereigns whether or not the world ever had a beginning. But to what end do I mention Holy Scripture? This Scripture you may happen to observe in religion, and the Christian religion itself: that is, that you concede as much authority to it as is granted by the civil sovereign, and otherwise you wish all such authority to be removed, as if it belonged to the sovereign, not only whether or not the world once had a beginning, but also whether or not the sacred Bible is the word of God. But you appear to be no more concerned with God Himself than with the divine word, since you appear, I think, easily ready to set Him aside. You take it to be ridiculous, and what could never be conceived by any imagination, that anything should ever exist that is not a body; and you also hold (as in the end of this treatise) that all "incorporeal substances" are to be dismissed as the "inane words of the Scholastics." Who does not see that thereby you not only deny (and not just in words *[tantum non ῥητῶς]*) angels and immortal souls, but the great and good God himself; and if you were not wary of the laws (which to you is the highest "rule of honoring and worshiping God") you would profess this openly. And however much you may mention God and the Holy Scriptures now and again (although I do not recall your mentioning the immortal soul), it is nevertheless to be doubted whether you do this ironically and for the sake of appearance rather than seriously and from conviction. (*Elenchus* 89–90)

These charges summarize essentially the whole case against Hobbes's religious opinions and it is worthwhile to examine them somewhat more deeply.

The accusation that Hobbes makes the question of the eternity of the world depend upon the will of the sovereign picks up on declarations in part 4 of *De Corpore* to the effect that, because man's knowledge is confined to what is finite, there can be no demonstrative knowledge of whether the world is eternal or whether it had a beginning.[10]

---

10. Hobbes reasons:

Whatever we men know we take from our phantasms; but there is no phantasm of an infinite, whether of magnitude or time; nor can a man or any other thing have any conception of the infinite, beyond that it is infinite; nor if someone

Hobbes concludes "questions of the magnitude and origin of the world are therefore not to be settled by philosophers, but by those who are lawfully responsible for regulating the worship of God" (*DCo* 4.26.1; *OL* 1:336). This is actually Hobbes's reply to an argument in Ward's *Philosophical Essay*. Ward had undertaken to prove that the past must be finite because the supposition of an infinite past yields a contradiction (specifically, the number of generations from the beginning of the world to Abraham is equal to that from the beginning of the world to his great-grandson Joseph).[11] Hobbes reconstructed Ward's reasoning in *De Corpore* (albeit without naming him) and compared it to that of someone who concludes that there must be as many even numbers as natural numbers because there are infinitely many of both; he then asked whether "those who thus do away with the eternity of the world do not by the same argument do away with the eternity of the creator of the world?" (*DCo* 4.26.1; *OL* 1:337). Although Wallis may be somewhat wide of the mark when he characterizes Hobbes as holding that the question of the world's beginning is to be resolved by the sovereign without reference to the Scriptures, it is evident that Hobbes's comments constitute a skeptical response to any "cosmological" argument for the existence of God, in which God is identified as the first cause of the world.[12]

---

should ascend by right reasoning from any effect to its immediate cause, and thence to a further cause, and so on perpetually, yet he will not be able to proceed eternally, but wearied will at some point give up, not knowing whether he could have gone further. Nor will anything absurd follow, whether the world is agreed to be finite or infinite; since whether the creator of the world had determined it to be one way or the other, the same things that now appear could still appear. (*DCo* 4.26.1; *OL* 1:336)

11. The argument appears at Ward 1652, 14–17. See Probst 1993 for a discussion of the argument and Hobbes's response.

12. This particular point will be of interest shortly, when we consider Hobbes's supposed proofs for the existence of God. It is also worth noting that in his reply to Wallis on this point, Hobbes steers away from the question whether God's existence can be known and back to questions of church government. "Lastly, what an absurd question it is to ask me whether it be in the Power of the Magistrate, whether the world be eternall or not? It were fit you knew tis in the Power of the Supreme Magistrate to make a Law for the punishment of them that shall pronounce publiquely of that question any thing contrary to what the Law hath once pronounced. The truth is, you are content that the Papall power be cut off, and declaimed against as much as any man will; but the Ecclesiasticall Power which of late was aimed at by the Clergy here, being a part thereof, every violence done to the Papall Power is sensible to them yet; like that which I have heard say of a man, whose leg being cut off for the prevention of a Gangrene that began in his Toe, would nevertheless complain of a pain in his Toe, when his leg was cut off" (*SL* 6; *EW* 7:352).

When Wallis refers to Hobbes's denial of God, angels, and immortal souls by rejecting the doctrine of immaterial substances, he has in mind some of the closing reflections of *De Corpore,* where (after duly acknowledging that natural philosophy must depend upon hypotheses, and is therefore not wholly demonstrable) Hobbes announces that

> nevertheless, since I have assumed no hypothesis that is not possible and easily comprehended and I have reasoned legitimately from these assumptions, I have demonstrated that they may be [the true causes of the phenomena], which is the end of physical contemplation. If, having assumed other hypotheses, someone else should demonstrate the same or greater things, we shall owe him more thanks than I judge due to myself, provided that the hypotheses employed are conceivable. For as for those who say that anything is moved or produced by itself, by species, by power, by substantial form, by incorporeal substance, by instinct, by antiperistasis, by antipathy, by sympathy, by occult quality, and the other inane words of the Scholastics, these are all said to no purpose. (*DCo* 4.30.15; *OL* 1:431)

The picture here of a physics that proceeds from purely materialistic principles, unencumbered by reference to God or immaterial beings, fits perfectly the seventeenth-century image of the scientific atheist. It is therefore no great wonder that Wallis should take these comments as reflecting Hobbes's irreligious outlook.

Publication of the *Elenchus* was hardly the end of the questions regarding Hobbes's religion. Ward pursued his own interpretation of the atheistic tendencies in Hobbesian philosophy, and Hobbes took pains to deny the allegations raised by both Wallis and Ward; the result is that there were several exchanges over the question of Hobbes's religion and his sincerity in professing Christianity. Ward's most extensive criticism of Hobbes's religion appeared in the sixth and final section of his *In Thomae Hobbii Philosophiam Exercitatio Epistolica* and follows the lines of attack used by Wallis. Ward set himself the task of inquiring whether Hobbes's theories are friendly or hostile to Christianity (*Exercitatio* 313–22) and then of addressing the broader issue of Hobbes's attitude toward religion in general (*Exercitatio* 332–40). He naturally concluded that Hobbesian political and metaphysical doctrines are inimical to Christianity in particular and hostile to religion in general. The details of this critique can be left aside, but a brief summary of Ward's arguments is useful for assessing the evidence for the charge that Hobbes was an atheist.

Ward bases his claim for the hostility of Hobbesian doctrine to Christianity on a specific understanding of what counts as Christian doctrine. True to his Anglican sympathies, the future bishop of Salisbury declares that "he is hostile to the Christian religion who perverts its rule, or who diminishes or destroys its authority, who disorders the churches of Christians, or who strives to root out the spiritual exercise of this religion from the souls of people. And all of these have a common and reciprocal force [ἐνεργεία], so where one of them is found to exist the others follow" (*Exercitatio* 317). Ward naturally found Hobbes guilty of all these varieties of anti-Christian activity. He was particularly concerned to argue that the biblical exegesis contained in *Leviathan* undermines the authority of the sacred Scriptures upon which Christianity is based and whose reliability Ward himself had defended in the third part of his *Philosophical Essay* of 1652.[13]

Furthermore, Ward sees anti-Christian tendencies in Hobbes's requirement that the civil sovereign be entrusted with the authority to determine which books of the Bible are canonical. The Hobbesian model of church polity amounts, in Ward's estimation, to the intrusion of the civil authority into the realm of the spiritual, with the result that the sovereign (rather than God) becomes the object of worship. "Religion," Ward declares "is a sacred and serious thing, belonging to the innermost reaches of souls; nor can it in any way consist with this opinion, which holds that religion should take its origin from the will of some man and from a civil institution" (*Exercitatio* 319). Hobbes's insistence on the absolute conformity of subjects to the religious dictates of the sovereign, whatever they might be, confirms Ward's opinion that the Malmesburian philosopher is fundamentally hostile to the Christian religion. Hobbes held that a subject is obliged to deny Christ if such denial should be commanded by his lawful sovereign, since "Profession with the tongue is but an externall thing, and no more than any other gesture whereby we signifie our obedience" (*L* 3.42, 271; *EW* 3:493). Hobbes even claims scriptural authority for this opinion, noting that in the Old Testament (2 Kings 5:17–19) the prophet Elisha allowed Naaman the Syrian to bow before the idol of Rimmon even though in his heart he accepts no other God than that of Israel; by such outward observance of a false god, says Hobbes, Naaman "de-

13. Probst 1993, 272, has a brief summary of this part of Ward's project, whose fundamental point is that "concerning the truth of those writings [i.e., the Scriptures] we have 1. The same Arguments, or as great, as for the truth of any writings in the world. Nay 2. We have arguments (Morall Arguments) to evince the truth of them such as no other writings can pretend to" (Ward 1652, 90).

nyed the true God in effect, as much as if he had done it with his lips" (*L* 3.42, 271; *EW* 3:493). Ward responded to all of this with apoplectic fury, likening Hobbes's theory to the outright denial of Christ and the martyrs, and assuring the reader that the only sane reaction to this theory was to lament that such odious nonsense had been printed.

The Hobbesian approach to Scripture left Ward with his "limbs numb with horror and indignation" (*Exercitatio* 324). Because *Leviathan* contains an extended argument to show that there is no scriptural authority for opinions that tend to diminish the power of the civil sovereign, Ward reads Hobbes as implying that neither Christ nor the Apostles could have done or demanded of their followers anything contrary to the will of the sovereign. In particular, Hobbes seems left with the implication that the saints and martyrs of the early church behaved in an un-Christian manner to the extent that their actions conflicted with the dictates of their respective sovereigns. Ward also takes issue with chapters 33 ("Of the Number, Antiquity, Scope, Authority, and Interpreters of the Books of Holy SCRIPTURE"), 36 ("Of the WORD of GOD, and of PROPHETS"), and 37 ("Of MIRACLES, and their Use") because he sees them as denying the authenticity and veracity of the Scriptures.

Like Wallis, Ward also detected atheistic tendencies in Hobbes's materialism. This is hardly an unexpected result because, as we have seen, materialism and atheism were closely identified in the seventeenth century. Hobbes's efforts to reconcile his materialism with Christian doctrine in *Leviathan* were unpersuasive to Ward, who declared that "nothing is more monstrous, nothing more unbecoming to a sane man than to wish to intrude this carnal and quasi-Muslim theory in place of Christianity" (*Exercitatio* 329–30). Ward even notes that, by denying all corporeal attributes (such as figure, motion, and place) to God while maintaining that the only substance is corporeal, Hobbes himself effectively makes it impossible to understand how God can exist (*Exercitatio* 340).

Hobbes's response to all such charges was a vehement denial. In *De Corpore,* for example, he attacks Ward ("Vindex") for suggesting in the *Vindicae Academiarum* that anything in the philosophy of *Leviathan* is contrary to religion.[14] Whenever pressed, Hobbes always

14. In particular, Hobbes denounces Vindex for having publicly charged him with being "irascible, a plagiarist, and an enemy to religion," on no further evidence than having heard rumors to this effect, which Hobbes considers as having been done "I should not say stupidly, but (although there is no vice without stupidity) rather with heinous wickedness [*scelerate*]" (Hobbes 1655, 174).

claimed scriptural warrant for his religious doctrines, and he accused both Ward and Wallis of opposing them for purely selfish motives, i.e., in the interest of increasing their own power as doctors of divinity at the expense of the civil sovereign.[15] In the *Six Lessons,* he addressed both of the Savilian professors and cautioned them to "[t]ake heed of calling them all Atheists that have read and approved my *Leviathan.* Do you think I can be an Atheist and not know it? Or knowing it durst have offered my Atheism to the Press? Or do you think him an Atheist, or a contemner of the Holy Scripture, that sayeth nothing of the Deity, but what he proveth by the Scripture?" (*SL* 6; *EW* 7:350). In *Mr. Hobbes Considered* of 1662, he went on to explain, referring to himself in the third person, that his conception of a material God is in no way contrary to Scripture:

It is by all Christians confest, that God is *incomprehensible;* that is to say, that there is nothing can arise in our fancy from the naming of him, to resemble him either in *shape, colour, stature,* or *nature;* there is no Idea of him; he is like nothing that we can think on. What then ought we to say of him? What Attributes are to be given him, not speaking otherwise than we think, nor otherwise than is fit, by those who mean to honour him? None but such as Mr. *Hobbes* hath set down, namely, expressions of reverence, such as are in use amongst men for signs of Honour, and consequently signifie *Goodness, Greatness,* and *Happiness.* . . . This is the Doctrine that Mr. *Hobbes* hath written, both in his *Leviathan,* and in his Book *de Cive,* and when occasion serves, maintains. What kind of Attribute I pray you is *immaterial,* or *incorporeal substance?* Where do you find it in the Scripture? Whence came it hither, but from *Plato* and *Aristotle* Heathens, who mistook those thin Inhabitants of the brain they see in sleep, for so many *incorporeal* men; and yet allowed them motion, which is proper only to things *corporeal?* Do you think it

15. In one illustrative passage, Hobbes claims that his "making the King Judge of Doctrines to be preach'd or published, hath offended you both; so has also his Attributing to the Civil Sovereign all Power *Sacerdotal.*" Further, he concludes that both Wallis and Ward "were angry also for [Hobbes's] blaming the Scholastical Philosophers," by rejecting many nonsensical notions common to school divinity and "*for detecting, further than you thought fit, the fraud of the Roman Clergy.* Your dislike of his Divinity, was the least cause of your calling him Atheist" (*MHC; EW* 4:434–35). The point here is Hobbes's familiar claim that school divines have attempted to mask their ignorance by the fraudulent use of meaningless terms, all with the intention of usurping the powers of the civil authority.

an honour to God to be one of these? And would you learn
Christianity from *Plato* and *Aristotle?* (*MHC; EW* 4:426–27)

The success of Hobbes's defense against any charge of atheism therefore depends in large measure upon the extent to which his interpretation of Scripture can be taken seriously. But, as we will see, this is a
matter of controversy.

As a further part of his defense against the charge of atheism,
Hobbes even tried to counterattack by turning the allegation of against
Wallis and Ward. He suggested (perhaps ironically) that the Savilian
professors were so ready to accuse others of closet atheism because
they themselves have grave doubts about the existence of God. As a
result, he asks whether "finding your doubts of the Deity more frequent than other men do, you are thereby the apter to fall upon that
kinde of reproach?" (*SL* 6; *EW* 7:353). Such a strategy had little effect
and need not be an object of our attention.

It should be evident that in claiming that the doctors of divinity
attack him for purely selfish motives Hobbes voiced a profound anticlericalism, or general hostility to the clergy. Although anticlerical atti-
tudes are certainly common among atheists, atheism and anticlericalism are conceptually distinct and it would be a mistake to conclude
that Hobbes's expressions of anticlerical sentiment in themselves constitute strong evidence of atheism. In Hobbes's day Protestant radicals
(who were certainly not atheists) frequently voiced anticlerical atti-
tudes, although these were typically based in the belief that the established church had departed from its true mission and its clergy had
become corrupted by various "ungodly" influences. Hobbes's anticlericalism differs from that of such radicals and derives from his political
theory, more specifically from his concern that the legitimate powers
of the sovereign may be undermined by ambitious doctors of divinity.
In any event, we cannot conclude that Hobbes was an atheist simply
from the fact that he was hostile to the clergy.

Thus far the textual basis for any imputation of atheism to Hobbes
is hardly overwhelming. He was certainly not the proponent of mainstream religious views, and his detailed and highly controversial read-
ings of Scripture were not the sort of thing one generally finds in the
theological literature of the period. It is evident that Hobbes rejected
much in the standard interpretation of the Christian Scriptures, but it
is not immediately obvious that he was attempting to signal his disbelief in the whole theological enterprise. When Hobbes reads such key
scriptural terms as *angel* or *spirit* as denoting bodies, he may genuinely

be trying to cast off the baleful influence of Greek philosophy and return Christian teaching to its roots in a decidedly materialistic worldview. Indeed, he can happily quote the early Church father Tertullian in defense of his materialism.[16] Similarly, in his elaborate attempt to harmonize his political theory with the scriptures by reducing the whole of Christian duty to obedience to the lawful sovereign, Hobbes may very well be engaged in an honest effort to redescribe Christian obligations in a way that fosters civil peace.[17] Further, Hobbes's employment of the cosmological argument to show the existence of God as "first cause" (*EL* 1.11.2, *EW* 4:59–60; and *L* 1.11, 51, *EW* 3:92–93) may be sincere endeavors to ground a rational belief in a benevolent creator, while his emphasis on the fact that we have no idea of God may be an expression of the orthodox view that God's nature is incomprehensible and utterly unlike anything we can grasp with the understanding. Finally, his detailed critical analysis of Scripture may be a genuine try to clarify difficulties in the received texts without intending to undermine their authority.

Then, again, things may not be so simple. Hobbes's project can be seen, and was seen by many of his contemporaries, as subversive of all religious belief, and particularly Christianity. We have already seen that Ward and Wallis had no doubts about the atheistic overtones in Hobbes's doctrine, but they were hardly alone in reading Hobbes this way. Literally dozens of Hobbes's contemporaries took him to be an atheist, and it is not difficult to see why: what Hobbes says on matters of religion leaves very little of established doctrine standing, and it is precisely for this reason that his religious writings were so controversial. In light of such facts, it is hardly a great interpretive leap to see the Malmesburian sage as engaged in a sly campaign to discredit the basis for any religious belief.

Hobbes's excursions into scriptural exegesis, for instance, have the effect of overturning much that had been accepted by the tradition. He argues that Moses could not have been the author of the Pentateuch

16. In *Mr. Hobbes Considered,* he rebukes Wallis on this point with the remark that "[t]here has hitherto appeared in Mr. *Hobbes* his Doctrine, no signe of Atheism; and whatsoever can be inferr'd from the denying of *Incorporeal Substances,* makes *Tertullian,* one of the ancientest of the Fathers, and most of the Doctors of the Greek Church, as much Atheists as he. For *Tertullian,* in his Treatise *De Carne Christi,* says plainly, *Omne quod est, corpus est sui generis. Nihil est incorporale, nisi quod non est.* That is to say, *Whatsoever is any thing, is a body of its kind. Nothing is Incorporeal, but that which has no Being*" (*EW* 4:429).

17. This is precisely how Lloyd 1992 reads Hobbes's political theory, although without attempting to characterize his doctrines as particularly orthodox.

(*L* 3.33, 200; *EW* 3:368), and that the books of the Old Testament were for the most part assembled long after the events they describe. Perhaps more tellingly, his account of the doctrine of the Trinity yields the remarkable result that Moses, Jesus, and the Apostles (together with their successors) are the three coequal persons of the Trinity. The doctrine depends upon Hobbes's understanding of the term *person*, which he defines as someone *"whose words or actions are considered, either as his own, or as representing the words and actions of another man, or of any other thing to whom they are attributed, whether Truly or by Fiction"* (*L* 1.16, 80; *EW* 3:147). God is therefore one person when represented by Moses, a distinct person when represented by Jesus, and still another when represented by the Apostles or their successors.[18] In the words of Edwin M. Curley, this theory "has certain disadvantages" when considered from the point of view of orthodox Christianity: it makes Moses, Jesus, and the Apostles coequal while denying the eternity of the three persons of the Trinity, and it hardly seems to restrict the trinity to just three members (Curley 1996a, 266).

Hobbes's accounts of prophecy and revelation are, if anything, more problematic for the Christian cause than his idiosyncratic theory of the Trinity. Someone skeptical of revealed religion typically argues that there can be no reliable criterion to determine whether God has communicated directly with any individual who claims to have had such communication. Hobbes explicitly endorses skepticism on this point in chapter 32 of *Leviathan*, when he observes that "How God speaketh to a man immediately, may be understood by those well enough, to whom he hath so spoken; but how the same should be understood by another, is hard, if not impossible to know. For if a man pretend to me, that God hath spoken to him supernaturally, and immediately, and I make doubt of it, I cannot easily perceive what argument he can produce to oblige me to beleeve it" (*L* 3.32, 196; *EW* 3:361). He then goes on to argue that such apparent revelatory vehicles as

18. Hobbes declares: "For as Moses, and the High Priests were Gods Representative in the Old Testament; and our Saviour himselfe, as a man, during his abode on earth: So the Holy Ghost, that is to say, the Apostles and their successors, in the Office of Preaching and Teaching, that had received the Holy Spirit, have Represented him ever since" (*L* 3.42, 267–68; *EW* 3:486–87). Martinich considers this part of Hobbes's philosophy to be a sincere attempt to make sense out of a doctrine that is beset with difficulties. He concludes that "[d]ue to the logical difficulties that the doctrine of the Trinity presents, almost any theory will be either inconsistent or heretical. What Hobbes has offered is a sophisticated attempt to render the doctrine both consistent and orthodox, and nothing more could be asked of either a philosopher or dogmatic theologian" (1992, 207–8).

dreams, visions, voices, or inspiration all fail to provide reliable evidence of actual revelation. He concludes that "though God Almighty can speak to a man, by Dreams, Visions, Voice, and Inspiration; yet he obliges no man to beleeve he hath done so to him that pretends it; who (being a man) may erre, and (which is more) may lie" (*L* 3.32, 196; *EW* 3:362). Such declarations plainly imply that nobody has ever had sufficient reason to believe claims of another's supernatural contact with God, and they should therefore apply as much to Moses, Jesus, or the Apostles as to any contemporary enthusiast. However, this skepticism is suddenly placed in abeyance in chapter 36 of *Leviathan,* when Hobbes discusses the Scriptures and their many claims of God's speaking to his prophets. Hobbes admits that it is a difficult matter to understand how God (who presumably lacks the usual apparatus of speech) can be said literally to speak to his prophets, and concludes that the prophets of the Old Testament must have received God's word in dreams and visions (*L* 3.36, 226–27; *EW* 3:417). Moses and the supreme prophets of Israel present a special case, because God is supposed to have spoken to them in a more familiar fashion. Hobbes determines that "in what manner God spake to those Soveraign Prophets of the Old Testament, whose office it was to enquire of him, as it is not declared, so also is it not intelligible, otherwise than by a voyce" (*L* 3.36, 229; *EW* 3:420). Prophecies in the New Testament are handled more easily, since at that time "there was no Soveraign Prophet but our Saviour; who was both God that spake, and the Prophet to whom he spake" (*L* 3.36, 229; *EW* 3:420). It strikes me as obvious that Hobbes has no warrant for this special treatment of the prophetic declarations recorded in the Scriptures. To put the matter bluntly, Hobbes's doctrines cannot be consistently maintained by anyone who takes revealed religion seriously. If claims of revelation based on dreams, visions, and inspiration are unreliable, and if there is no intelligible sense in which God can literally speak to man, then Hobbes's own principles demand that the extraordinary claims made by the Old Testament prophets, Moses, and Jesus himself should not be taken seriously today. Moreover, such claims should not have been so taken when they were first propounded centuries ago.

Even Hobbes's proofs of the existence of God can be seen as undermining the cause of religion. The sort of argument he seems to favor involves reasoning back to a "first cause." Since the investigation of causes "must come to this thought at last, that there is some cause, whereof there is no former cause, but is eternall; which is it men call God" (*L* 1.11, 51; *EW* 3:92). But, read in the context of his claim that there is no way to know whether the world is eternal, this sort of argument seems at best inept, or perhaps even an ironic parody of traditional arguments for God's existence.[19] Finally, when he insists (as he does in many places) that we can have no idea of God, or that God is utterly unlike anything we can understand or reason about, Hobbes might well be inviting the reader to conclude that the concept of God is another senseless notion imported from the vain philosophy of the schools.

Edwin M. Curley has made the most extensive case for reading Hobbes as atheist, and I find myself very much in sympathy with his argumentation.[20] Curley bases part of his case on the testimony of Aubrey, who reports that upon reading Spinoza's *Tractatus Theologico-Politicus,* Hobbes declared that Spinoza "had outthrown him a bar's length, for he durst not write so boldly" (Aubrey 1898, 1:357). This declaration by itself is hardly sufficient to establish Hobbes's atheism, although it suggests that Hobbes was attracted to the evident skepticism about Scripture and its interpretation that characterizes Spinoza's *Tractatus.* Elsewhere Aubrey reports such anecdotes as the occasion "[w]hen Mr T. Hobbes was sick in France, the divines came to him, and tormented him (both Roman Catholic, Church of England and Geneva). Said he to them 'Let me alone, or else I will detect all your cheats from Aaron to yourselves!' I think I have heard him speak something to this purpose" (Aubrey 1898 1:160). These anecdotes are admittedly balanced by others where Aubrey reports that Hobbes's "writings and vertuous life testifie against" his reputation for atheism, while the philosopher himself is reported to have expressed a prefer-

---

19. As we saw above, Wallis was prepared to draw a conclusion much like this, when he complained that what Hobbes had written in *De Corpore* made the eternity of the world (and hence the principal argument for God's existence) depend upon the will of the sovereign. Wallis continued this interpretation in *HHT,* when he insinuated that *Leviathan* had failed to win the approval of the reading public "Unless with such, as thought it *a piece of Wit* to pretend to *Atheism. . . .* For, one while they find him affirming, That, beside the Creation of the World, there is no Argument to prove a Deity; Another while, That it cannot be evinced by any Argument, that the World had a beginning; and, That, whether it had or no, is to be decided not by Argument, but by the Magistrates Authority" (*HHT* 6). Then there is the further difficulty that Hobbes's *Anti-White* (probably written in late 1642) declares that "those who declare that they will show that God exists . . . act unphilosophically" (Hobbes 1976, 305). This pronouncement comes from the same author who in the *Elements of Law* (1640) and *Leviathan* (1651) claims to have philosophically convincing arguments for the existence of God. At the very least, Hobbes seems to have been ambivalent about the prospects for rational arguments for a deity.

20. See Curley 1992, which is supplemented by Curley 1996a and 1996b.

ence for the Church of England. Nevertheless, Aubrey's testimony has at least some force in making the case for Hobbes's religious skepticism, if not outright atheism.

Curley's complex and intriguing reading need not be reproduced in detail here, but the question of ironic intentions (if any) in Hobbes's religious writings does need to be taken up. The heart of his interpretation consists of five theses, namely:

> 1) that if we compare Spinoza's Theological-Political Treatise with Hobbes's theological-political treatise, i.e., with his Leviathan, on a variety of topics which they both discuss in the theological portions of their works—specifically on the topics of prophecy, miracles, and the authority of Scripture—we shall find quite a lot in Spinoza's work which Hobbes might have found to be bolder than what he had written on the same topics; 2) that where Spinoza's position is bolder, Hobbes' less radical position is often stated in a way suggesting irony; 3) that since irony can function both as a protective device and as a way of hinting at views one would hesitate to express openly, Hobbes' use of it is evidence that he would have gone further than he did in the direction of unorthodoxy, if the political situation had permitted him to do so safely; 4) that it is entirely credible that Hobbes said to Aubrey what Aubrey says he said; and finally 5) that Hobbes is properly viewed as a precursor of such Enlightenment figures as Voltaire and Hume, that in spite of the deference he often shows to orthodox Christian doctrines, he is essentially a secular thinker, whose religious views are subversive of those held by most Europeans of his time. (Curley 1992, 511–12)

A crucial piece of this approach is Curley's claim that Hobbes uses a rhetorical device he calls "suggestion by disavowal," wherein an author marshals evidence or argument that tend toward a certain conclusion, but then disavows the seemingly obvious conclusion. The purpose of suggestion by disavowal is clear enough: it provides a kind of cover in case one is later charged with expressing unsavory opinions, while at the same time it indicates the author's real intention. To take a simple but apposite example, consider Hobbes's claim that "If I should say I have heard that *Dr. Wallis* is esteemed at *Oxford* for a simple fellow, and much inferior to his fellow-professor *Dr. Ward* (as indeed I have heard, but do not believe it) though this be no great disgrace to *Dr. Wallis,* yet he would think I did him injury" (*SL* 6; *EW* 7:354). The obvious point of this remark is to disparage Wallis's intellectual abilities by repeating some hearsay; yet Hobbes disavows such an intention and, if challenged, he could always claim that he did not literally endorse such a judgment and therefore did not behave improperly. If Curley is right and Hobbes is using such a rhetorical device in his discussion of religious matters, there seem to be strong grounds for thinking that the Monster of Malmesbury was an atheist.

Curley's argumentation has been opposed by A. P. Martinich, who reads Hobbes as an orthodox Calvinist. The guiding principle in Martinich's reading is that, for the most part, Hobbes meant what he said and should be taken literally. Thus, absent any compelling evidence to read him ironically, Martinich concludes that Hobbes's professions of belief should be taken at face value. Curley's imputation of suggestion by disavowal is, according to Martinich, "a defective idea of how to recognize irony" (Martinich 1992, 351). Although he is willing to concede that there are passages where this device may be at work, Martinich thinks that the "vast amount of what Hobbes wrote about religion is on the face of it favorable to religion or presupposes it. Given this general appearance of approval, a few allegedly dubious passages cannot be used to drive an interpretation" (Martinich 1992, 351).

The matter of Hobbes's religious belief can thus be reduced to the question of whether he was sincere or ironic in his many discussions of religion. I take it that if an "ironist" reading of Hobbes can be shown to be more plausible than a "literalist" interpretation, then there are good grounds for concluding that Hobbes was probably an atheist. The issue to resolve is therefore: can we justify taking Hobbes's pronouncements on religion as ironic? It is obviously not sufficient (and perhaps not even necessary) to show that Hobbes contradicts himself on certain key points. If self-contradiction were sufficient for irony, all philosophers worth reading would probably end up in the ironist camp, and Hobbes's most ironic pieces would be his circle quadratures, which are veritable storehouses of self-contradiction. Still, I think that there is adequate evidence to support the ironist reading of Hobbes's discussion of religion, and thereby to justify the conclusion that he was an atheist. Quentin Skinner has made the point that in his treatment of religion "Hobbes makes systematic use of the various devices specifically recommended by the theorists of eloquence for contriving a tone of irony and ridicule" (Skinner 1996, 24). More significantly, Hobbes's contemporary critics were unanimous in their conviction that he used irony in his writings on religious issues. We saw, for example, that Wallis accused Hobbes of treating theological matters "ironically" and "sarcastically," and he was convinced that

Hobbes would have professed his disbelief openly, had he not feared legal sanction. Furthermore, numerous other contemporaries complained of Hobbes's use of irony to make his opponents and their views look ridiculous (Skinner 1996, 394–95). I suppose it is possible that seventeenth-century readers were not in a position to appreciate Hobbes's intentions, but it seems much more likely that his contemporaries could detect irony more readily than we can. In light of these facts, I take it that the ironist interpretation of Hobbes's religious writings is the more plausible, and I conclude that Hobbes was probably an atheist.

I admit that any reading of Hobbes as an atheist must take a very large amount of what he wrote as disingenuous—he does, after all, devote a great deal of time and effort to the discussion of religious issues, and almost none of these pronouncements could be taken literally if he was a convinced atheist. The hypothesis of disingenuity is generally a last resort in the interpretation of a philosophical text, but there is reason to think that this is the place for such a hypothesis. One must remember that, aside from simple considerations of prudence, Hobbes's own political theory demands that he never openly profess atheism, regardless of what he actually believed. Hobbes never lived under a sovereign who permitted the open expression of atheistic views, so he would have been obliged by his own political theory to keep any atheistic thoughts to himself; in such circumstances he could at most have resorted to an ironic mode that does not literally profess disbelief. Because Hobbes's published work and public actions always manifest conformity, any evidence for his hidden atheism must be circumstantial and depend upon the supposition that his real intellectual commitment is to a theory of the world where matter and motion account for all the phenomena of nature and there is no room for a God. Although the very nature of the case dictates that we can never have conclusive evidence one way or the other, I prefer to think of Hobbes as a sly and interesting atheist rather than a confused and bizarre Christian.

7.2 HARSH LANGUAGE AND LEARNED CRITICISM: THE RHETORICAL SIDE OF THE DISPUTE

A significant number of the exchanges between Hobbes and Wallis were devoted to what we might call the rhetoric of controversy because they focus on the language in which the dispute itself was conducted. Eloquence, or the power to use language to persuade an audience, was a prized virtue in Hobbes's day, all the more so in the context of a heated "battle of the books." In order to dispute effectively a seventeenth-century English author had to manifest a command of the language of controversy and, wherever possible, show that his opponent's usage fell below the standard of gentlemanly refinement consistent with the virtue of eloquence. It goes without saying that moderation and civility are also virtues, and a central problem for the disputant is to attack his opponent as fiercely as possible without lapsing into "mere raillery," unduly harsh abuse, or excessively coarse language. The result of this linguistic balancing act is that two rhetorical strategies commonly come into play as the disputants trade charges and countercharges. The first strategy is to rebuke one's opponent for bad style and thereby show that he fails the tests of education and refinement appropriate to one who undertakes to conduct a public controversy. The second is a kind of "linguistic brinksmanship" that involves attacking one's opponent as sharply and wittily as possible without actually breaking the boundaries of polite usage.

A preliminary point to make in an examination of the rhetorical aspects of the dispute is that there was a well-established rhetorical tradition in seventeenth-century England, and educated people were assumed to be familiar with a standard repertoire of tropes and figures that could be deployed to embellish spoken or written discourse (Skinner 1996, 211). Moreover, the educated gentleman was supposed to be familiar enough with classical Latin and Greek authors that, at need, his prose could be liberally sprinkled with quotations from or allusions to classical texts. For disputes carried out in Latin (as were parts of that between Hobbes and Wallis), it was also necessary to observe the norms of Latin grammar and style, and one of the easiest charges to level against an opponent was that he had used poor Latin, typically by lapsing into Anglicisms that violate the principles of Latin eloquence.

Hobbes and Wallis traded accusations of poor Latin throughout their dispute, with the result that numerous pages in their respective polemics are given over to the consideration of grammatical minutiae or the citation of classical authorities on technical questions of style or usage. The specific points discussed are of no interest in themselves (I assume, for instance, that nothing of philosophical, mathematical, or historical significance hangs on whether Hobbes uses the ablative case correctly). Consequently there is no need to investigate this aspect of the dispute in any detail. The flavor of these grammatical exchanges can be gathered from one case, namely Hobbes's critique of Wallis's use of the Latin verb *adduco.* In the *Elenchus,* Wallis had objected that Hobbes's treatment of angles of incidence and reflection in the eighth article of chapter 19 (*DCo* 3.19.8; *OL* 1:238–40) was needlessly long and confusing. Commenting on what he saw as wasted effort in the solution of a simple problem (namely, given two points on the circumference of a circle, to draw two right lines to them, so that their reflected lines are parallel, or contain any given angle), Wallis remarked "malleum adducis quo occidas muscam" (*Elenchus* 94). In English, roughly, "you bring a hammer to kill a fly." Hobbes responded that the objection was mathematically irrelevant, and that in any case it was expressed in poor Latin.

> I must put you in mind that these words of yours, *Adducis malleum, ut occidas muscam,* are not good Latine, *Malleum affers, Malleum adhibes, Malleo uteris,* are good. When you speak of *bringing* bodies animate, *Ducere* and *Adducere* are good, for there *to bring* is *to guide or lead.* And of Bodies inanimate, *Adducere* is good for *Attrahere,* which is to draw to. But when you bring a hammer, will you say *Adduco malleum, I lead a hammer?* A man may lead another man, and a ninny may be said to lead another ninny, but not a hammer. Neverthelesse, I should not have thought fit to reprehend this fault upon this occasion in an Englishman, nor to take notice of it, but that I finde you in some places nibbling (but causelessly) at my Latine. (*SL* 5; *EW* 7:322)

Wallis rebutted the accusation by first alleging that Hobbes had initially mistaken the case and had even boasted to friends that he could show up the false learning of his enemy because *duco* and *adduco* could never take inanimate bodies as direct objects: "You were, I heare, of opinion, when you first made braggs of this notion, (or else your friends belye you) that they were not to be used but of bodies Animate: But, that being notoriously false, some body it seems had rectified that mistake, and informed you better, and therefore you dare not say so now" (*Due Correction* 15). The point of this response is obviously to draw a parallel between Hobbes's efforts at Latin criticism and his quadratures—both seeming to start off with a false opinion that friends tried to dissuade him from making public. Wallis proceeded to argue that "*duco,* with its compounds, is a word of as great variety and latitude of signification, as almost any the Latine tongue affords," and its use with inanimate direct objects is sanctioned by the authority of over a hundred classical loci that he took the trouble to provide.

Hobbes was bound to counter these objections to his Latin competence. He replied in Στίγμαι that no more than two of the passages cited by Wallis were germane to the case, and that neither of these was a proper construction involving *adduco* (Στίγμαι 2; *EW* 7:391–93). To support his philological criticisms, Hobbes enlisted the help of Henry Stubbe, whose extensive classical learning was ideally suited to this kind of polemical encounter. He solicited a letter from Stubbe attacking Wallis's philological conclusions, and then printed it anonymously at the end of Στίγμαι under the pretense that it had been sent to him by an unnamed third party.[21] After an extensive (if utterly one-sided) analysis of the many grammatical and philological points in the dispute, Stubbe declared:

> And now I conceive enough hath been said to vindicate Mr. Hobbes, and to shew the insufferable ignorance of the puny professor, and unlearned Critick. If any more shall be thought necessary, I shall take the paines to collect more examples and Authorities, though I confess I had rather spend time otherwise then in matter of so little moment. As for some other passages in his book, I am no competent judge of Symbolick Stenography. (Στίγμαι appendix; *EW* 7:426–27)

Stubbe made good on his threat to "collect more examples and Authorities," and published them in 1657 under the title *Clamor, Rixa, Joci, Mendacia, Furta, Cachiny; or, a Severe Enquiry into the late Oneirocritica Published by John Wallis, Grammar-Reader in Oxon.* This piece was a rebuttal to Wallis's *Dispunctio* (which contained some disparaging remarks about Stubbe)[22] and included the text of the origi-

21. After some preliminary skirmishing on the propriety of *adducis malleum,* Hobbes writes, "Being come thus far I found a friend that hath eased me of this dispute; for he shewed me a letter written to himself from a learned man, that hath out of very good Authors collected enough to decide all the Grammatical questions between you and me both Greek and Latin. He would not let me know his name, nor any thing of him but only this, that he had better ornaments then to be willing to go clad abroad in the habit of a Grammarian" (Στίγμαι 2; *EW* 7:393). The pretense of the letter's being written to a friend of Hobbes is, of course, a sham, as is evident from Stubbe's letter to Hobbes of 26 November/6 December 1656. Stubbe there writes of an enclosure, "I here send yoᵘ a long letter; I doubt not but you can forgiue such faults . . . as are occasioned by excesse of zeale to serue yoᵘ. . . . I leaue yᵉ busynesse wholly to yoʳ management" (*CTH* 1:378).

22. In the course of the philological wrangling in *Dispunctio,* Wallis writes, "Now, enter your second, (with a Rod at his back,) what says hee? Hee says 'Tis ridiculous. That's easyly sayd: But saying so, doth not make it so. Hee says that, *Hee had, he is sure, been whipped soundly in Westminster School, by the learned Master thereof at present.* . . . That the *present master* is a *learned* man . . . I shall easily grant; . . . But I perceive hee hath had the hap sometimes to breed up sawcy boys. But leaving this Epistoler to

nal letter as printed in Στίγμαι, together with Wallis's replies, and Stubbe's further responses. The emphasis the disputants placed on such seemingly trivial points of usage testifies to the powerful role seventeenth-century authors accorded to matters of language in the conduct of controversies such as this one. The war between Hobbes and Wallis involved weapons beside the logic and close argumentation characteristic of geometric proof, and one of the most important was that of eloquence.

Subtleties of classical grammar and style were not the only focus of attention in that part of the dispute that focused on questions of language. Each disputant also tried to show that his opponent's efforts were too crude to pass the test of true eloquence, either because his language was excessively harsh and vulgar or because his attempted witticisms fell flat. Thus, when Hobbes devoted the last of his *Six Lessons* to the manners of the Savilian professors, he found it useful to accuse them of not knowing how to conduct themselves in a public dispute: "It cannot be expected that there should be much Science of any kinde in a man that wanteth Judgement; nor Judgement in a man that knoweth not the Manners due to a publique disputation in writing" (*SL* 6; *EW* 7:331). Hobbes also had to argue that his own harsh language was consistent with good manners, which he did by invoking "Vespasian's Law," or the principle that good manners forbid one from initiating the use of bad language in a dispute, but there is no harm in replying in kind. Wallis responded with a catalogue of Hobbes's ill language and suggested that the reason he had switched from the Latin of *De Corpore* to the English of the *Six Lessons* was that "when ever you have thought it convenient to repaire to Billingsgate, to learn the art of Well-speaking, for the perfecting of your naturall Rhetorick; you have not found that any of the Oister-women could teach you to raile in Latin, and therefore it was requisite that you apply your selfe to such language as they could teach you" (*Due Correction* 2).[23] Such mutual accusations of bad language are of little interest in themselves, except to the extent that they testify to the importance the disputants attached to rhetorical form, but they are a recurring theme throughout the exchanges between Hobbes and Wallis.

Real eloquence requires wit as well as proper words, and Wallis attempted to evince both the breadth of his learning and the sharpness

---

ruminate when and where he was whipped last, and, for what offence (whether *Ignorance* or *Impudence*) let's hear what he hath to say to the businesse" (*Dispunctio* 20).

23. Billingsgate is that part of London where the fish markets were located. It was synonymous with rude language.

of his wit with a number of clever rhetorical tricks. One of these involves the rhetorical figure known as *paranomasia*, which is an extended wordplay based on the similar sounds of different words. Wallis uses this figure in the *Elenchus* as a joking way of showing that Hobbes himself is really the specter Empusa. He does this by concocting a derivation of *Hobbes* from *Empusa*, working his way through from Greek to English by construing *Empusa* as derived from ἕν (one) and ποῦς (foot), which he then connects to the child's game called "ludus Empusae" in ancient Rome. The game requires one child to hop about on one foot as long as possible but be beaten by his playmates if the other foot should touch the ground; thus we get *hop* from *Empusa*, which quickly turns into *Hobbes* as the hobgoblin Empusa is identified with the philosopher from Malmesbury.[24]

Hobbes denounced such attempted witticisms as "Levity and Scurrility" that comport poorly with the "gravity and sanctity requisite to the calling of the Ministry" (*SL* 6: *EW* 7:354–55). In particular, he dismisses the paranomasia on Empusa by asking, "When a stranger shall read this, and hoping to finde therein some witty conceit, shall with much adoe have gotten it interpreted and explained to him, what will he think of our Doctors of Divinity at *Oxford*, that will take so much pains as to go out of the language they set forth in, for so ridiculous a purpose?" (*SL* 6; *EW* 7:355). Stubbe lent his talents to the defense of Hobbes in this matter and examined at great length the philological basis for Wallis's "silly clinch, which will not passe for wit either at *Oxford*, or at *Cambridge*; no nor at *Westminster*" (Στίγμαι appendix; *EW* 7:410).

A final instance of Wallis's efforts to evince great wit through wordplay and questionable taste will close our study of the linguistic aspects of the dispute. This episode arose in connection with Wallis's criticisms of Hobbes's application of his "method of motions" in chapter 16 of *De Corpore*, which deals with accelerated motion (*DCo* 3.16.2; *OL* 1:185–86). Among many other criticisms, Wallis had complained that

24. In the original (which doesn't really translate well): "Erat enim & *Empusa tua*, Daemonium illud Atheniense, ex pede dignoscenda, quippe erat (uti ais) *pedibus altero aeneo, alter Asinino*, (& utro horum tu dignosci malis, penes te sit optio,) sed utut (uti videtur) duos pedes habuerit non tamen nisi uno incedebat, ut aliunde si opus sit discas (unde liquet ex eorum lemurum numero fuisse quos nos Anglicè dicimus *Hob-goblins*,) nempe ab ἕν & ποῦς factum est nomen illud: unde & puerorum ludus ille, *Empusae* ludus dictus (Anglicè *Fox, fox, come out of your hole*, hoc est, *heus vulpes de foveâ prodi*) nomen sortitur, quo puer ille qui *vulpes* audit, alter suspenso pede, altero subsultans incedit, (quod est Anglicè *to Hop*,) à reliquis flagellandus si utroque terram tetigerit" (*Elenchus* 4).

the formulation "in every uniform motion" should be taken as a plural "in all uniform motions," because the doctrine "does not concern single motions taken singly, but many motions taken together and compared with one another" (*Elenchus* 40). Expanding on this criticism in the *Due Correction,* he comments that "there must be at least two Motions, because two Times; unlesse you will say, that one and the *same motion* may be *now, and anon too*" (*Due Correction* 96). Hobbes correctly guessed that the expression "now, and anon too" was intended as some kind of joke. He wrote to Stubbe asking for help in tracking down the source of the quotation. Stubbe replied that "[i]t is not taken out of any ballad, but referres to Obadiah Sedgewicke, who haueing married for his first wife a chambermayde, presently after yᵉ marryage dinner was ended, hee took her aside into the draweing roome, and desired to anticipate those nuptiall pleasures hee was otherwise not to partake of till night. Yᵉ mayde desired him to forbeare till night; but he replyed, now and anon too. and thus you haue yᵉ story as it was related by Sedgewickes brother in lawe to mee yᵉ other day" (Stubbe to Hobbes, 29 November/9 December 1656; *CTH* 1:379). Armed with this information Hobbes then counterattacked against Wallis's charge that "the ribauldry in your obscene Poem *De Mirabilibus Pecci*" indicates the Malmesburian philosopher's utter lack of sophistication (*Due Correction* 3). Hobbes responded by charging that, at least in the matter of comparative obscenity, Wallis must be judged the more vulgar of the two disputants.[25]

7.3 LOYALTY, DUPLICITY, AND THE POLITICS OF THE RESTORATION

The final extramathematical topic at issue in the dispute is that of the political loyalties of the two principal disputants. We have already cov-

---

25. Hobbes writes: "For my verses of the Peak, though they be as ill in my opinion as I beleeve they are in yours, and made long since, yet are they not so obscene, as that they ought to be blamed by Dr. *Wallis*. I pray you Sir, whereas you have these words in your *Schoole-Discipline* page 96. *unlesse you will say that one and the same motion may be now, and anon too;* what was the reason you put these words *now and anon too* in a different Character, that makes them to be the more taken notice of; Do you think that the story of the Minister that uttered his affection (if it be not a slander) not unlawfully but unseasonably, is not known to others as well as to you? what need you then (when there was nothing that I had said could give the occasion) to use those words; there is nothing in my verses that do *olere hircum,* so much as this of yours. I know what good you can receive by ruminating on such Ideas, or cherishing of such thoughts" (Στίγμαι 2; *EW* 7:389).

ered much of this ground in section 7.1.1, when we investigated their differing opinions on the question of church government. Wallis's Presbyterian principles put him at odds with Hobbes's political absolutism, and on any construal of what counts as politics, it is evident that they had serious political differences. The political situation in England underwent several dramatic changes over the course of the Hobbes-Wallis dispute, most especially with the restoration of the monarchy in 1660, and questions of loyalty made their way into the dispute at the Restoration.

When Charles II took his place as England's monarch, those who had gained much under the Commonwealth and Protectorate had reason to feel vulnerable. Shortly before his return Charles II issued the "Declaration of Breda," wherein he announced a general pardon for all who "shall lay hold upon this our grace and favor, and shall, by any public act, declare their doing so, and that they return to the loyalty and obedience of good subjects; excepting only such persons as shall hereafter be excepted by Parliament, those only to be excepted" (Prall 1968, 282). The Parliament to which Charles II referred in the declaration was not a purged group of radicals characteristic of the Parliaments that had sat from 1649, but the reinstitution of the old Long Parliament.[26] This group had a guaranteed majority of Royalists, many of whom had scores to settle with the revolutionary regime. Any who hoped to retain some measure of their gains during the interregnum therefore sought to distance themselves from their former allies and recast themselves as moderates. Most tried to argue that they had never wholeheartedly supported the abolition of monarchy and disowned the more extreme actions taken in support of the Parliamentary cause, most notably the 1649 execution of Charles I. The result of this scramble for political cover was a flood of obsequious declarations of high principle following the Declaration of Breda, all intended to curry favor with the restored monarchy.

Wallis certainly had reason to worry about his future at the Restoration. He was intruded into Oxford's Savilian Professorship of Geometry in 1649 at the behest of the Parliamentary Board of Visitors, and in 1657 he acquired the office of "keeper of the archives" through a highly irregular appointment that was apparently engineered by his po-

---

26. The Long Parliament was purged by army radicals in December of 1648, leaving the Rump Parliament, which executed Charles I and abolished the monarchy. The Rump was replaced by the "Barebones" Parliament in 1653, which was dissolved with the establishment of the Protectorate. Three different Protectorate Parliaments then followed.

litical connections, notwithstanding the fact that his holding of the office conflicted with the Savilian statutes.[27] Moreover, as a member of the Westminster Assembly and an active supporter of the Presbyterian cause, Wallis had taken the Solemn League and Covenant in 1643; he had also subscribed to the engagement in 1650, and—most damningly—he had assisted the Parliamentary forces during the Civil War by deciphering the king's correspondence captured at the Battle of Naseby. The king's pardon was readily forthcoming, however, and Parliament confirmed Wallis's two positions in the university. He was even admitted as one of the king's chaplains in ordinary and appointed in 1661 to the group of divines charged with revising the Book of Common Prayer.

Hobbes undoubtedly welcomed the Restoration because the reestablishment of the Stuart house suited his professed preference for monarchy over other forms of government. Nevertheless, the events of 1660 did not leave him altogether without grounds for concern. His return to England from France in 1651 was interpreted by many as an abandonment of the Royalist cause he had once supported. Furthermore, his reputation for atheism and his pronounced anticlericalism made him a favorite target for the denunciations of Anglican divines.[28] Shortly after the restoration of Charles II, Wallis publicly charged Hobbes with disloyalty to the monarchy. In 1662 the Savilian professor of geometry claimed that the events of recent years had made *Leviathan* "somewhat out of season," since the monarchy had been restored.

27. Aubrey reports that in 1657 Wallis "gott himselfe to be chosen (by unjust means) to be Custos Archivorum of the University of Oxon, at which time Dr. Zouch had the majority of voices, but because Dr. Zouch was a malignant, (as Dr. Wallis openly protested, and that he had talked against Oliver) he was putt aside. Now, for the Savilian Professor to hold another place besides, is so downeright against Sr. Hen. Savile's Statutes that nothing can be imagined more, and if he does, he is downright perjured. Yet the Dr. is allowed to keepe the other place still" (Aubrey 1898, 2:569). Anthony à Wood gives an equally damning assessment in his *Athenae Oxonienses:* "The famous Dr. Rich. Zouch, who had been an Assessor in the Chancellours Court for thirty years or more, and was well vers'd in the Statutes, Liberties, and Privileges of the University, did, upon great intreaties stand for the said place of Antiquary or Custos Archivorum thereof, but he being esteemed a Royalist, Dr. J. W. was put up and stood against him, tho altogether incapable of that place, because he was one of the Savilian Professors, a Cambridge man, and a stranger to the usages of the University. At length by some corruption, or at least connivance of the Vice Chancellour, and perjury of the Senior Proctor (Byfield), W. was pronounced elected" (Wood [1813] 1967, 2:cols. 414–15). This incident also led to Stubbe's rebuke of Wallis in *The Savilian Professours Case Stated* (Stubbe 1658).

28. As we saw in chapter 6, Hobbes had to face the possibility of being tried for heresy when the Restoration Parliament ordered an investigation of his works in 1666, an investigation eventually blocked by his patron Arlington.

More specifically, Wallis charged that *Leviathan* had been "written in Defense of *Olivers Title* (or whoever by whatsoever means can get to be upmost)," and that Hobbes's return to England in 1651 amounted to "deserting his *Royal Master* in distresse" (*HHT* 5).

The first of these two accusations of disloyalty is obvious nonsense and was easily shown to be so by Hobbes. When *Leviathan* was published in 1651 Oliver Cromwell was a general in the Parliamentary army. The office of Lord Protector was not established until late 1653, so (unless Hobbes had a rare prophetic gift) there is no plausible sense in which he could have written *Leviathan* in defense of Cromwell's claim to power.[29] Furthermore, because his political philosophy places obedience to lawful authority at the center of a citizen's obligations, Hobbes could hardly be charged with encouraging rebellion.

The accusation that Hobbes "deserted" the Royalist cause by returning to England in 1651 is somewhat more difficult to assess. Scholars have often puzzled over the question of why Hobbes should have returned, given that his previous association with the Royalist cause would have left him with few prospects under the new regime. Some have argued that Hobbes's actions show him to have been a supporter of the Commonwealth, and to have been an active participant in the engagement controversy, which centered on the question of whether and by what means submission to the newly established Commonwealth could be justified.[30] The "Review and Conclusion" to *Leviathan* contains a fairly direct acknowledgment of the question of the difficult political situation in England, and has been read as Hobbes's declaration of support for the "engagers" who advocated submission to the new government. Hobbes declares that *Leviathan* was "occasioned by the disorders of the present time," and that his intention was "to set before mens eyes the mutuall Relation between Protection and Obedience; of which the condition of Humane Nature, and the Laws Divine, (both Naturall and Positive) require an inviolable observation" (*L*, "Review and Conclusion," 395–96; *EW* 3:713).

The formula of "mutual relation between Protection and Obedi-

29. Hobbes makes this point as follows: "What was *Oliver* when that Book came forth? It was in 1650, and Mr. *Hobbes* returned before 1651. *Oliver* was then but General under your Masters of the Parliament, nor had yet cheated them of their usurped Power. For that was not done till two or three years after, in 1653, which neither he nor you could foresee" (*MHC; EW* 4:420).

30. See Metzger 1991, 131–35, for an overview of the controversy and Hobbes's place in it. The engagement that stands at the center of the controversy reads, "I do declare and promise that I will be true and faithful to the commonwealth of England, as it is now established, without a King or House of Lords" (Prall 1968, 238).

ence" encapsulates the principal argument of the engagers, who sought to justify submission to the Commonwealth on the grounds that the de facto power of the state could provide protection for its citizens only in exchange for their obedience. Skinner sees in this formulation evidence that Hobbes's *Leviathan* is a "somewhat belated though highly important contribution to the lay defence of 'engagement'" (Skinner 1972, 96). Thus, rather than reading Hobbes as standing aloof from the political issues posed by the engagement, Skinner sees him as participating in that debate, and siding with those who sought legitimacy for the new regime. Indeed, because Hobbes only uses the language of "mutual relation" in the "Review and Conclusion" of *Leviathan,* Skinner takes this as evidence that "Hobbes actually adapted his own earlier presentations of his theory to bring it into line with these other theorists of *de facto* powers. The formula of 'mutual relation' might be said to be implied by the doctrine of *De Cive.* But it is nowhere explicitly stated in that work and although it virtually surfaces in the 1650 version of the *Elements of Law,* it is not stated absolutely explicitly until the Review and Conclusion of *Leviathan* in 1651" (Skinner 1972, 97).

This reading of Hobbes's intentions at the time of his return to England is nevertheless beset with important difficulties. In particular, it conflicts with Hobbes's own account of his reasons for returning and the intentions in publishing *Leviathan.* In reply to Wallis's accusation of deserting the king, Hobbes insisted "[n]or did he desert His Majesty, as you falsely accuse him, as His Majesty Himself knows. . . . It is true, that Mr. *Hobbes* came home; but it was because he would not trust his safety with the French Clergy" (*MHC; EW* 4:415). A similar account appears in his Latin prose *Vita* (*OL* 1:xvii). These, of course, may be nothing more than post hoc rationalizations penned years after the event, but the portrayal of Hobbes as compelled to leave France finds support from other sources. A very similar account is recorded by Loedwijk Huygens (brother of the mathematician Christiaan Huygens), whose report of a visit to Hobbes in 1652 contains the remark that "[I] rode . . . to the renowned philosopher Hobbius, who, upon having been exiled from France for the strange notions in the book which he entitled Leviathan, has come back to live here" (Huygens 1982, 75). Further evidence of Hobbes's unwilling return to England comes from the fact that Hobbes was banned from the court in exile as the result of political intrigue involving his former friend Edward Hyde, later earl of Clarendon. Hyde had become hostile to Hobbes's political philosophy (which he regarded as based on a depraved view of human motivation and a cynical, self-serving theory of obligation) and took credit for having the philosopher driven from court (Metzger 1991, 92–97; Sommerville 1992, 24–25). Thus, there seems to be little basis for thinking that Hobbes abandoned the Royalist cause and sided with its enemies.

Hobbes's intentions in writing *Leviathan* (so far as they can be determined) also seem to avoid the charge of pure opportunism. It is true, as Skinner notes, that Hobbes boasted in 1656 that his doctrine "hath framed the minds of a thousand Gentlemen to a conscientious obedience to present Government, which otherwise would have wavered in that Point" (*SL* 6; *EW* 7:336). But this fact alone does not justify the inference that Hobbes wrote *Leviathan* for the purpose of legitimating the authority of the Commonwealth. Hobbes's own account was that during his exile he "staid about *Paris,* and had neither encouragement nor desire to return into *England,*" and he "wrote and published his *Leviathan,* far from the intention either of disadvantage to His Majesty, or to flatter *Oliver* (who was not made Protector til three or four years after) or purpose to make way for his return" (*MHC; EW* 4:415).

Royalist hopes were by no means extinguished during the period when Hobbes was writing *Leviathan* (1650–51). On the contrary, there was every expectation in the exiled court that Charles II could gain the throne by military means. The remaining Royalist forces were joined in an alliance with Scotland in 1650, which was concluded after the king acceded to Scottish demands that he commit himself to the establishment of Presbyterianism. In the course of events Charles II's attempted invasion of England came to grief in September of 1651 with his crushing defeat at the Battle of Worcester. At that point even the most optimistic Royalists were prepared to concede that the cause was lost, at least for the foreseeable future. In this context there is no reason to think that Hobbes wrote *Leviathan* with the intention of supporting the Commonwealth, since its main argument could just as easily have been intended to show the reluctant supporters of the vanquished Parliamentary cause that they owed allegiance and obedience to the crown. Hobbes himself suggests that *Leviathan* was a work that could have been put to use to defend either the king or his enemies. In the dedicatory epistle of the *Problemata Physica,* which was addressed to Charles II, Hobbes remarks that his opponents should not "turn it to a fault, if fighting against your enemies, and snatching up whatever weapons I could, I made use of a double-edged sword" (*PP* epistle; *OL* 4:303).

Hobbes actually welcomed Wallis's accusations of disloyalty, as they gave him the opportunity to counterattack and try to salvage something from a dispute he was evidently losing badly. We saw in chapter 6 that by the time that Wallis's *Hobbius Heauton-timorumenos* appeared in 1662, Hobbes's standing as a man of science had been demolished by the continued refutation of his mathematical works. When Wallis began to raise questions of Hobbes's political loyalties, the philosopher from Malmesbury gladly compared his record with that of Wallis. He exulted that "Mr. *Hobbes* could long for nothing more than such an occasion to tell the world his own and your little stories, during the time of the late Rebellion" (*MHC; EW* 4:413). Hobbes assembled a catalog of Wallis's misdeeds in *Mr. Hobbes Considered,* which included siding with the Parliamentary forces against the king, deciphering the royal correspondence, acting with the Westminster Assembly to alter the form of church government without consent of the king, currying favor with Cromwell and his associates by dedicating his *Elenchus* to vice-chancellor Owen, and preaching principles that inspired rebellion. Hobbes concludes:

> Therefore of all the Crimes (the Great Crime not excepted) done in that Rebellion, you were guilty; you, I say, *Dr. Wallis,* (how little force or wit soever you contributed) for your good will to their cause. The King was hunted as a Partridge in the Mountains; and though the Hounds have been hang'd, yet the Hunters were as guilty as they, and deserved no less punishment. And the Decypherers, and all that blew the horn, are to be reckoned amongst the Hunters. Perhaps you would not have had the prey killed, but rather have kept it tame. And yet who can tell? I have read of few Kings deprived of their Power by their own Subjects, that have lived any long time after it, for reasons that every man is able to conjecture. (*MHC; EW* 4:419)

In actual fact neither Wallis nor Hobbes seems to have suffered from this airing of mutual charges of disloyalty. Wallis was not damaged by his Cromwellian past and continued to enjoy the favor of Charles II while he became a zealous Conformist to the Church of England. For his part, Hobbes was not abandoned by his friends or allies, and although he always had enemies at the court, Charles II seems to have regarded him affectionately. These charges of political intrigue and disloyalty thus function as a side issue in the dispute rather than a focal point.

CHAPTER EIGHT

Persistence in Error

Why Was Hobbes So Resolutely Wrong?

For all men by nature Reason alike, and well, when they have good principles. For who is so stupid, as both to mistake in Geometry, and also to persist in it, when another detects his error to him?
—Hobbes, *Leviathan*

The account assembled in the preceding chapters leads fairly naturally to a single conclusion: Hobbes was led by a misplaced faith in the efficacy of his materialistic foundations for geometry to think that he could quickly dispatch all the great problems of that science. For his part, Wallis undertook the refutation of Hobbesian geometry primarily for the purpose of discrediting Hobbes's "dangerous" metaphysical, theological, and political theories. The result was a bitter and very public dispute that dragged on for decades and encompassed a host of issues beyond technical details of circle quadrature.

Yet if this line of thought is pursued, there remains the problem of explaining why Hobbes should have been so persistent in his geometric errors. Why, after all, did he not simply admit that he was mistaken, cut his losses, and spare himself the profound humiliation that inevitably followed his repeated forays into the murky waters of circle quadrature? In short, there is a fundamental question that must loom large in any account of Hobbes's geometric endeavors: how did a man whose mathematical abilities were once ranked highly enough to earn him a reputation as one of Europe's mathematical cognoscenti end up filling hundreds of pages with miserably failed attempts at the solution of great geometric problems?

Making the minimal assumptions that most of what agents do makes sense to them as they do it and that they generally regard their beliefs as rationally grounded, we are left with the problem of explaining how Hobbes could have been engaged in such an apparently irrational project for such a long time.

I believe that Hobbes's conduct in the dispute with Wallis can be largely explained by appealing to the status of geometry within his system. Having insisted that all of philosophy must be grounded in the metaphysics of matter and motion, Hobbes readily accepted the consequence that geometry is an integral part of philosophy and that continued failure in geometric matters must be symptomatic of philosophical ineptitude. In other words, Wallis was right to characterize Hobbes as having "set such store by geometry as to hold that without it there is hardly anything sound that could be expected in philosophy" (*Elenchus* 108). He was also correct in his judgment that the most effective way to oppose Hobbes's philosophical principles was to "show how little he understands this mathematics (from which he takes his courage)" (Wallis to Huygens, 1/11 January 1659; *HOC* 2:296). Early on in the dispute, Hobbes was sensitive to the damage that Wallis could do to his reputation as a man of learning, and he tried to downplay the significance of his purely technical errors by insisting that

> [i]t is in Sciences as in Plants; Growth and Branching is but the Generation of the Root continued; nor is the Invention of Theoremes any thing else but the knowledge of the Construction of the Subject prosecuted. The unsoundness of the Branches are no prejudice to the Roots; nor the Faults of Theoremes to the Principles. And Active Principles will correct false Theoremes if the Reasoning be good; but no Logique in the world is good enough to draw evidence out of false or unactive Principles. (*SL* epistle; *EW* 7:188)

This pretense could not last. Hobbes had staked too much on his geometry to pretend that his failings were so many minor faults to be corrected by the rigorous application of his true principles; and even so he was still committed to rectifying his errors through the application of his "active principles" and thereby making good his claims to preeminence in geometry, i.e., actually squaring the circle.

Hobbes ultimately saw his entire philosophical program threatened by the prospect of his inability to deliver the great mathematical results he had promised. He concluded that it was no use to attempt to salvage a modicum of respectability by insisting that the intrinsic difficulty of the problems could excuse his failures. Hobbes's geometric errors were generally errors of ignorance (although he would never publicly admit to such a fault), and he had in any case proclaimed that the superiority of his methodology should greatly advance the science of geometry by solving previously unsolved problems. To find the area of a circle is, on Hobbes's account of the matter, no more than to deduce the properties of something we construct. Thus, to suggest that a problem so simple to pose might nevertheless escape our efforts at solution would be tantamount to saying that the proper science of the commonwealth (itself an "artificial body" constructed by human agreement) might nevertheless fail to settle questions of right and obligation. Hobbes could simply not abide such a prospect.

After some hesitation in the 1650s and 1660s, Hobbes concluded that he had nothing to lose by rejecting the geometry of his opponents and maintaining the essential correctness of his procedures, no matter how wildly they may conflict with the received view of the matter. In this Hobbes's actions were much like those of a man who "throweth his goods into the Sea for *feare* the ship should sink" (*L* 2.21, 108; *EW* 3:197). Throwing classical geometry overboard was a serious price to pay, but Hobbes came to see it as the only way to avoid an even worse fate: the outright refutation of his philosophy as a whole. Whether this was ultimately a rational decision on Hobbes's part is a question I will leave to the judgment of the reader.

If anything else is clear about this dispute it is the fact that the claims of neither Hobbes nor Wallis can be properly characterized as a dispassionate search for the truth. Both participants' motives for pursuing the quarrel are doubtless complex, but it is beyond question that Wallis placed a high priority on discrediting Hobbes's metaphysical and theological doctrines, which he regarded as utterly pernicious. Furthermore, I take it that Hobbes's bitter enmity toward Wallis was the product of both personal and political factors: Wallis had done more damage to Hobbes's intellectual reputation than any other of his critics, and in any case the Savilian professor's standing as a "school divine" placed him squarely in opposition to Hobbesian political principles. There is consequently no need to doubt Hobbes's declaration that "I had never answered your Elenchus as proceeding from Dr. *Wallis,* if I had not considered you also as the Minister to execute the malice of that sort of people that are offended with my *Leviathan*" (Στίγμαι 1; *EW* 7:381).

8.1 HOBBESIAN GEOMETRY AND THE SOCIOLOGY OF SCIENTIFIC KNOWLEDGE

For all that there is an undeniable sociopolitical element to the controversy, I am convinced that it was not driven by strictly social or extra-mathematical factors. Hobbes was grievously mistaken about matters of great mathematical importance, and his mistakes in this science produced catastrophic consequences. There is, however, another way of thinking about the course of the dispute. Proponents of the "sociology of scientific knowledge" hold that scientific or mathematical controversy is merely symptomatic of "deeper" sociopolitical differences, and that the resolution of such controversy involves the triumph of one set of social interests over another. Thus, a sociologist of knowledge interested in the struggle between Hobbes and Wallis would hold that the *real* issues between them were those concerning religion and politics, and that mathematics mattered in the course of the quarrel only to the extent that it provided the contestants a convenient arena. Since the publication of Shapin and Schaffer's *Leviathan and the Air-Pump* (1985), this sort of analysis has become quite commonplace in Hobbes scholarship. Nevertheless, a consideration of the details of the present case shows that there is something quite seriously wrong with the attempt to see the controversy as driven purely by social factors. It will therefore be worthwhile to consider, by way of conclusion, the case against a purely sociological reading of Hobbes's dispute with Wallis.

Shapin and Schaffer enthusiastically endorse the thesis that scientific controversy is nothing more than a cover for more fundamental and important differences of opinion on how best to achieve social order. According to their analysis, scientific disputes are typically clashes between different "forms of life" or structures of rules and social conventions that dictate what knowledge is and how competing claims to knowledge are to be adjudicated. Moreover, these rival forms of life are linked to different political programs, and the resolution of a scientific controversy is achieved when one political program gains ascendancy and carries its affiliated scientific form of life in its wake. They inform us

(1) that the solution to the problem of knowledge is political; it is predicated upon laying down rules and conventions of relations between men in the intellectual polity; (2) that the knowledge thus produced and authenticated becomes an element in political

action in the wider polity; it is impossible that we should come to understand the nature of political action in the state without referring to the products of the intellectual polity; (3) that the contest among alternative forms of life and their characteristic forms of intellectual product depends upon the political success of the various candidates in insinuating themselves into the activities of other institutions and other interest groups. He who has the most, and the most powerful, allies wins. (Shapin and Schaffer 1985, 342)

On Shapin and Schaffer's account, the conflict between Hobbes and Boyle was not really a dispute about scientific methodology or the existence of a vacuum, but was instead a clash between two different political programs. Hobbes's rejection of experimental science in favor of a priori deductions from mechanistic first principles is the consequence of his conception of sovereignty as absolute and undivided, while the Royal Society's preference for an experimental methodology is merely the symptom of its members' desire for a polity midway between strict absolutism and the dreaded mob rule of democracy.[1] This analysis of the dispute between Hobbes and Boyle concludes that the ultimate failure of Hobbesian physics was the result of Hobbes's inability to secure sufficiently powerful allies in the political world of the Restoration. It is important to observe that such an account makes the success or failure of a scientific research program entirely independent of the truth or falsehood of the program's central claims: the history of science is a random walk driven exclusively by social and political forces.

The prima facie difficulty with such an account is its strong element of sociological reductionism. It is gratuitous simply to assume that the only real conflicts are political conflicts, and it is a silly exercise in sociological positivism to maintain without argument that the only possible motivation for one's acceptance or rejection of a scientific the-

---

1. As Shapin and Schaffer put it, "We see that both games proposed for natural philosophers assumed a causal connection between the political structure of the philosophical community and the genuineness of the knowledge produced. Hobbes's philosophical truth was to be generated and sustained by absolutism. Boyle and his colleagues lacked a precise vocabulary for the polity they were attempting to erect. Almost all of the terms they used were highly contested in the early Restoration: 'civil society,' a 'balance of powers,' a 'commonwealth.' The experimental community was to be neither tyranny nor democracy. The 'middle wayes' were to be taken" (1985, 339). Later, they conclude that "[t]he experimental philosophers aimed to show those who looked at their community an idealized reflection of the Restoration settlement. Here as a functioning example of how to organize and sustain a peaceable society between the extremes of tyranny and radical individualism" (1985, 341).

ory is a perceived connection between the theory and one's "deeper" sociopolitical interests. If interests are defined quite broadly, so that whatever a person seeks or desires is something in which he has an interest, then the "interest model" is vacuously true: people are always motivated by their interests. Sociological reductionists, however, take interests in a much narrower sense than this. They countenance only such genuinely social interests as concern for the maintenance of the social order, the desire for political power, or the aspiration for advancement in the social structure. The sociological positivist's thesis that historical figures are motivated only by such social interests is not obviously true, and it stands in need of at least some defense. It is worthwhile to inquire briefly into the theoretical justification advanced for sociological reductionism.

The principal (and, to my knowledge, the only serious) argument for the explanatory primacy of social factors proceeds by appeal to the authority of Wittgenstein. The argument reasons that any agent's actions or claims to knowledge must be understood as deriving from his adherence to the dictates of a form of life, so that it is by reference to the social facts embodied in a form of life that all actions are to be explained. David Bloor, a principal champion of the sociology of scientific knowledge, is explicit in holding such a view:

> There can no longer be any excuse for offering the meaning of an actor's beliefs as an explanation of his behavior, or of his future beliefs. Verbalized principles, rules and values must be seen as endlessly problematic in their interpretation, and in the implications that are imputed to them. They are the phenomena to be explained. They are dependent, not independent variables. The independent variable is the substratum of conventional behavior that underlies meaning and implication. As Wittgenstein put it: "what has to be accepted, the given, is—so one could say—*forms of life.* (Bloor 1983, 137)

Shapin and Schaffer toe the same line and claim to base their account on a "liberal, but informal, use of Wittgenstein's notions of a 'language game' and a 'form of life'" (1985, 15). As we have seen, they regard a scientific community as a form of life, and take scientific disputes to be clashes between different forms of life. They further hold that there can be no external standard of truth or rationality against which the claims of the disputants can be judged, because the form of life itself dictates what is to count as true or rational. Reason, evidence, and argument thus fail, and the controversy can be resolved only by other (i.e., political) means, including the formation of strategic alliances with other social interest groups, Machiavellian strategies for outmaneuvering the enemy, brute force, and similar ploys.

This is not the place to pronounce upon the worth of Wittgenstein's later philosophy, but it is important to see just what follows from the equation of scientific (and mathematical) communities with forms of life. In particular, it is worthwhile asking whether this account of forms of life can underwrite an interpretation of the Hobbes-Wallis controversy grounded in strict sociological reductionism. What, then, is a form of life? As with most Wittgensteinian concepts, it is not easy to answer this question succinctly. A form of life is fundamentally a collective or communal concept: it is a system of rules and practices upon which members of a community agree, and conformity to which is required if one is to be a member of the community. A form of life is also the community's standard for what patterns of behavior and response will count as intelligible or coherent. As Kripke defines it in his discussion of rule-following: "The set of responses in which we agree, and the way they interweave with our activities, is our *form of life*. Beings who agreed in consistently giving bizarre . . . responses would share in another form of life. By definition, such another form of life would be bizarre and incomprehensible to us" (Kripke 1982, 96). The dictates of different forms of life thus constitute boundaries between mutually unintelligible communities, and there is no possibility of meaningful interaction across these frontiers. Thus, to employ the Wittgensteinian notion of a form of life in the analysis of scientific controversy, it must be shown that adherents of one research program find their opponents' claims not merely false but downright incomprehensible. To use the terminology derived from Kuhn's *Structure of Scientific Revolutions* (1970), the research programs would have to be incommensurable.[2] More specifically with regard to the Hobbes-Wallis dispute, the program of sociological reductionism would have to show that Hobbes and Wallis literally could not understand one another, and that they regarded each other's claims as flatly unintelligible rather than true or false.

2. This connection between Kuhn and the sociologists of scientific knowledge should not be taken to suggest that Kuhn himself was sympathetic to their project. In fact, he voiced serious reservations about the "historical philosophy of science" (Kuhn 1992). Further, as Friedman (1998) argues, the interpretation of Wittgenstein favored by sociologists of knowledge has little basis in the writings of Wittgenstein and has led to its embrace of an utterly implausible philosophical agenda.

The sociological reduction of mathematics and its history requires more than simply an appeal to forms of life as a way of characterizing mathematical communities. The sociologist of mathematical knowledge also wants to replace the usual account of mathematical objects and mathematical truth. A purely sociological approach to mathematics (especially that of the sort offered by Bloor) locates the objects of mathematical investigation in human social convention—they are literally "social constructs" brought into being by the system of human conventional behavior that underlies all meaning and implication. As Bloor puts it, the way to "give content" to the ordinary notion of mathematical objects and objectivity "is to equate it with the social" (Bloor 1991, 98). The sociological positivist's account of mathematical truth does not evaluate the truth of mathematical assertions "in terms of our practices 'corresponding' to some mysterious mathematical reality," but rather by reference to purely social factors (Barnes, Bloor, and Henry 1996, 185).

From the equation of mathematics and "the social" it follows fairly readily that there can be different mathematics, just as there are different patterns of social organization. It is no news to be told that languages, customs, political systems, religions, and other obviously social entities vary widely from culture to culture. Since he identifies the object of mathematics with institutionalized belief, Bloor's account requires that there be at least the possibility of such variation in mathematics. It is important to recognize that an alternative mathematics must consist of something deeper than just a different notation or a different way of developing familiar material. The alternative must literally contain different truths. For example, it is obvious and uninteresting that our system of arithmetical notation could have been different; in particular, we could have used the symbol '3' to designate the number four and '4' to designate the number three; then, assuming that the rest of our notation was unaltered, the expression '3 + 1 = 5' could have expressed a truth in arithmetic, namely the truth that four plus one is five. This kind of trivial semantic conventionalism is compatible with any view on the nature of mathematics. The sociological reductionist's commitment to the possibility of an alternative mathematics yields a much stronger consequence. It requires that the addition of the numbers three and one could yield the number five. Bloor readily admits this consequence, observing that an alternative mathematics would have to look like "error and inadequacy" from the standpoint of our mathematics:

The "errors" in an alternative mathematics would have to be systematic, stubborn, and basic. Those features which we deem error would perhaps all be seen to cohere and meaningfully relate to one another by the practitioners of the alternative mathematics. They would agree with one another about how to respond to them; about how to develop them; about how to interpret them; and how to transmit their style of thinking to subsequent generations. The practitioners would have to proceed in what was, to them, a natural and compelling way. (Bloor 1991, 108)

If one grants the possibility of a genuinely alternative mathematics in the relevant sense, it follows quite easily that alternative systems should be studied impartially and "symmetrically," in the sense that no alternative should be accorded the privileged status of the incontestable truth about mathematics. Bloor's requirement of symmetry holds that a properly developed sociology of knowledge "would be symmetrical in its style of explanation. The same types of cause would explain, say, true and false beliefs" (Bloor 1991, 7).[3] Shapin and Schaffer employ considerations of symmetry when they consciously leave the category of "misunderstanding" out of their account of the controversy between Hobbes and Boyle: Hobbes's departures from Boyle's program in pneumatics were not a consequence of his *misunderstanding* experimental science; rather, he developed an alternative to it (Shapin and Schaffer 1985, 11–12). In the sociology of mathematics, the symmetry requirement bids the investigator not dismiss alternatives to "our" mathematics as incoherent, erroneous, or inconsistent. Applied specifically to the case of Hobbes's mathematical adventures, the sociologist of knowledge will have to describe Hobbes's deviation from the traditional view of the subject as the exploration of an alternative mathematics rather than a thoroughly confused lapse into incoherence and self-contradiction.

A sociological reductionist's account of Hobbes's mathematical career would, presumably, go something like this: Hobbes began as a member of the form of life known as the mathematicians' community, but he eventually embraced mathematical principles that were contrary to those of the group. These principles were a causal consequence of

---

3. Thus formulated, the symmetry requirement is utterly trivial. Inasmuch as they are all beliefs, both true and false beliefs have the same kind of cause, viz., whatever kind of cause it is that produces belief. I assume that Bloor intends a more finely grained causal typology than this, but will leave the issue aside here.

his commitments to a particular theory of social organization and they reflected his absolutist views on the proper ordering of the commonwealth and the production of knowledge within it. His opponents (principally those in the Royal Society) had a different political agenda, and sought to silence Hobbes by denying him a forum in which to air his views. Nevertheless, Hobbes persevered and used his new doctrines to explore an alternative to traditional mathematics. Having broken with the traditional form of life, Hobbes could not make himself understood by those who labored under the old paradigm. Eventually, his opponents were so successful in their Machiavellian strategies for marginalizing Hobbes that his program disappeared from the mathematical landscape.

8.2 THE INADEQUACY OF A PURELY SOCIOLOGICAL EXPLANATION

Whatever its merits may be in regard to other scientific or mathematical controversies, the project of sociological reductionism can shed little light on the dispute between Hobbes and Wallis. This may come as a surprise, considering that the Shapin-Schaffer account of Hobbes's scientific battles is generally hailed as a paradigmatic example of how the sociology of scientific knowledge can and should be done.[4] Nevertheless, there is little hope for a purely sociological interpretation of Hobbes's mathematical career. The reasons for this are many, but they can be outlined fairly quickly.

In the first place, the appeal to some version of the Wittgensteinian notion of "forms of life" can hardly contribute much to our understanding of the dispute. As I have indicated, the sociology of knowledge would require that Hobbes and Wallis adhere to fundamentally different forms of life, and such adherence requires that there be an impenetrable barrier to mutual understanding between them. But this is simply not the case. As I showed in some detail in chapter 6, Hobbes clearly understood most (but not all) of Wallis's criticisms, and he ad-

---

4. Barnes, Bloor, and Henry wax ecstatic in their evaluation of Shapin and Schaffer's work, concluding that through the "magnificent and marvellously well-substantiated insights of their book" they "offer a model for the historical and sociological study of boundary drawing in science which deserves to be extended to other places at other times" (1996, 154). Lynch sees in their work "a convincing argument for an intimate connection between political issues and the natural philosophies of Thomas Hobbes and his opponents" (1991, 295).

mitted error on quite a number of points. Indeed, as the debate between them raged on over the decades, it was obvious that they understood each other all too well. On any reasonable definition of what
counts as a form of life, they were incontestably part of the same form
of life: seventeenth-century intellectuals engaged in a "battle of the
books" with conventions as explicit and well-understood as those of
chess. Even in their disagreements about very general matters in the
philosophy of mathematics (such as the nature of ratios or the relation
between mathematics and natural philosophy), they understood what
was being debated, the reasons advanced for each position, and the
consequences attending to such reasons. Although one or the other
disputant might dismiss his opponent's claims as "nonsense" or "unintelligible," even this does not show a significant failure of understanding. Wallis, for instance, understood perfectly well what Hobbes meant
when he claimed that time and a line are homogeneous; but he took
the claim to be false and dismissed it as "absurd," not thereby intending to indicate his inability to understand what was asserted, but
rather to show that he took it to be simply false.

It is also critical to note that there is simply no interesting set of
social factors that can put Hobbes's mathematical opponents into a
single camp. By religion, class background, political affiliation, profes-
sional training, patronage, membership in scientific societies, or any
other social category, those who criticized Hobbes's mathematics are
a very diverse group indeed. These critics include Wallis, a Presbyterian
divine and professor at Oxford; Huygens, a Dutch Calvinist of independent means who was elected to the Parisian Académie Royale des
Sciences and granted a royal pension; André Tacquet, a Belgian Jesuit
who taught at the universities of Bruges, Louvain, and Antwerp; Roberval, a French Catholic professor at the Collège Royal; de Sluse, a
Belgian Catholic priest who held a variety of high offices in the church;
Mylon, the son of Louis XIII's *contrôleur-général des finances* and an
advocate at the *Parlement* of Paris; Pierre de Carcavi, a moderately
wealthy French nobleman with no teaching position; John Pell, an
English Protestant professor and clergyman who also held important
government positions under the Protectorate of Cromwell; Viscount
Brouncker, an Anglo-Irish aristocrat; and Seth Ward, an Anglican Royalist who nevertheless managed to hold the Savilian Professorship of
Astronomy during the interregnum. It is only by reference to mathematical (as opposed to sociological) facts that one might hope to explain why such a diverse group could make common cause against

Hobbes's mathematics.[5] All of this shows that, although there was certainly a political and social dimension to the dispute, the role of such factors is necessarily limited.

There is the further difficulty that Hobbes clearly began as "one of the guys" in the 1640s—a regular and fully paid-up member of the mathematical community, but there is no easy way to explain how he could end up constructing a whole new form of mathematical life all by himself. A form of life is supposed to be a collective achievement, something that determines how particular "games" like mathematics are to be played, and decidedly not the sort of thing that can be constructed by a solitary individual. Just as there is supposed to be no private language, there can be no such thing as a private mathematics. But the record of Hobbes's increasing hostility toward the mathematics of his opponents shows him to have been quite alone on his trek into isolation and self-contradiction, because even friends like du Verdus would not follow him in his rejection of traditional geometry. Hobbesian mathematics circa 1670 was a decidedly private affair, and the sociology of knowledge must have difficulty accounting for this fact.

A perhaps more serious difficulty confronting the sociologist of knowledge is the fact that Hobbes has no coherent program of his own. Taking his statements at face value and subjecting them to the standard interpretation of such terms as *more, less, magnitude, root,* and *multiply,* Hobbes's pronouncements at the end of his career yield such self-contradictions as $1.9997 = 2$, or $3.2 = \pi$. Hobbes rejects the reasoning that leads to these absurdities, but there is no stable doctrine that seems to underlie his objections. Consider, for example, his claim that "the root of 100 stones is 10 stones" (Hobbes 1669b, 2; OL 4:508). He elsewhere defines the root of any quantity as a quantity that, when multiplied by itself, yields the quantity of which it is the root. But what is the product of 10 stones by 10 stones, or, more properly, how do you multiply a stone by a stone? The very notion seems incapable of serious consideration. I conclude that the Hobbesian dictum "in every extraction of a numerical square root the numbered thing is the same as in the number from which the root is extracted" lacks any plausible sense even on Hobbes's own terms. Furthermore, when Hobbes insisted that his geometric constructions could not be refuted by alge-

5. I suppose one might argue that the relevant social factor is membership in the community of recognized mathematicians, but such a criterion simply begs the question, and in any case, Hobbes certainly had a sufficient reputation to be counted as a member of such a community in the 1640s and 1650s.

braic means, others (notably Huygens and de Sluse) provided purely geometric arguments that showed the failure of his constructions. Thus, we need not think that Hobbes was somehow working out a nonalgebraic alternative to traditional mathematics, which alternative must be judged according to some sui generis standard.

Another problem for sociological positivism in this case concerns the remarkable fact that the sociological analysis of Hobbesian mathematics must, it seems, take a line diametrically opposed to the sociological reconstruction of Hobbes's dispute with Boyle. Shapin and Schaffer imagine that Hobbes's insistence that natural philosophy proceed by a priori deductions from mechanical first principles is a causal consequence of his political absolutism. Just as the sovereign must lay down absolute and unchallengeable dictates to govern the commonwealth, so the properly developed natural philosophy must start with abstract hypotheses rather than messy empirical data.[6] Boyle, in contrast, seeks a "commonwealth" or "balance of powers" guaranteed by publicly certified agreements, and his insistence upon the role of experiment is supposed to be linked to this political agenda. But in the realm of mathematics, it is Hobbes who appeals to the empirical adequacy of his results while his opponents insist upon the unchallenged (and unchallengable) axioms of traditional geometry. Hobbes claimed, for example, that two lines in a construction "are equal, at least so closely that the difference cannot be discovered either by the senses or by reasoning" (*CTH* 2:609). The rest of the mathematical world greeted such appeals to approximation with derision, arguing that the issue is not whether two lines have any discernible difference in length, but whether their equality has been demonstrated on the basis of axiomatic first principles. Further, Wallis attacked Hobbes's geometric principles for importing questionable physical principles into the realm of mathematics and thereby failing to respect the abstract, universal, and unchallengable character of geometric reasoning.

6. Shapin and Schaffer insist that "[t]he force by which Leviathan lays down and executes the laws of the commonwealth is therefore the same force that lies behind geometrical inferences" (1985, 153). Later they claim that Hobbes's dispute with Boyle "was a contest about power and assent. Geometry was normative for social relations because it was consistent with the Hobbesian model of assent" (1985, 328). Barnes, Bloor, and Henry agree that "[t]he model of knowledge for Hobbes was geometrical reasoning, which could secure total and irrevocable agreement. . . . Hobbes's method in natural philosophy, as in his political theory, ultimately depended upon unquestioning obedience to an absolute authority" (1996, 153). It may be useful to point out that by the time of *PRG*, Hobbes seems to have given up on the idea that his (or any other) geometrical principles could garner universal assent.

Ought we therefore to conclude Hobbes was a closet ally of the broad popular masses while his opponents in the Royal Society were crypto-absolutists, whose political agenda can be glimpsed in their unswerving allegiance to the authority of Archimedes?

A further reason for skepticism about a purely sociological approach to this controversy is the fact that Wallis was involved in bitter disputes throughout his career, and there is no way to construct an analysis of these battles that could reduce them to some unifying set of social factors. Richard S. Westfall described Wallis as a "bellicose character engaged in endless quarrels and controversies" (1958, 18), and this opinion can be upheld by a even the most cursory look at his published works. The catalog of Wallis's controversial writings begins with *Truth Tried*, a 1642 critique of Lord Robert Brooke's treatise *The Nature of Truth*. It continues with the *Commercium Epistolicum* of 1658, which is a record of the exchange of letters between the French mathematicians Fermat and Bernard Frenicle de Bessey on the one side and Wallis and Brouncker on the other. This exchange began when Fermat posed some number-theoretic challenge problems to the English mathematical community through the mediation of Thomas White and Kenelm Digby; in the course of time, however, Wallis and Fermat traded hostile comments on the relative merits of Wallis's method of inductions as well as priority claims for French and English mathematicians in the discovery of important results.[7] Wallis pursued his hostility toward French mathematicians in his *Treatise of Algebra* of 1685, which contained a fanciful history of algebra that attributed essentially all of the significant developments in that subject to English authors, and particularly to Thomas Harriot.[8] Another of Wallis's mathematical polemics was his 1657 *Adversus Meibomii, de Proportionibus Dialogus*—a piece whose level of vituperation led Barrow (who held no high opinion of Meibom's efforts) to call it a "diatribe" (*LM* 18, 293).

Mathematical matters were not the only source of Wallis's contributions to the literature of controversy. Theological issues, such as the

---

7. This dispute is examined in Mahoney 1994, 335–47. It is also summarized in Scott 1938, chap. 5.

8. In the preface to the *Treatise of Algebra,* he declares that Harriot "hath taught (in a manner) all that which hath since passed for the *Cartesian* method of *Algebra;* there being scarce any thing of (pure) *Algebra* in *Des Cartes,* which was not before in *Harriot;* from whom *Des Cartes* seems to have taken what he hath (that is purely *Algebra*) but without naming him" (*Treatise of Algebra* preface, sig. a2). French mathematicians did not find this amusing.

nature of the Trinity, the propriety of infant baptism, or the proper account of the Christian Sabbath provided him with the chance to engage in public controversy with a variety of authors that produced no fewer than eight published letters on the Trinity and numerous other minor pieces of polemical theology (Wallis 1692a, 1692b, 1696, 1697). Furthermore, Wallis became embroiled in a bitter dispute with William Holder concerning the credit for teaching a deaf-mute to pronounce some words of English. The result was his *Defence of the Royal Society* (Wallis 1678), which sharply rebuked Holder for the accusations in his *Supplement to the Philosophical Transactions of July 1670* (Holder 1678). Nor were Wallis's quarrels restricted to his published works; his letters clearly show him to have been a man eager for a fight. Scott reports that "Wallis was of a highly contentious disposition. His correspondence, unhappily, leaves no room whatever for doubt on that point. No man ever scorned personal popularity more completely than he" (1938, 88). It is far from credible that all of these quarrels could be correlated with some social or political interests, and the natural interpretation of Wallis's penchant for controversy must be in terms of individual psychological factors rather than some fanciful sociological just-so story in which he appears as the defender of a form of life. The relevance of this to the dispute with Hobbes should be obvious, as it shows that we need not seek for some underlying set of social interests to explain why Wallis pursued his battle with Hobbes so vigorously.

A final reason for treating the social factors as secondary in the dispute between Hobbes and Wallis is that they are not a constant. We saw in chapter 2 that the original impetus to quarrel involved the perceived threat that Hobbes posed to the universities. Yet the question of the status of English universities was settled well before the Restoration and disappeared from the exchanges between Wallis and Hobbes, while mathematical questions remained at center stage for decades. On the other hand, questions of political loyalty appeared relatively late in the dispute, and only after the mathematical terrain had been thoroughly worked over. Moreover, the key mathematical questions on which the dispute centered were not decided on the basis of social or political factors, but on straightforward mathematical grounds. Hobbes's numerous attempted quadratures and cube duplications won no adherents even among his friends and patrons, and the reason for this is evident: he was simply and spectacularly wrong. This is not to say that Hobbes failed in every aspect of his mathematics; after all (as we saw in chapter 4) he could pinpoint weaknesses in Wallis's mathematical work and his approach to questions in the philosophy of mathematics has some important strengths. Nevertheless, in his claims to have squared the circle or to have solved other great problems, Hobbes's efforts consistently fell short of the mark. This is perhaps best shown by the fact that his many quadratures all yield different results and thus cannot even be reconciled with one another, to say nothing of reconciling them with the truth.

8.3 HOBBES AND THE BETRAYAL OF RIGHT REASON

The absurdities to which Hobbes committed himself in defense of his geometric claims make it tempting to see him as driven by blind and irrational forces unconnected to mathematics, and considerations of explanatory symmetry might well make it seem that the same must be true of his opponents. After all, if we seek to understand the controversy from the point of view of the participants, and if we grant that Hobbes, at least, saw some significant merit in his claims to have squared the circle, we seem compelled to find some standpoint from which Hobbes's mathematical writings make sense. Wittgensteinian sociologists of knowledge will urge that it is only from within a form of life that utterances or other actions can be regarded as meaningful, and then conclude that Hobbes must have been part of an alternative (and apparently very exclusive) form of mathematical life, whose conflict with the prevailing norms must be examined symmetrically, i.e., without recourse to concepts like truth or error.

I think, however, that Hobbes himself outlines a better account of the matter. He possessed a keen psychological insight, particularly in matters of human motivation, and what he says about mankind can be taken to apply to his own case. In *Leviathan* he famously claimed to "put for a generall inclination of all mankind, a perpetuall and restlesse desire of Power after power, that ceaseth onely in Death" (*L* 1.11, 47; *EW* 3:85–86), while he also described ambitious men as "setting themselves against reason, as oft as reason is against them" (*L* 1.11, 50; *EW* 3:91). Hobbes plainly intended for his mathematical work to establish his standing at the forefront of European mathematics, and his mathematical ambitions were nothing less than a restless desire for the kind of power that comes with the reputation as a great savant. To have been mocked and humiliated at the hands of Wallis was, as Hobbes himself confessed, almost more than he could bear. His failures eventually turned him against the mathematics he had once declared as the pinnacle of human reason. His refusal to yield ground was the product of shattered ambition and wounded pride, as well as his sense that he had nothing further to lose if his geometry were to go down to defeat. In fact, when Hobbes described the role of reason in controversy in *Leviathan,* his words would unwittingly be quite appropriate to his conflict with Wallis:

> And as in Arithmetique, unpractised men must, and Professors themselves may often erre, and cast up false; so also in any other subject of Reasoning, the ablest, most attentive, and most practised men, may deceive themselves, and inferre false Conclusions; Not but that Reason it selfe is always Right Reason, as well as Arithmetique is a certain and infallible Art: But no one mans Reason, nor the Reason of any one number of men, makes the certaintie; no more than an account is therefore well cast up, because a great many men have unanimously approved it. And therefore, as when there is a controversy in an account, the parties must by their own accord, set up for right Reason, the Reason of some Arbitrator, or Judge, to whose sentence they will both stand, or their controviersie must either come to blowes, or be undecided, for want of a right Reason constituted by Nature; so is it in all debates of what kind soever: And when men that think themselves wiser than all others, clamor and demand right Reason for judge; yet seek no more, but that things should be determined, by no other mens reason but their own, it is as intolerable in the society of men, as it is in play after trump is turned, to use for trump on every occasion, that suite whereof they have most in their hand. For they do nothing els, that will have every of their passions, as it comes to bear and sway them, to be taken for right Reason, and that in their own controversies: bewraying their want of right reason, by the claym they lay to it. (*L* 1.5, 18–19; *EW* 3:30–31)

There is probably no better description of Hobbes's own predicament. Convinced that he had delivered principles that could make short work of any mathematical problem, blinded by his passionate desire to defeat Wallis and reap his share of mathematical glory, and ultimately embittered by his failures, Hobbes himself betrayed his want of right reason by his claim to it.

Selections from Hobbes's
Mathematical Writings

Hobbes's mathematical writings are a varied and voluminous lot produced over the course of more than three decades. The following selections are intended to provide a relatively detailed account of Hobbes's mathematics that can illustrate some of the points made in the course of the book. There can be no question of attempting a complete treatment of all Hobbes's mathematical efforts, not least because the frequency of technical errors and outright blunders make it an oeuvre that does not handsomely repay close attention. I have included the following mathematical items: Hobbes's contribution to Pell's 1647 refutation of Longomontanus; the 1656 and 1668 versions of part 3, chapter 17, article 2 of *De Corpore,* which present alternative demonstrations of the fundamental theorem for the quadrature of "deficient figures"; the original circle quadrature intended for part 3, chapter 20, article 1 of *De Corpore* (but not actually published) and the first of three quadratures printed in the twentieth chapter of the 1656 version of *De Corpore;* the comparison of the Archimedean spiral and the parabola from part 3, chapter 20, article 5 of *De Corpore;* the 1661 cube duplication published anonymously in Paris; and the first proposition from Hobbes's 1669 *Quadratura Circuli, Cubatio Sphaerae, Duplicatio Cubi breviter demonstrata,* which purports to square the circle.

A.I HOBBES'S CONTRIBUTION TO PELL'S REFUTATION OF LONGOMONTANUS

This theorem was originally published in John Pell's *Controversiae de verâ circuli mensurâ . . . Prima pars* (Pell 1647, 50–51) as one of several alternative proofs of the result that can be stated as follows: if a is the tangent to an arc less than a quadrant of a circle, and β is the tangent to one-half of the same arc, and the circle has radius r, then $a{:}\beta :: 2r^2{:}(r^2 - \beta^2)$. The primary interest in Hobbes's proof is that it shows his commitment to the classical theory of proportions. Where others who contributed to Pell's campaign against Longomontanus used more compressed "analytic" techniques and employed re-

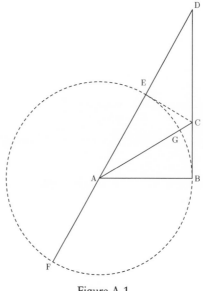

Figure A.1

latively short algebraic arguments, Hobbes's proof employs a rather prolix classical style characteristic of the "synthetic" demonstrations in Euclid's *Elements;* this is especially apparent in Hobbes's use of the various techniques for manipulating ratios and proportions, which he takes from the fifth book of Euclid. A manuscript version of Hobbes's proof survives as British Library MS. Add. 4278, f. 200^r, from which it is possible to date the work to June of 1645. I have made minor changes to Hobbes's notation (these include using 'AB^2' where he uses 'ABq', expressing proportions by '$A{:}B :: C{:}D$' rather than Hobbes's preferred symbolism '$A.B :: C.D$', using the symbol '×' for multiplication, and adding parentheses where necessary to identify the factors in a product).

The tangent to an arc less than a quadrant is to the tangent of half that arc as twice the square of the radius to the square of the radius minus the square of the tangent of one half the arc.

Let *EBF* [in figure A.1] be a circle with center *A* and radius *AB,* and let the arc less than a quadrant *AB* be *BE,* half of which is *BG.* The tangent of the whole arc *BE* is the right line *BD.* The tangent of the arc *BG* is the right line *BC.*

I say $BD{:}BC :: 2AB^2{:}AB^2 - BC^2$.

Let the points *E, C* be connected. Then because the right lines *AB, AC* of the triangle *ABC* are equal to the right lines *AE, AC* of the triangle *AEC,* and the angle BAC is equal to the angle EAC, then by Euclid (*Elements* 1, prop. 4), the base EC is equal to the base BC, and the angle AEC is equal to the angle ABC, which is a right angle. Because the triangles DEC, DBA have equal angles at E and B, namely right angles, and they have the angle at D in common, they will be similar. Therefore the proportion $DB:AB :: DE:EC$ holds, and since EC and BC are equal, the proportion $DB:AB :: DE:BC$ also holds.[1] Permuting the terms of this proportion yields $DB:DE :: AB:BC$, and squaring the terms gives $DB^2:DE^2 :: AB^2:BC^2$. Now let DA be produced to F, and the rectangle contained by FD and DE (that is by $2AB + DE$ and DE) will be equal to DB^2. Thus DB is a mean proportional between $2AB + DE$ and DE. And therefore DB^2 is to DE^2 as $2AB + DE$ to DE. But it was shown that $DB^2:DE^2 :: AB^2:BC^2$. Therefore the proportion $AB^2:BC^2 :: 2AB + DE:DE$ holds. Further, by separating the ratio, the proportion $AB^2:AB^2 - BC^2 :: 2AB + DE:2AB$ is established.[2] And multiplying by the line DE, we have $AB^2:AB^2 - BC^2 :: (2AB + DE) \times DE:2(AB \times DE)$. But $(2AB + DE) \times DE$ is DB^2. Also, $2(AB \times DE)$ is equal to $2(DB \times BC)$ (since $DB:DE :: AB:BC$, as was shown above, and thus the rectangle contained by DB and BC is equal to the rectangle contained by AB and DE, and so $2(AB \times DE)$, the double of the second rectangle, is equal to $2(DB \times BC)$, the double of the first). Therefore $AB^2:AB^2 - BC^2 :: DB^2:2(DB \times BC)$ or $AB^2:AB^2 - BC^2 :: DB^2:2(BC - BD)$. And removing the common altitude DB, the proportion $AB^2:AB^2 - BC^2 :: DB:2BC$ arises. But because $2BC:BC :: 2AB^2:AB^2$, there will arise (*ex aequali* by a perturbed proportion, as in the scheme set out below),[3] the proportion $2AB^2:AB^2 - BC^2 :: DB:BC$. Which was to be demonstrated.

1. I use "the proportion $DB:AB :: DE:BC$ holds" and the like for Hobbes's rather awkward Latin "*DB.AB :: DE.BC* sunt proportionales." At this point in the demonstration, Hobbes embarks on an argument based on the techniques for transforming ratios and proportions that are set out in book 5 of Euclid.

2. Euclid defines "separation of a ratio" in the fifteenth definition of book 5: "**Separation of a ratio** means taking the excess by which the antecedent exceeds the consequent in relation to the consequent itself." In other words, given the ratio A:B, the separation of the ratio is A − B:B (assuming that A is greater than B). The theorem that justifies the maneuver in Hobbes's demonstration is proposition 17 in book 5 of the *Elements*: "If magnitudes be proportional componendo, they will also be proportional seperando"; expressed symbolically, the theorem asserts that if A:B :: C:D, then (A − B):B :: (C − D):D.

3. In *Elements* (5, def. 17), Euclid declares "the ratio **ex æquali** arises when, there being several magnitudes and another set equal to them in multitude which taken two and two are in the same proportion, as the first is to the last among the first magnitudes, so is the first to the last among the second magnitudes; Or, in other words, it means taking the extreme terms by virtue of the intermediate terms." The eighteenth definition in book 5 reads: "A **perturbed proportion** arises when, there being three magnitudes and another set equal to them in multitude, as antecedent is to consequent among the first magnitudes, so is antecedent to consequent among the second magnitudes, while, as the

$$\left\{ \begin{array}{l} 2AB^2 \\ AB^2 \quad \ldots \ldots \quad DB \\ AB^2 - BC^2 \quad \ldots \ldots \quad 2BC \\ BC \end{array} \right\}$$

A.2 THE PRINCIPAL THEOREM IN THE QUADRATURE OF DEFICIENT FIGURES

The seventeenth chapter of *De Corpore* is devoted to the study of deficient figures produced by the motion of a line that diminishes as it moves through a given space. The determination of the area of such figures is the main goal of the chapter, and the key theorem asserts that if the ratio by which the line diminishes is in the *n*th power of the distance it has moved through the figure, then the ratio of the deficient figure to its complement is 1:*n*. Expressed in more modern notation, this is a version of the result that $\int_0^a x^n dx = a^{n+1}/(n + 1)$. The remaining articles in chapter 17 of *De Corpore* contain a series of tables calculating the areas of various deficient figures and comparing them to one another. Wallis remarked that this chapter contains quite a few true propositions, notwithstanding the fact that the proof of the principal theorem is a failure. He concluded that, although he could not say where Hobbes had gotten them, "it is to be suspected that they are not yours, because they are true, while things of yours are wont to be false" (*Elenchus* 84).

Hobbes's presentation of this material has a very strong similarity to Cavalieri's method of indivisibles, most specifically in the fourth of his *Exercitationes Geometricae,* but it may also derive from Roberval's approach to the theory of indivisibles. One key point of similarity between Hobbes and the "indivisiblist" mathematicians is reflected in Hobbes's use of the language of indivisibles: he speaks of lines being drawn through "every possible part of a right line," refers to the latitude of small spaces as "indivisible," and describes a figure as "made up of so many indivisible spaces." The basic theorem is presented here in the version printed in the 1656 English translation of *De Corpore* and then in the version from the 1668 edition of *De Corpore.* The 1656 version contains two arguments for the result, the first of which was new to the English *De Corpore,* while the second is a reworked version of the

---

consequent is to a third among the first magnitudes, so is a third to the antecedent among the second magnitudes." Hobbes's use of these concepts is relatively straightforward: the previous argumentation has established two sequences of ratios that can be compared to one another, and the middle terms of both sequences can be removed by using the Euclidean definitions.

argument in the original 1655 Latin *De Corpore*. The 1668 version gives a rather different version of the argument.

Although the various proofs differ in interesting ways, their similarities are also quite striking.[4] Hobbes attempts to establish his general result concerning the ratios of areas between a deficient figure and its complement, and in doing so he first attempts to establish that a ratio holds between pairs of lines in the deficient figure and its complement, after which he concludes that the same ratio will hold between the areas of the figure and its complement. This overall style of argument parallels the procedures of Cavalieri, but Hobbes's argumentation ultimately runs into serious difficulties in the details. The fact that the central result is given three distinct attempted proofs shows fairly clearly that Hobbes was made aware of the inadequacies in his argumentation, but (as inspection of the following material shows) the final version is hardly an improvement over those that preceded it. One important feature of Hobbes's argumentation is the fact that, although he speaks of figures as "made up" of indivisible elements, he does not attempt to represent the area of the figure as an infinite sum in the style of Roberval's *Traité des indivisibles* or Wallis's *Arithmetica Infinitorum*. Instead, he tries to work out a means of comparing the ratios of areas without using the sort of "algebraic" means he condemned as essentially ungeometrical.

A.2.1 De Corpore, *Part 3, Chapter 17, Article 2 (1656 Version)*[5]

A Deficient Figure, which is made by a Quantity continually decreasing to nothing by proportions every where proportionall and commensurable,[6] is to

4. Wallis took the differences between the Latin and English versions of chapter 17 of *De Corpore* as Hobbes's admission of error: "As to the Demonstration, you keep a vapouring (nothing to the purpose,) as if it were a good demonstration and not confuted. Yet, when you have done, (because you knew it to be naught) you leave it quite out in the English, and give us another (as bad) instead of it. That is, you confess the charge. Your fundamentall Proposition was not demonstrated; and so this whole chapter comes to nothing" (*Due Correction* 119). This is a rather uncharitable reading of the evidence, since Hobbes did include the original argument (slightly modified in response to Wallis's criticisms) in the English version of *De Corpore*. Nevertheless, Wallis is correct in regarding Hobbes's argumentation as ultimately unconvincing.

5. I have corrected numerous typographical errors in the original and have made slight alterations to the diagram to make it fit the text. The frequency of typographical errors is particularly high in the second half of the article, which contains Hobbes's reworked version of the argument from the 1655 version *of De Corpore*. Almost all of these errors are in the labels for lines in the diagram. This suggests that in revising the argument to meet Wallis's criticisms Hobbes was either undecided about how best to proceed or making revisions hastily.

6. It should be remembered that Hobbes uses the English term *proportion* as a translation for the Latin *ratio*—a fact that leads to some significant linguistic confusions in

its Complement, as the proportion of the whole altitude, to an altitude diminished in any time, is to the proportion of the whole Quantity which describes the Figure, to the same Quantity diminished in the same time.

Let the quantity AB [in figure A.2.1], by its motion through the altitude AC, describe the Complete Figure AD; and againe, let the same quantity, by decreasing continually to nothing in C, describe the Deficient Figure ABEFC, whose Complement will be the Figure BDCFE. Now let AB be supposed to be moved till it lie in GK, so that the altitude diminished be GC, and AB diminished be GE; and let the proportion of the whole altitude AC to the diminished altitude GC, be (for example) triplicate to the proportion of the whole quantity AB or GK, to the diminished quantity GE.[7] And in like manner, let HI be taken equal to GE, & let it be diminished to HF; and let the proportion of GC to HC be triplicate to that of HI to HF; & let the same be done in as many parts of the straight line AC as is possible; and a line be drawn through the points B, E, F and C. I say the deficient figure ABEFC, is to its Complement BDCFE as 3 to 1, or as the proportion of AC to GC is to the proportion of AB, that is, of GK to GE.

For (by the second Article of the 15. Chap.) the proportion of the complement BEFCD to the deficient figure ABEFC is all the proportions of DB to OE, and of DB to QF, and of all the lines parallel to DB terminated in the line BEFC, to all the parallels to AB terminated in the same points of the line BEFC.[8] And seeing the proportions of DB to OE, and of DB to QF &c. are

---

his account of the theory of ratios. Hobbes defines proportionality and commensurability of ratios at DCo 3.17.1; EW 1:247. The relevant definitions read as follows: "Four Proportions are said to be *Proportionall*, when the first of them is to the second, as the third is to the fourth. For example, if the first proportion be duplicate to the second; and again the third be duplicate to the fourth, those Proportions are said to be *Proportionall*. And Commensurable Proportions are those, which are to one another as number to number. As when to a proportion given, one proportion is duplicate, another triplicate, the duplicate proportion will be to the triplicate proportion as 2 to 3; but to the given proportion it will be as 2 to 1; and therefore I call those three proportions *Commensurable*." Despite some obscurity in the exposition, Hobbes's intent is fairly clear— when the ratio $\alpha{:}\beta$ is duplicate of $\gamma{:}\delta$, while the ratio $\varphi{:}\psi$ is likewise duplicate of $\chi{:}\omega$, then the pairs of ratios are proportional, which is essentially a generalization of the definition of four quantities standing in a proportion, i.e., in the same ratio. Similarly, commensurability is a matter of ratios being expressible in terms of integers.

7. The sixteenth article of chapter 13 of *De Corpore* (in the 1656 version) contains Hobbes's definition of multiplication of ratios: "A Proportion is said to be multiplied by a Number when it is so often taken as there be Unities in that Number" (*DeC* 2.13.16; *EW* 1:164). Thus, the ratio 1:3 taken twice is the ratio 1:9, or the ratio 5:4 taken twice is the ratio of 25:16; and in general if the ratio $\alpha{:}\beta$ is multiplied by n the result is $\alpha^n{:}\beta^n$. Thus, Hobbes is here assuming that $AC{:}GC = AB^3{:}GE^3$.

8. Hobbes here refers to a proposition added to the English version of chapter 15 that was not present in the 1655 Latin original. The proposition considers the velocities

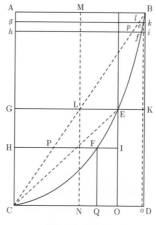

Figure A.2.1

every where triplicate of the proportions of *AB* to *GE*, and of *AB* to *HF* &c. the proportions of *HF* to *AB*, and of *GE* to *AB* &c. (by the 16 Article of the 13 Chap.)[9] are triplicate of the proportions of *QF* to *DB*, and of *OE* to *DB* &c. and therefore the deficient figure *ABEFC* which is the aggregate of all the lines *HF, GE, AB*, &c. is triple to the complement *BEFCD* made of all the lines *QF, OE, DB*, &c.; which was to be proved.[10]

It follows from hence, That the same complement *BEFCD* is 1/4 of the whole Parallelogram. And by the same method may be calculated in all other

---

with which deficient figures are described, and it makes reference to the first figure of chapter 17 (figure A.2.1 here). The key portion of the proposition asserts that "as the proportions of the Swiftnesses wherewith *QF, OE, DB*, and all the rest supposed to be drawn parallel to *DB*, and terminated in the Line *BEFC*, are to the proportions of their several Times designed by the several parallels *HF, GE, AB* and all the rest supposed to be drawn parallel to the Line of time *CD*, and terminated in the Line *BEFC* (the aggregate to the aggregate) so is the Area or Plain *DBEFC* to the Area or Plain *ACFEB*" (*DCo* 3.15.2; *EW* 1:208–9). The point of the proposition is that throughout the curve *BEFC* the ratio between the parallels in the deficient figure to the parallels in its complement is constant, and indeed equal to the ratio of the areas of the deficient figure to its complement.

9. As mentioned above, this article is Hobbes's definition of multiplication of ratios.

10. The conclusion fails to follow, since from the fact that the lines are in triplicate proportion (i.e., in the ratio of cubes), nothing Hobbes has said thus far establishes that the areas are in the ratio of one to three. As Wallis remarked "this is but the same Bull that hath been baited so often. viz. *because the diameters* (DB, OE, QF, &c. that is CA, CG, CH,) *are in the triplicate proportion of the Ordinates* (AB, GE, HF,) *therefore the ordinates are in the triplicate proportion of the diameters. . . . But how doe you prove this consequence? Nay, not a word of proof. We must take your word for it*" (*Due Correction* 120).

Deficient Figures generated as above declared, the proportion of the Parallelogram to either of its parts; as that when the parallels encrease from a point in the same proportion, the Parallelogram will be divided into two equal Triangles; when one encrease is double to the other, it will be divided into a Semiparabola and its Complement, or into 2 and 1.

The same construction standing, the same conclusion may otherwise be demonstrated, thus.[11]

Let the straight line *CB* be drawn cutting *GK* in *L*, & through *L* let *MN* be drawn parallel to the straight line *AC*; wherefore the Parallelograms *GM* and *LD* will be equal. Then let *LK* be divided into three equal parts, so that it may be to one of those parts in the same proportion which the proportion of *AC* to *GC* or of *GK* to *GL* hath to the proportion of *GK* to *GE*. Therefore *LK* will be to one of those three parts as the Arithmetical proportion between *GK* and *GL* is to the Arithmetical proportion between *GK* and the same *GK* want the third part of *LK*; and *KE* will be somewhat greater then a third of *LK*.[12] Seeing now the altitude *AG* or *ML* is by reason of the continual decrease, to be supposed less than any quantity that can be given; *LK* (which is intercepted between the Diagonal *BC* and the side *BD*) will also be less then any quantity that can be given; and consequently, if *G* be put so neer to *A* in *g*, as that the difference between *Cg* and *CA* be less then any quantity that can be assigned, the difference also between *Cl* (removing *L* to *l*) and *CB*, will be less then any quantity that can be assigned; and the line *gl* being drawn & produced to the line *BD* in *k* cutting the crooked line in *e*, the proportion of *AC* to *gC* will still be triplicate to the proportion of *gk* to *ge*, and the difference between *k* and *e* and the third part of *kl* will be less then any quantity that can be given; and therefore the Parallelogram *eD* will differ from a third part of the Parallelogram *Ae* by a less difference then any quantity that can be assigned. Again, let *HI* be drawn parallel and equal to *GE*, cutting *CB* in *P*, the crooked line in *F*, and *EO* in *I*, and the proportion of *Cg*, to *CH* will be triplicate to the proportion of *ge* to *HF*, and *IF* will be greater then the third part of *PI*. But again,

---

11. What follows is a reworked version of the argument in the original 1655 *De Corpore*.

12. In the original version of this argument, Hobbes claimed that *KE* would be one-third of *LK* (Hobbes 1655, 145). Wallis objected that, since the point *G* is taken arbitrarily, the ratio between *KE* and *LK* could take on any desired value and the argument therefore fails (*Elenchus* 70). In the *Six Lessons*, Hobbes replied that "you did not then observe, that I make *the Altitude AG, less then any Quantity given*, and by consequence *EK* to differ from a third part by a less difference then any Quantity that can be given" (*SL* 5; *EW* 7:300). Hobbes therefore modified the argument by admitting that the ratio differs from the desired result, but that the difference can be made as small as desired.

setting H in h so neer to g, as that the difference between Ch and Cg may be as but a point, the point P will also in p be so neer to l, as the difference between Cp and Cl will be but as a point; and drawing hp till it meet with BD in i, cutting the crooked line in f, and having drawn eo parallel to BD, cutting DC in o, the Parallelogram fo will differ less from the third part of the Parallelogram gf, then by any quantity that can be given. And so it will be in all other Spaces generated in the same manner. Wherefore the differences of the Arithmetical and Geometrical Means,[13] which are but as so many points B, e, f. &c. (seeing the whole Figure is made up of so many indivisible Spaces) will constitute a certain line, such as the line $BEFC$, which will divide the complete Figure AD into two parts, whereof one, namely $ABEFC$, which I call a Deficient Figure, is triple to the other, namely $BDCFE$, which I call the Complement thereof, And whereas the proportion of the altitudes to one another, is in this case everywhere triplicate to that of the decreasing quantities to one another; in the same manner if the proportion of the altitudes had been every where quadruplicate to that of the decreasing quantities it might have been demonstrated, that the Deficient Figure had been quadruple to its Complement; and so in any other proportion, Wherefore, a Deficient Figure, which is made, &c. Which was to be demonstrated.

A.2.2 De Corpore, *Part 3, Chapter 17, Article 2 (1668 Version)*

A deficient figure made by a quantity continually decreasing by ratios everywhere proportional and commensurable until it evanesces is to its complement as the ratio of the whole altitude to an altitude diminished in any time is to ratio of the whole quantity which describes the figure to the same quantity diminished in the same time.

Let the parallelogram $ABCD$ (figure A.2.2) be described, and let the base AB be understood to be moved parallel to CD, so that as it is moved it perpetually decreases until it evanesces in point C; and let the ratio of AB diminished to the same whole AB be everywhere the same as the ratio of AC to AG, either everywhere duplicate, or triplicate, or in any other ratio of a ratio to a ratio. While AB decreases in this way the point B describes a certain line, say $BEFC$. Now I say that if the ratio of AC to AG is the same as the ratio of AB to GE,

13. Hobbes claims in the corollary to *DCo* 2.13.28 that "if any quantity be supposed to be divided into equal parts infinite in number, the difference between the Arithmetical and Geometrical Means will be infinitely little, that is, none at all. And upon this foundation chiefly, the Art of making those Numbers which are called *Logarithmes* seems to have been built." This principle was also invoked in the 1655 version of the argument. It is unclear to me just how Hobbes thinks that this principle can assist in his argument.

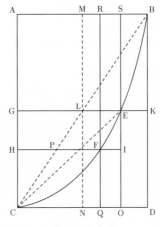

Figure A.2.2

then the space *ABEFC* is to the space *DCFEB* as one to one; but if the ratio of *AC* to *AG* is the duplicate of the ratio of *AB* to *GE,* then the space *ABEFC* is to the space *DBEFC* as two to one; if triplicate, as three to one, and so on.

LEMMA 1

By whatever ratio the velocity of a moved point is increased, in the same ratio are also increased the spaces passed over by the point in the same or equal times.

LEMMA 2

If between two right lines there are interposed an infinite number of both arithmetical and geometrical means, these do not differ in magnitude.[14]

In the parallelogram *ABCD* let the side *AB* be understood to be moved parallel to the side *CD,* and in moving to decrease until at last it evanesces in the point *C;* and by such motion the figure *ABEFC* is described, leaving the complement whose line *BEFC* is described by the endpoint *B* of the decreasing line *AB.* And in the same time let the side *AC* be understood to move uniformly to *BD.* Thus *CD* can be taken for the measure of time. But the right lines parallel to the line *CD,* terminated on one side in the line *BEFC,* and on the other in the right line *AC,* will be measures of the parts of time in which *AB* is moved to *CD* and *AC* to *BD.*

14. This is just the corollary to *DCo* 2.13.28 that Hobbes appealed to in his original argument, as well as in the revised version of that argument from 1656. The first lemma recapitulates principles from *DCo* 3.16, articles 1–6.

Now let there be taken arbitrarily in the right line *CD* a point *O*, and let *OS* be drawn parallel to the side *BD*, cutting the line *BEFC* in *E* and the right line *AB* in *S*. Again, from the point *Q* taken arbitrarily in *CD* let *QR* be drawn parallel to the same side *BD*, cutting *BEFC* in *F* and *AB* in *R*. And let *EG*, *FH* also be drawn, parallel to *CD* and cutting *AC* in *G* and *H*. Finally, let the same be supposed to be done in every point of the line *BEFC*.

I say that as the aggregate of all the velocities by which the right lines *QF*, *OE*, *DB*, and all the rest generated in the same manner is to the aggregate of the times designated by the right lines *HF*, *GE*, *AB*, and the rest, so the plane surface *DCFEB* is to the plane surface *ABEFC*. Just as *AB* by decreasing through the line *BEFC* in the time *CD* evanesces in point *C*, so *CD* (equal to *AB* itself) by decreasing through the same line *CFEB* in the same time evanesces in the point *B*, having described the right line *DB* equal to *AC*. Therefore the velocities with which *AC* and *DB* are described are equal to one another. On the other hand, seeing that in the same time in which the point *O* describes the right line *OE* the point *S* describes the right line *SE*, then *OE* will be to *SE* as the velocity with which *OE* is described to the velocity with which *SE* is described. And for the same reason *QF* will be to *RF* as the velocity with which *QF* is described to the velocity with which *RF* is described, and thus for all the other parallels. Therefore as the right lines that are parallel to the side *AB* and terminated in the line *BEFC* are the measures of the times, so the right lines that are parallel to the side *BD* (and terminated in the same line *BEFC*) are the measures of the velocities. Now (by lemma 1) in whatever ratio the velocities are increased, in the same ratio the right lines passed over in the same times are increased, namely *QF*, *OE*, *DB*, etc.

Now all the lines *QF*, *OE*, *DB*, etc. constitute the plane surface *DBEFC*; and all the lines *HF*, *GE*, *AB*, etc.—that is, all of *ES*, *FR*, *CA*, etc.—constitute the plane surface *ACFEB*. The former of these are the aggregate of the velocities, the latter the aggregate of times. Thus as the aggregate of the velocities is to the aggregate of the times, so the complement *DBEFC* is to the figure *AB-EFC*. Therefore if indeed the ratios of *DB* to *OE* and *OE* to *QF* should be (for example) triplicate, then vice versa the ratios of *OE* to *DB*, and *QF* to *OE* will be subtriplicate of the ratios of *GE* to *AB* and *HF* to *GE*. Thus the aggregate of all of *QF*, *OE*, *BD*, etc. will be to the aggregate of all *HF*, *GE*, *AB*, etc. (by lemma 2) subtriple.[15] Therefore as the aggregate of the velocities is to the aggregate of the times by which the deficient figure is described, so will the complement of the figure to the deficient figure itself, that is to say the complement *DBEFC* to the figure *ABEFC*. Which was to be demonstrated.

---

15. Even the most charitable reading of Hobbes's principle leaves it obscure why this consequence should follow.

A.3 TWO OF HOBBES'S QUADRATURES FROM DE CORPORE, PART 3, CHAPTER 20

I here present two versions of the several attempted quadratures that Hobbes at various times intended for chapter 20 of *De Corpore*. The first is the original effort that Hobbes planned to include, but he abandoned when *De Corpore* was in proofs. Wallis acquired a copy of this from Hobbes's printer and quoted from it at great length (*Elenchus* 97–107). I have reconstructed the proof from Wallis's quotation of the unpublished sheets. The second is the first of three attempted quadratures published in the 1656 English version of *De Corpore*. Its construction is the same as that of the unpublished first quadrature, but Hobbes no longer claims it as an exact result. Instead, he declares that in his twentieth chapter "I have let stand there that which I did before condemn, not that I think it exact, but partly because the Division of Angles may be more exactly performed by it then by any organicall way whatsoever" (*SL* epistle; *EW* 7:186). Hobbes's indecision and frustration are palpable in the second half of the 1656 version of the argument: he grants that his construction conflicts with established values for π, but he cannot quite see the flaw in his procedures and is unwilling to let the matter rest. I find these two demonstrations more interesting than many other Hobbesian attempts at quadrature because they give a better picture of the "method of motions" he used to approach the problem and from which he expected such great results. This method attempts to rectify curvilinear arcs by imagining them to be straightened as points on the arc move uniformly through specified points in the construction. As is so often the case with Hobbes's mathematics, his constructions in both cases are quite complex (with a number of otiose lines cluttering the diagram) and the argumentation ultimately fails to achieve the desired result. Nevertheless, it is instructive to see just what sort of argument Hobbes had first found persuasive as he prepared chapter 20 of *De Corpore*.

A.3.1 The Original Quadrature Intended for De Corpore, *Part 3, Chapter 20, Article 1*

To find a right line equal to the perimeter of a circle.

Let *ABD* [in figure A.3.1] be the quadrant of a circle, about which is circumscribed the square *ABCD*. Let the sides of the square be bisected at *E, F, G,* and *H*. Then the lines *FH, EG* will divide $\overset{\frown}{BD}$ into three equal parts at *I* and *K*. Let *IM* (the sine of $\overset{\frown}{BI}$) be drawn, and *IM* will be half of the radius *BC*. Let $\overset{\frown}{BI}$ be divided into four equal parts at *L, N,* and *O,* and let the right line *MI* be so divided at *P, Q,* and *R*. Let *PL, QN,* and *RO* be connected, and to *SN* (the sine of $\overset{\frown}{BN}$) let *NT* be added, which is equal to *SN* itself. Finally, let *IT* be drawn and produced to meet *BC* in *V*. I say the right line *BV* is equal to

$\widehat{BI}$, and thus three times BV (which is Be) is equal to the arc BD, and twelve times the same BV is equal to the perimeter of the circle of which ABD is a quadrant.[16]

Because both MQ and QI, as well as SN and NT, are equal, then as ST is to SN, so MI is to QI. Let the right line TI be produced to meet BA produced in q. Then NQ produced falls on the same point q. And qP, qQ, qR joined and produced to BV will cut BV in four equal parts at X, Y, and Z.

On the right lines MQ, QI let two equilateral triangles MgQ, QhI be constructed, and with centers g and h let $\widehat{MQ}$, $\widehat{QI}$ be drawn. Either of these is equal to either of $\widehat{BN}$, $\widehat{NI}$.[17] Again, on the right lines MP, PQ, QR, RI let there be constructed as many equilateral triangles MkP, PlQ, QpR, RbI. And with centers k, l, p, b let $\widehat{MP}$, $\widehat{PQ}$, $\widehat{QR}$, $\widehat{RI}$ be drawn, any one of which is equal to any of $\widehat{BL}$, $\widehat{LN}$, $\widehat{NO}$, $\widehat{OI}$.

Now because $\widehat{IQ}$, $\widehat{IN}$ are equal, rectilinear motion through qQ places the point Q in N; and $\widehat{IN}$, $\widehat{IQ}$ will coincide, of course $\widehat{IQ}$ (which is slightly more curved than $\widehat{IN}$) being slightly straightened. And because $\widehat{IR}$, $\widehat{RQ}$ are both together equal to $\widehat{IN}$, these are straightened by the same motion and are placed in $\widehat{IN}$, with which they coincide. And thus by the rectilinear motion through qR the midpoint R, which is brought to the middle of the right line YV, will be brought to the middle of $\widehat{IN}$, that is through O.[18]

Similarly, because motion through qQ places the point Q in N, and motion through qM places the point M in B, and $\widehat{MQ}$, $\widehat{BN}$ are equal, $\widehat{MQ}$ coincides with $\widehat{BN}$. And by the same motion $\widehat{MP}$, $\widehat{PQ}$ will be placed in the same $\widehat{BN}$, with which they coincide. Therefore, the motion through qP places the midpoint P in the middle of $\widehat{BN}$, that is in L.[19]

In the same manner, by the perpetual bisection of the right line MI, and by constructing equilateral triangles on the parts thus produced, there will arise an infinity of arcs, that is, as many as one might wish, equal to each other

16. Elementary trigonemetric calculation shows that this assertion is false. By construction (taking the radius AB as a unit), $ST = 2\sin(15°)$, $MI = 1/2$, and $SM = \cos(15°) - \cos(30°)$. BV therefore has a length of approximately .5236539, and Hobbes requires a value for π at approximately 3.1419234.

17. More precisely, taking AD as unit, $\widehat{BN} = \widehat{NI} = \pi/12 = \widehat{MQ} = \widehat{QI}$.

18. Wallis observes that this is the crucial misstep in the argument: "In no way [will the point R pass through O], for the right line qRZ does not pass through O. Although the points q, Q, N lie in the same right line, as also do q, M, B, nevertheless q, R, O do not, nor do q, P, L lie in the same line. And if you should contend that they do, it remains for you to prove it" (*Elenchus* 98). From here forward, the argument proceeds from a false supposition and delivers the inevitable false result.

19. Hobbes's "method of motions" fails here again. He has no guarantee that qP continued will pass through L, and calculation shows that, indeed, it does not. In effect, he is assuming what he needs to prove, namely that the fourth part of the line BV is equal to a fourth part of $\widehat{BI}$.

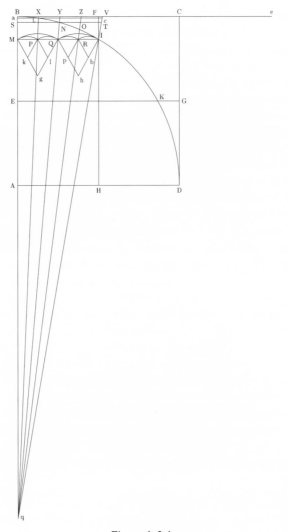

Figure A.3.1

and all taken together equal to $\widehat{BI}$. And by motion through the right lines drawn from q through the several parts of division of the right line *MI*, these arcs will be placed in as many equal parts of the right line *BV*. But the same right lines drawn from q will cut the right lines *BV* and *MI*, and $\widehat{BI}$ in the same ratios.[20]

20. Had the previous argumentation succeded, the demonstration could have ended here because the determination of the arbitrary arc lengths for the arc *BI* would suffice

Let *aL* (the sine of $\overset{\frown}{BL}$) be drawn, and let it be produced until it cuts *IV* at *c*. Then, because *MP* is the fourth part of *MI*, *aL* will be the fourth part of *ac*. And because *qB* is greater than *qa*, *BV* will also be greater than *ac*. Therefore *BV* is greater than four sines of $\overset{\frown}{BL}$, which arc is one fourth of $\overset{\frown}{BI}$. In the same way it can be shown that if $\overset{\frown}{BI}$ were divided into any number of equal parts (so that the difference between the arc itself and the aggregate of as many sines of one of these smallest parts as there are parts in the division is less than any given quantity), the right line *BV* would still be greater than all of these sines taken together. Therefore the right line *BV* is not less than the arc *BI*. But it cannot be greater either, because if this point *B* itself is taken for the sine of the smallest part of the arc, then the aggregate of all the sines considered as points is the right line *BV* itself, and it is equal to the arc *BI*.[21] Therefore the right line *BV* is equal to the arc *BI*, and *Be* is equal to the arc *BD*, and four times *Be* is equal to the perimeter of the circle of which *ABD* is a quadrant. Therefore a right line has been found equal to the perimeter of the circle, which was to be done.

A.3.2 *The First Quadrature from the 1656* De Corpore

Let the Square *ABCD* [in figure A.3.2] be described, and with the *Radii AB*, *BC* and *DC* the three Arches *BD, CA* and *AC;* of which let the two *BD* and *CA* cut one another in *E*, and the two *BD* and *AC* in *F*. The Diagonals therefore *BD* and *AC* being drawn will cut one another in the center of the Square *G*, and the two Arches *BD* and *CA* into two equal parts in *H* and *Y;* and the Arch *BHD* will be trisected in *F* and *E*. Through the Center *G* let the two Straight Lines *KGL* and *MGN* be drawn parallel and equal to the sides of the Square *AB* and *AD*, cutting the four sides of the same Square in the points *K, L, M* and *N;* which being done, *KL* will pass through *F*, and *MN* through *E*. Then let *OP* be drawn parallel and equal to the side *BC*, cutting the Arch *BFD* in *F*, and the sides *AB* and *DC* in *O* and *P*. Therefore *OF* will be the sine of the arch *BF*, which is an arch of 30 degrees; and the same *OF* will be equal to half the Radius. Lastly, dividing the arch *BF* in the middle in *Q*, let *RQ* the Sine of the

---

for the quadrature of the circle. Hobbes continues, however, and the remaining argumentation resembles the kind of double reductio ad absurdum argument found in Archimedean exhaustion proofs. His idea is to show that the line *BV* can be neither greater nor less than the arc *BI*. Unfortunately for Hobbes, his argument begs the crucial question.

21. This part of the argument begs the question, as Hobbes later realized. In the printed version of the 1655 *De Corpore* he admitted, "Seeing that it is possible that the line *qP* produced to the perpendicular *Li* may cut it beyond *L*, it is also possible that *aL* is greater than one-fourth of the right line *ac*. Therefore it is not demonstrated that the right line *BV* is equal to the arc *BI*" (Hobbes 1655, 171).

arch *BQ* be drawn and produced to *S*, so that *QS* be equal to *RQ*, and consequently *RS* be equal to the chord of the arch *BF*; and let *FS* be drawn and produced to *T* in the side *BC*. I say, the Straight Line *BT* is equal to the Arch *BF*; and consequently that *BV* the triple of *BT* is equal to the Arch of the Quadrant *BFED*.[22]

Let *TF* be produced till it meet the side *BA* produced in *X*; and dividing *OF* in the middle in *Z*, let *QZ* be drawn and produced till it meet with the side *BA* produced. Seeing therefore the Straight Lines *RS* and *OF* are parallel and divided in the midst in *Q* and *Z*, *QZ* produced will fall upon *X*, and *XZQ* produced to the side *BC* will cut *BT* in the midst in *α*.

Upon the Straight line *FZ* the fourth part of the Radius *AB* let the equilateral triangle *aZF* be constructed; & upon the center *a*, with the Radius *aZ* let the arch *ZF* be drawn, which arch *ZF* will therefore be equal to the arch *QF* the half of the arch *BF*.[23] Again, let the straight line *ZO* be cut in the midst in *b*, and the straight line *bO* in the midst in *c*; and let the bisection be committed in this manner till the last part *Oc* be the least that can possibly be taken; and upon it, and all the rest of the parts equal to it into which the straight line *OF* may be cut, let so many equilateral triangles be understood to be constructed; of which let the last be *dOc*. If therefore upon the center *d*, with radius *dO* be drawn the arch *Oc*, and upon the rest of the equal parts of the straight line *OF* be drawn in like manner so many equal arches, all of those arches together taken will be equal to the whole arch *BF*; & the half of them, namely, those that are comprehended between *O* & *Z*, or between *Z* & *F* will be equal to the arch *BQ* or *QF* and in summe, what part soever the straight line *Oc* be of the straight line *OF*, the same part will the arch *Oc* be of the arch *BF*, though both the arch and the chord be infinitely bisected. Now seeing the arch *Oc* is more crooked then that part of the arch *BF* which is equal to it; and seeing also that the more the straight line *Xc* is produced the more it diverges from the straight line *XO*, if the points *O* and *c* be understood to be moved forwards with straight motion in *XO* and *Xc*, the arch *Oc* will thereby be extended by a little and little, till at the last it come some where to have the same crookedness with that part of the arch *BF* which is equal to it. In like manner, if the straight line *Xb* be drawn, and that point *b* be understood to be moved forwards at the same time, the arch *cb* will also by little and little be extended, till its crookedness come to be equal to the crookedness of that part of the arch *BF* which is equal to it. And the same will happen in all those small equal arches which are described upon so many equal parts of the straight line *OF*.

22. This construction is the same as the previous attempted quadrature, although the labels are different. The same trigonometric calculation set forth in note 16 suffices to refute the claim.

23. Indeed, the arcs both have the value $\pi/12$, taking *AD* as unit.

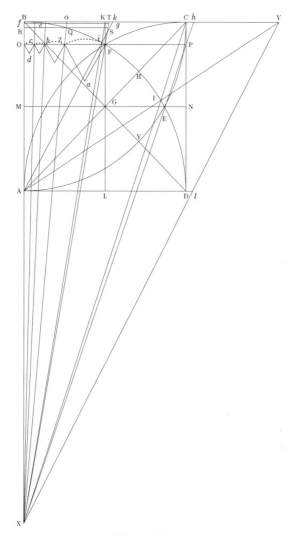

Figure A.3.2

It is also manifest, that by straight motion in *XO* and *XZ* all those small arches will lie in the arch *BF* in the points *B*, *Q* and *F*.[24] And though the same small equall arches should not be coincident with the equall parts of the arch *BF* in all other points thereof, yet certainly they will constitute two crooked lines, not onely equall to the two arches *BQ* and *QF* and equally crooked; but also

24. What Hobbes claims to be "manifest" here is essentially the same falsehood that destroyed the quadrature in A.3.1. Although the rectilinear motion Hobbes de-

having their cavity towards the same parts; which how it should be, unlesse all those small arches should be coincident with the arch *BF* in all its points, is not imaginable.[25] They are therefore coincident, and all the straight lines drawne from *X* & passing through the points of division of the straight line *OF*, will also divide the arch *BF* into the same proportions into which *OF* is divided.[26]

Now seeing *Xb* cuts off from the point *B* the fourth part of the arch *BF*, let that fourth part be *Be;* and let the Sine thereof *fe* be produced to *FT* in *g*, for so *fe* will be the fourth part of the straight line *fg*, because as *Ob* is to *OF*, so is *fe* to *fg*. But *BT* is greater then *fg;* and therefore the same *BT* is greater then four Sines of the fourth part of the arch *BF*. And in like manner, if the arch *BF* be subdivided into any number of equal parts whatsoever, it may be proved that the straight line *BT* is greater then the Sine of one of those small arches so many times taken as there be parts made of the whole arch *BF*. Wherefore the Straight line *BT* is not lesse then the Arch *BF*. But neither can it be greater, because if any straight line whatsoever, lesse then *BT*, be drawn below *BT* parallel to it and terminated in the straight lines *XB* and *XT*, it would cut the arch *BF;* and so the Sine of some one of the parts of the arch *BF* taken so often as that small arch is found in the whole arch *BF*, would be greater then so many of the same arches; which is absurd.[27] Wherefore the Straight line *BT* is equal to the arch *BF*, & the Straight line *BV* equal to the Arch of the Quadrant *BFD;* and *BV* four times taken, equal to the Perimeter of the Circle described with the Radius *AB*. Also the Arch *BF* and the Straight line *BT* are every where divided into the same proportions; and consequently any given Angle, whether greater or less then *BAF* may be divided into any proportion given.

But the straight line *BV* (though its magnitude fall within the terms assigned by *Archimedes*) is found, if computed by the Canon of Sines, to be somwhat

---

scribes here will undoubtedly straighten the small arcs, he has no guarantee that they will be brought into coincidence with $\overset{\frown}{BF}$, and this fact is independent of his rather jarring assumption that the line *OF* can be infinitely subdivided and an infinity of small arcs produced.

25. Hobbes evidently felt some unease about the soundness of this argumentation, but he could not give up the idea that he had found the way to rectify $\overset{\frown}{BF}$. His assertion that the two "crooked lines" are equal to $\overset{\frown}{BQ}$ and $\overset{\frown}{QF}$ is, however, in error and begs the very question at issue.

26. As with the previous attempt at quadrature, the demonstration could have ended here, since this result would be equivalent to the quadrature of the circle. Hobbes nevertheless continues with the same kind of flawed attempt at double reductio ad absurdum he had earlier undertaken.

27. The "absurdity" arises from the fact that Hobbes in effect assumes that his construction divides the arc into equal parts, which is exactly the result he needed to prove. In fact, *BT* will exceed the arc $\overset{\frown}{BF}$.

greater then that wch is exhibited by the *Ludolphine* numbers.[28] Nevertheless, if in the place of *BT,* another straight line, though never so little less, be substituted, the division of Angles is immediately lost, as may by any man be demonstrated by this very Scheme.[29]

Howsoever, if any man think this my Straight line *BV* to be too great, yet, seeing the Arch and all the Parallels are every where so exactly divided, and *BV* comes so neer to the truth, I desire he would search out the reason, Why (granting *BV* to be precisely true) the Arches cut off should not be equal.

But some man may yet ask the reason why the straight lines drawn from *X* through the equal parts of the arch *BF* should cut off in the Tangent *BV* so many straight lines equal to them, seeing the connected straight line *XV* passes not through the point *D,* but cuts the straight line *AD* produced in *l;* and consequently require some determination of this Probleme,[30] Concerning which, I will say what I think to be the reason, namely, that whilest the magnitude of the Arch doth not exceed the magnitude of the Radius, that is, the magnitude of the Tangent *BC,* both the Arch and the Tangent are cut alike by the straight lines drawn from *X;* otherwise not. For *AV* being connected, cutting the arch *BHD* in *I,* if *XC* being drawn should cut the same arch in the same point *I,* it would be as true that the Arch *BI* is equal to the Radius *BC,* as it is true that the Arch *BF* is equal to the straight line *BT,* and drawing *XK* it would cut the arch *BI* in the midst in *i;* Also drawing *Ai* and producing it to the Tangent *BC* in *k,* the straight line *Bk* will be the Tangent of the Arch *Bi,* (which arch is equal to half the Radius) and the same straight line *Bk* will be equal to the straight line *kI.* I say all this is true, if the preceding demonstration be true; and consequently the proportional section of the Arch and its Tangent proceeds hitherto. But it is manifest by the Golden Rule,[31] that taking *Bh* dou-

28. Ludolph van Ceulen (1539–1610) calculated π to thirty-five decimal places while Archimedes found the inequalities $3.14084 < \pi < 3.142858$. Hobbes's construction makes π approximately 3.1419234.

29. Again, Hobbes seems unable to resist begging the question. He assumes that he has found the means to divide any angle into any given number of equal parts, and that this result is "lost" if his construction should fail. But the division is correct only if he has already squared the circle.

30. This is presumably the kind of argument used by one of Hobbes's friends to persuade him of the inadequacy of his quadrature. If the construction had been successful, the line *XD* produced should intersect *BC* produced at *V,* but it does not.

31. The "Golden Rule" or "Rule of Three" is the elementary principle that allows the fourth term of a proportion to be computed if the first three terms are given. As Wallis states it in *Mathesis Universalis:* "Let the third be multiplied by the second, and the product be divided by the first. The quotient will exhibit the fourth term sought" (*MU* 38; *OM* 1:196). Hobbes's point here is that elementary trigonometric calculation shows that his construction fails to solve the problem posed.

ble to *BT*, the line *Xh* shall not cut off the arch *BE* which is double to the arch *BF*, but a much greater. For the magnitude of the straight lines *XM*, *XB* and *ME* being known (in numbers) the magnitude of the straight line cut off in the Tangent by the straight line *XE* produced to the Tangent may also be known; and it will be found to be less then *Bh;* wherefore the straight line *Xh* being drawn will cut off a part of the arch of the Quadrant greater then the arch *BE*. But I shall speak more fully in the next Article concerning the magnitude of the arch *BI*.

And let this be the first attempt for the finding out of the dimension of a Circle by the Section of the arch *BF*.

A.4 THE COMPARISON OF THE SPIRAL OF ARCHIMEDES WITH THE PARABOLA

This result is one of the more intriguing pieces of mathematics to make its way into *De Corpore*. It is Hobbes's account of the rectification of the Archimedean spiral in part 3, chapter 20, article 5. As I argued in chapter 3, the reasoning Hobbes employs here is closely connected with Roberval's analysis of the same problem, as well as making use of Galileo's analysis of the construction of the parabola from uniformly accelerated motion. The argument is essentially sound, the only flaw being Hobbes's assumption that he has rectified arc of the quadrant (in the preceding sections) and that he had found the means to rectify the parabola in the eighteenth chapter of *De Corpore*. This led Hobbes to declare that

> From the known Length of the Arch of a Quadrant, and from the pro-
> portional Division of the Arch and of the Tangent *BC*, may be deduced
> the Section of an Angle into any given proportion; as also the Squaring
> of the Circle, the Squaring of a given Sector, and many the like proposi-
> tions, which it is not necessary here to demonstrate. I will therefore
> onely exhibit a Straight line equal to the Spiral of Archimedes, and so
> dismiss this speculation. (*DCo* 3.20.4; *EW* 1:307)

The argument remained unaltered between its first appearance in the 1655 and 1656 versions of *De Corpore,* but it was removed from the 1668 version. I present the English translation of 1656.

The length of the Perimeter of a Circle being found, that Straight line is also found, which touches a Spiral at the end of its first conversion. For upon the center *A* [in figure A.4] let the circle *BCDE* be described; and in it let *Archimedes* his Spiral *AFGHB* be drawn, beginning at *A* and ending at *B*. Through the center *A* let the straight line *CE* be drawn, cutting the Diameter *BD* at

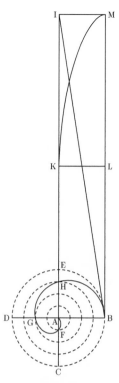

Figure A.4

right angles; and let it be produced to *I;* so, that *AI* be equal to the Perimeter *BCDEB.* Therefore *IB* being drawn will touch the Spiral *AFGHB* in *B;* which is demonstrated by *Archimedes* in his book *de Spiralibus.*[32]

And for a Straight Line equal to the given Spiral *AFGHB,* it may be found thus.

Let the straight line *AI* (which is equal to the Perimeter *BCDE*) be bisected in *K;* and taking *KL* equal to the Radius *AB,* let the rectangle *IL* be completed. Let *ML* be understood to be the axis, and *KL* the base of a Parabola, and let *MK* be the crooked line thereof. Now if the point *M* be conceived to be so moved by the concourse of two movements, the one from *IM* to *KL* with velocity increasing continually in the same proportion with the Times, the other from *ML* to *IK* uniformly, that both those motions begin together in *M* and end in *K; Galilaeus* has demonstrated that by such motion of the point *M,* the

32. The relevant results are propositions 18–20 of Archimedes' "On Spirals." See Archimedes 1919, 171–76, and Dijksterhuis [1956] 1987, 268–74.

crooked line of a Parabola will be described.[33] Again, if the point *A* be conceived to be moved uniformly in the straight line *AB,* and in the same time to be carried round upon the center *A* by the circular motion of all the points between *A* and *B, Archimedes* has demonstrated that by such motion will be described a Spiral line. And seeing the circles of all these motions are concentrick in *A;* and the interiour circle is alwayes lesse then the exteriour in the proportion of the times in which *AB* is passed over with uniform motion; the velocity also of the circular motion of the point *A,* will continually encrease proportionally to the times. And thus far the generations of the Parabolical line *MK,* and of the Spiral line *AFGHB,* are like. But the Uniform motion in *AB* concurring with circular motion in the Perimeters of all the concentrick circles, describes that circle, whose center is *A,* and Perimeter *BCDE;* and therefore that circle is (by the *Coroll.* of the first article of the 16 Chapter) the aggregate of all the Velocities together taken of the point *A* whilst it describes the Spiral *AFGHB.*[34] Also the rectangle *IKLM* is the aggregate of all the Velocities together taken of the point *M,* whilest it describes the crooked line *MK.* And therefore the whole velocity, by which the Parabolicall line *MK* is described, is to the whole velocity with which the Spiral line *AFGHB* is described in the same time, as the rectangle *IKLM,* is to the Circle *BCDE,* that is to the triangle *AIB.* But because *AI* is bisected in *K* & the straight lines *IM* & *AB* are equal, therefore the rectangle *IKLM* and the triangle *AIB* are also equal. Wherefore the Spiral line *AFGHB,* and the Parabolical line *MK,* being described with equal velocity and in equal times, are equal to one another. Now

33. This is the familiar result from the "fourth day" of Galileo's "Two New Sciences" (Galileo 1974, 221).

34. The corollary in question is Hobbes's version of the "mean speed theorem" and asserts that "[i]f the *Impetus* be the same in every point, any straight line representing it may be taken for the measure of Time; and the Quicknesses or *Impetus* applied ordinately to any straight line making and Angle with it, and representing the way of the Bodies motion, will designe a parallelogram which shall represent the velocity of the whole Motion. But if the *Impetus* or Quickness of Motion begin from Rest, and increase Uniformly, that is, in the same proportion continually with the times which are passed, the whole Velocity of the Motion shall be represented by a Triangle, one side whereof is the whole time, and the other the greatest *Impetus* acquired in that time; or else by a parallelogram, one of whose sides is the whole time of Motion, and the other, half the greatest *Impetus;* or lastly by a parallelogram having for one side a mean proportional between the whole time & half of that time, & for the other side the half of the greatest *Impetus.* For both these parallelograms are equal to one another, & severally equal to the triangle which is made of the whole line of time, and the greatest acquired *Impetus;* as is demonstrated in the Elements of Geometry" (*DCo* 3.16.1, corollary 2; *EW* 1:219). Hobbes's use of it here is to argue that the increasing velocity of the point describing the spiral can be analyzed as a right triangle, having one side equal to the radius of the circle and the other equal to the circumference. Such a triangle has an area equal to that of the circle.

in the first article of the 18 Chapter a straight line is found out equal to any Parabolical line. Wherefore also a Straight line is found out, equal to a given Spiral line of the first revolution described by *Archimedes;* which was to be done.

A.5 HOBBES'S 1661 CUBE DUPLICATION

This is the original version of Hobbes's efforts to duplicate the cube, which stand at the center of his unsuccessful campaign for membership in the Royal Society in 1661–62. It was written in French and published anonymously in Paris in 1661. Subsequent versions of the same basic construction were published in Hobbes's *Dialogus Physicus* (1661) and *Problemata Physica* (1662). In addition, Hobbes circulated several other variations of the argument among fellows of the Royal Society in 1662, and Charles II requested that the society deliver its verdict on the validity of Hobbes's argumentation in the same year. The most striking feature of this particular version of the argument is the clutter of extraneous lines in the construction. The inadequacy of the construction can be shown quite easily: after the stipulation that the line *AS* is to be taken equal to the semidiagonal *BI*, this fact is never appealed to in the further course of the argument, which means that the proof would proceed just as correctly if *AS* had been of any length whatever.[35]

The Duplication of the Cube
By V.A.Q.R.

A RIGHT LINE BEING GIVEN, to find two mean proportionals between it and its half.

Let the right line *AB* [in figure A.5] be given, whose square is *ABCD*, and let this be divided into four equal squares by the right lines *EF, GH,* which intersect in the center of the square *ABCD*, at point *I*. In this way, the four sides will be divided in two equal parts at the four points *E, F, G, H.* Thus, it is required to find two mean proportionals between *DC* and *DF.*

I draw the diagonals *AC, BD,* and describe the four circle quadrants *ABD, BCA, CDB, DAC,* whose arcs cut the said diagonals at *K, L, M, N.* At these points the arcs are each divided into two equal parts, as is well known.

I produce *BA, CD* to the points *O* and *P,* so that *AB* is equal to *AO* and *DP* is equal to *DC.* And having described the circle quadrant *ADO* and drawn the diagonal *AP* (which will cut the arc *DO* into two equal parts at point *Q*),

---

35. This is essentially Huygens's argument as relayed to the Royal Society in his evaluation of Hobbes's duplication (*CTH* 2:538–39).

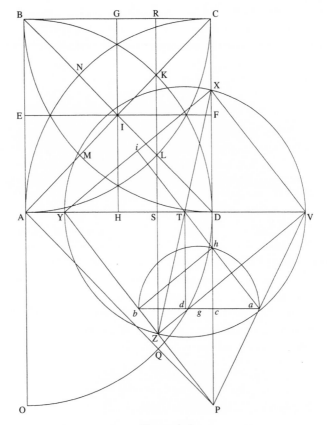

Figure A.5

I further take R in BC so that BR will be equal to the sine of 45 degrees, that is to say to the semidiagonal BI.[36] Consequently, SD is the excess of the side AD over the semidiagonal AS.

I cut this SD in two equal parts at T. In AD produced I take DV equal to DF, and making T the center and TV the semidiameter, I describe the circle $VXYZ$, cutting DC at X, DA at Y, and the right line RS produced at Z. And I say that the two right lines DY, DX are the two mean proportionals demanded between DP (equal to AB) and DV (equal to its half, DF).

For, drawing the right lines VX, XY, the angle VXY (in the semicircle) will be a right angle. And the right line XT, being drawn and produced to the con-

36. The original French reads, "Et ayant décrit le quart de cercle ADO, & tiré la diagonale AP (qui coupera l'arc DO en deux parties égales, au point Q.) Et étant produite de l'autre part en R, marquera BR égale au sinus droit de 45 degrés, c'est à dire à la semidiagonale BI."

cavity of the circle $VXYZ$ will fall on Z, because ST, TD are equal. Consequently, SZ will be equal to DX, and XZ will be a diameter of the circle $VXYZ$. Therefore, the angle XYZ in the semicircle will be a right angle, and in drawing the right lines YZ, VZ we make the rectangle $VXYZ$, whose sides VX, YZ are parallel.[37]

Now, if the right line YZ produced falls on P, the whole line PZY will be right and parallel to VX, and the alternate angles YPX, VXP will be equal. The angles YPX and XYD will also be equal, and the three right triangles PDY, YDX, and XDV will be similar. Consequently, the four right lines PD, DY, DX, DV will be in the same continued ratio.

It is therefore required to demonstrate that the right line YZ produced falls on P.

Let PV be drawn and divided in two equal parts at a. Let the right line ab also be drawn parallel to AV and cutting PD at c. Further, let Td be drawn parallel to PD, cutting ab at d, and let dc be divided in two equal parts at g. On the center g with distance ga let the semicircle ahb be described, cutting PD at h and ab at b.

This being done, the two right lines ah, bh being drawn will form a right angle at h. Now ac is half of DV, and because dg, gc are equal, db will also be equal to half of DV, and ab will be half of YV.

Therefore, as PD is to DY, that is to say to the line composed of DS and SY, so also is Pc (the half of PD) to cb, the line composed of the halves of DS and SY, and consequently Pb being produced will fall on Y. And the right lines hb, ha will be the halves of the right lines XY, XV; and XY being divided in two equal parts at i, the figure $Yihb$ will be a rectangle, and Yb will be parallel to XV.[38] But YZ is parallel to XV. Therefore, YZ produced will fall on P. And (because of what was demonstrated) the four right lines PD, DY, DX, DV are in one and the same continued ratio. I have therefore found two mean proportionals between a given right line and its half. Which was required to be done.

Consectary. A cube that has the lesser of these two means as an edge is double of a cube that has the half of the greater extreme for an edge.[39] Because the ratio of a cube to a cube is triplicate of the ratio of the edge to an edge. And the ratio of PD to DX is triplicate of the ratio of PD to DY.

---

37. Reading "tirant les doites YZ, VZ" for "tirant la doite YZ."

38. This consequence fails to follow. The prior argumentation establishes that $Yihb$ is a parallelogram, but not that it is a rectangle. From this point forward, the demonstration proceeds from a false assumption to a false result.

39. Reading, "Vn Cube qui a pour côté la moindre de ces deux moyennes" for "Vn Cube qui a pour côté la plus grande de ces deux moyennes," which misstates the intended result.

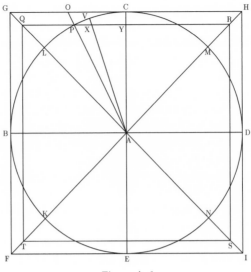

Figure A.6

A.6 THE 1669 QUADRATURA

This Hobbesian quadrature is a vastly different kind of argument from what we saw in *De Corpore*. This argument makes no appeal to the "method of motions," which Hobbes seems by this time to have largely abandoned, nor does it attempt to determine the area of the circle by considering indivisible "least elements" of lines or surfaces. Instead, Hobbes relies upon a very simple geometric construction and a short (but ultimately fallacious) argument about the relationship between the areas of finite parts of circle sectors. More surprisingly, the theorem implies a value of 3.2 for π, which is quite significantly further off the mark than Hobbes's earlier efforts.

Proposition I: To Find a Circle Equal to a Given Square

Let the given circle be *BCDE* [in figure A.6], whose center is *A*, and let it be divided into four parts by the diameters *BD*, *CE*. Let the square *FGHI* be circumscribed about this circle, which touches the circle at points *B*, *C*, *D*, *E*. Let the diagonals *GI*, *HF* be drawn, cutting the circle at the points *K*, *L*, *M*, *N*. Let the half-side *CG* be bisected at *O*, and let *AO* be drawn cutting the circle at *P*. Through the point *P* let the right line *QR* be drawn parallel to *GH*, cutting *AG*, *AH* at *Q* and *R*, and *AC* at *Y*. And let the square *QRST* be completed. I say that the square *QRST* is equal to the given circle *BCDE*.

Because the right line *CG* is bisected at *O*, and the bases *CG*, *YQ* of the triangles ACG, AYQ are parallel, the base YQ is also bisected at P, and thus the triangles AYP, APQ are equal.

In $\overarc{LC}$ let $\overarc{LV}$ be taken equal to $\overarc{CP}$, and let AV be drawn, cutting YP at X.

Now $APL + PQL + CYP = AVL = ACP$ (because $APL + PQL = AYP$). Also, $ACV + AVP = ACP = AVL$.

Thus $APL + PQL + CYP = ACV + AVP$.

Subtracting the equals APL, ACV from both sides, there remains $PQL + CYP = AVP$.

Therefore because the sector AVP added to the equal sectors ACV, ALP makes the whole sector ACL, so also the two trilinear figures PQL, CYP added to the same equal sectors ACV, ALP make two equal triangles equal to the same sector ACL.[40] Now the trilinear figure PQL added to the sector ALP makes the triangle APQ. And (because the sectors ALP, ACV are equal and triangles AYP, APQ are equal) the same trilinear figure PQL added to the sector ACV makes the triangle AYP.

Therefore if PQL, CYP are equal, the whole triangle AYQ will be equal to the whole sector ACL. But if PQL is greater or less than CYP, the triangle AYQ will be greater or less than the sector ACL. Therefore either no right triangle can be taken with vertex A and equal to the sector ACL, or PQL and CYP are equal.[41] But the first is absurd. Therefore, PQL, CYP are equal, of which the first *(PQL)* extends wholly outside the sector ACL, while the second *(CYP)* is wholly contained within the sector ACL.

Thus the triangles AYP, APQ taken together (that is an eighth part of the whole square $QRST$) are equal to the two sectors ACP, APL taken together (that is to an eighth part of the whole given circle $BCDE$), and the whole square $QRST$ is equal to the whole circle $BCDE$.

Therefore a square has been found equal to a given circle.[42]

40. As Wallis observes (1669a, 2), this misstates the case slightly since "these will not *make two triangles* (even though they can be equal to two triangles)."

41. Sadly for Hobbes, both disjuncts of this claim are false.

42. Elementary calculation reduces this conclusion to absurdity. Since $AO{:}AP :: AC{:}AY :: OC{:}PY$, it follows that $AO^2{:}AP^2 :: ACO{:}AYP$ (since similar figures with proportional sides are in duplicate ratio). Then, setting the radius AP or AC equal to R, we get $CO = R/2$, and $AP^2 = R^2$. Further, $AO^2 = R^2 + R^2/4 = 5R^2/4$ and (substituting and simplifying the proportion $AO^2{:}AP^2 :: ACO{:}AYP$) we obtain $ACO = R^2/4$. Thus, $AYP = R^2/5$ and $AYQ = 2R^2/5$. In consequence $QRST$ will be $16R^2/5$, which results in a value of 3.2 for π.

References

Andersen, Kirsti. 1985. Cavalieri's Method of Indivisibles. *Archive for History of the Exact Sciences.* 28:292–367.

Apostle, Hippocrates George. 1952. *Aristotle's Philosophy of Mathematics.* Chicago: University of Chicago Press.

Aquinas, Thomas. 1964. *Summa Theologica: Latin Text and English Translation, Introduction, Notes, and Appendices.* 60 vols. Cambridge: Blackfriars; New York: Macmillan.

Archimedes. 1912. *The Works of Archimedes.* Ed. and trans. Thomas L. Heath. 2d ed. Cambridge: Cambridge University Press.

Ariew, Roger, and Marjorie Grene, eds. 1995. *Descartes and His Contemporaries: Meditations, Objections, and Replies.* Chicago: University of Chicago Press.

Aristotle. 1984. *The Complete Works of Aristotle: The Revised Oxford Translation.* Ed. Jonathan Barnes. 2 vols. Princeton: Princeton University Press.

Arndt, Hans Werner. 1971. *Methodo scientifica pertractatum: Mos geometricus und Kalkülbegriff in der philosophischen Theorienbildung des 17. und 18. Jahrhunderts.* Quellen und Studien zur Philosophie, ed. Günter Patzig, Erhard Scheibe, and Wolfgang Wieland, vol. 4. Berlin and New York: Walter de Gruyter.

Ashworth, E. J. 1974. *Language and Logic in the Post-medieval Period.* Synthese Historical Library, vol. 12. Dordrecht and Boston: Reidel.

Aubrey, John. 1898. *"Brief Lives," chiefly of Contemporaries, set down by John Aubrey, between the Years 1669 & 1696.* Ed. Andrew Clark. 2 vols. Oxford: Oxford University Press, Clarendon Press.

Auger, Léon. 1962. *Un savant méconnu: Gilles Personne de Roberval (1602–1675); Son activité intellectuelle dans les domaines mathématique, physique, méchanique, et philosophique.* Paris: A. Blanchard.

Aylmer, G. E. 1978. Unbelief in Seventeenth-Century England. In *Puritans and Revolutionaries: Essays in Seventeenth-Century History presented to Christopher Hill,* ed. Donald Pennington and Keith Thomas, 22–46. Oxford: Oxford University Press, Clarendon Press.

Barnes, Barry, David Bloor, and John Henry. 1996. *Scientific Knowledge: A Sociological Analysis.* London: Athlone.

Barnouw, Jeffrey. 1979. Bacon and Hobbes: The Conception of Experience in the Scientific Revolution. *Science, Technology, and the Humanities* 2:92–110.

————. 1980. Vico and the Continuity of Science: The Relation of His Epistemology to Bacon and Hobbes. *Isis* 71:609–20.

————. 1990. Prudence et science chez Hobbes. In Zarka and Bernhardt 1990, 107–18.

————. 1992. Le vocabulaire du *conatus*. In Zarka 1992, 103–24.

Baroncelli, Giovanna. 1992. Bonaventura Cavalieri tra matematica e fisica. In Bucciantini and Torrini 1992, 67–102.

Barrow, Isaac. 1685. *Isaaci Barrow Lectiones Mathematicae XXIII; In quibus Principia Matheseôs generalia exponuntur: Habitae Cantabrigiae* A.D. *1664, 1665, 1666. Accesserunt ejusdem Lectiones IV. In quibus Theoremata & Problemata Archimedis De Sphaerâ et Cylindro, Methodo Analyticâ eruuntur.* London: J. Playford for George Wells.

————. 1860. *The Mathematical Works of Isaac Barrow, D.D.* Ed. William Whewell. 2 vols. bound as one. Cambridge: Cambridge University Press.

Beaulieu, Armand. 1989. Torricelli et Mersenne. In *L'Oeuvre de Torricelli: Science galiléenne et nouvelle géométrie,* ed. François De Gandt, 39–52. Publications de la Faculté des Lettres et Sciences Humaines de Nice, no. 32. Paris: Les Belles Lettres.

————. 1992. Le group de Mersenne. Ce que l'Italie lui a donné. Ce qu'il a donné a l'Italie. In Bucciantini and Torrini 1992, 17–34.

Becker, Oskar. 1954. *Grundlagen der Mathematik in geschichtlicher Entwicklung.* Orbis Academicus: Problemgeschichten der Wissenschaft in Dokumenten und Darstellungen, ed. Fritz Wagner and Richard Brodführer. Munich: Karl Alber.

Beeley, Philip. 1996. *Kontinuität und Mechanismus: Zur Philosophie des jungen Leibniz in ihrem Ideengeschichtlichen Kontext.* Studia Leibnitiana Supplementa, vol. 30. Stuttgart: Franz Steiner Verlag.

Benoit, Paul, Karine Chemla, and Jim Ritter, eds. 1992. *Histoire de fractions, fractions d'histoire.* Basel, Boston, and Berlin: Birkhäuser Verlag.

Bernhardt, Jean. 1975. L'Anti-White de Hobbes. *Archives internationales d'histoire des sciences* 25:104–15.

————. 1978. Genèse et limites du matérialisme de Hobbes. *Raison présente* 47:41–61.

————. 1979. La polémique de Hobbes contre la Dioptrique de Descartes dans le Tractatus Opticus II (1644). *Revue internationale de philosophie* 33:432–42.

————. 1985a. Nominalisme et mécanisme dans la penseé de Hobbes. *Archives de philosophie* 48:235–249.

————. 1985b. Sur le passage de F. Bacon a Th. Hobbes. *Les études philosophiques* 57:449–57.

————. 1986. Témoinage direct de Hobbes sur son "Illumination Euclidienne." *Revue philosophique* 2:281–82.

————. 1988. Nominalisme et mécanisme dans la pensée de Hobbes (II). *Archives de philosophie* 51:579–96.

————. 1989. *Hobbes.* Paris: Presses Universitaires de France.

————. 1993. Empirisme rationell et statut des *universalia:* le problème de la théorie de la science chez Hobbes. *Revue d'histoire des sciences* 46:131–52.

Bernstein, Howard R. 1980. Conatus, Hobbes, and the Young Leibniz. *Studies in History and Philosophy of Science* 11:167–81.

Bertman, Martin. 1973. Hobbes: Philosophy and Method. *Scientia* 108: 769–80.

Bickley, Francis. 1911. *The Cavendish Family.* London: Constable.

Birch, Thomas. [1756–57] 1968. *The History of the Royal Society of London for Improving of Natural Knowledge from its First Rise.* 4 vols. Reprint, ed. A. Rupert Hall and Marie Boas Hall. Sources of Science, no. 44. New York: Johnson Reprint.

Bird, Alexander. 1996. Squaring the Circle: Hobbes on Philosophy and Geometry. *Journal of the History of Ideas* 57:217–31.

Blay, Michel. 1992. Deux exemples de l'influence le l'école galiléenne sur les premiers travaux de l'Académie Royale des Sciences de Paris. In Bucciantini and Torrini 1992, 49–66.

Bloor, David. 1978. Polyhedra and the Abominations of Leviticus. *British Journal for the History of Science* 11:243–72.

———. 1983. *Wittgenstein: A Social Theory of Knowledge.* New York: Columbia University Press.

———. 1988. Rationalism, Supernaturalism, and the Sociology of Knowledge. In *Scientific Knowledge Socialized,* ed. Imre Hronszky, Márta Fehér, and Balázs Dajka, 59–74. Budapest: Akadémia Kiadó.

———. 1991. *Knowledge and Social Imagery.* 2d ed. Chicago: University of Chicago Press.

Borrelli, Gianfranco, ed. 1990. *Thomas Hobbes: La ragioni del moderno tra teologia e politica.* Naples: Morani.

Bos, Henk J. M. 1981. On the Representation of Curves in Descartes' Géométrie. *Archive for History of the Exact Sciences* 24:295–338.

———. 1984. Arguments on Motivation in the Rise and Decline of a Mathematical Theory; the "Construction of Equations," 1637–ca. 1750. *Archive for History of the Exact Sciences* 30:331–80.

———. 1991. *Lectures in the History of Mathematics.* History of Mathematics, vol. 7. Providence, R.I.: American Mathematical Society; London: London Mathematical Society.

———. 1993. On the Interpretation of Exactness. In *Philosophy of Mathematics: Proceedings of the 15th International Wittgenstein-Symposium,* ed. Klaus Puhl and Johannes Czermak, 2 vols., 1:23–44. Vienna: Verlag Hölder-Pichler-Tempsky.

Bosman, H. 1927. André Tacquet (S.J.) et son traité d' «Arithmétique théorique et pratique». *Isis* 9:66–82.

Bowle, John. [1951] 1969. *Hobbes and His Critics: A Study in Seventeenth Century Constitutionalism.* London: Johnathan Cape. Reprint, London: Frank Cass.

Boyer, Carl B. 1946. Proportion, Equation, Function: Three Steps in the Development of a Concept. *Scripta Mathematica* 12:5–13.

————. 1956. *A History of Analytic Geometry.* New York: Scripta Mathematica.

Boyle, Robert. 1992. *An Examen of Mr. T. Hobbes his Dialogus Physicus de Natura Aëris.* Oxford: H. Hall for Thomas Robinson.

Brandt, Frithiof. 1928. *Thomas Hobbes' Mechanical Conception of Nature.* Copenhagen: Levin & Munksgaard; London: Librairie Hachette.

Breger, Herbert. 1986. Leibniz's Einführung des Transzendenten. In *300 Jahre "Nova Methodus" von G. W. Leibniz (1684–1984),* ed. Albert Heinekamp, 119–32. Studia Leibnitiana, special vol. 14. Stuttgart: Franz Steiner Verlag.

Breidert, Wolfgang. 1979. Les mathématiques et la méthode mathématique chez Hobbes. *Revue internationale de philosophie* 129:415–31.

Brockdorff, Cay von. 1932. Ein Verehrer Thomas Hobbes' als Interpret des Aristoteles. *Veröffentlichungen des Hobbes-Gesellschaft* 1:8–18.

————. 1934. *Des Sir Charles Cavendish Bericht für Joachim Jungius über die Grundzüge der Hobbes'schen Naturphilosophie.* Kiel: Hobbes-Gesellschaft.

Brooke, Robert Grevile. 1641. *The Nature of Truth, its Union and Unity with the Soule.* London: R. Bishop for Samuel Cartwright.

Brunchvicg, Léon. 1947. *Les étapes de la philosophie mathématique.* 3d ed. Paris: Presses Universitaires de France.

Bucciantini, Massimo, and Maurizio Torrini, eds. 1992. *Geometria e Atomismo nella Scuola Galileiana.* Biblioteca di «Nuncius», no. 10. Florence: Leo S. Olschki.

Burgess, Glenn. 1990. Contexts for the Writing and Publication of Hobbes's *Leviathan. History of Political Thought* 9:675–702.

Cajori, Florian. 1929. Controversies between Wallis, Hobbes, and Barrow. *Mathematics Teacher* 2:146–51.

Cantor, Moritz. [1900–1901] 1965. *Vorlesungen über die Geschichte der Mathematik.* 4 vols. Leipzig: Teubner. Reprint, New York: Johnson Reprint.

Cassinet, Jean. 1992. La relation d'ordre entre rapports dans les Éléments d'Euclide: développments au XVIIe siècle. In Benoit, Chemla, and Ritter 1992, 341–50.

Cavalieri, Bonaventura. 1635. *Geometria indivisibilibus continuorum nova quadam ratione promota.* Bologna: Jacobi Montij.

————. 1647. *Exercitationes Geometricae Sex . . .* Bologna: Jacobi Montij.

————. 1987. *Carteggio.* Ed. Giovanna Baroncelli. Archivo della Corrispondenza degli Scienziati Italiani, no. 3. Florence: Leo S. Olschki.

Cawdrey, Daniel. 1657. *Independencie a Great Schism; proved against Dr. Owen in his Apology in his Tract of Schism.* London: J. S. for John Wright.

————. 1658. *Independencie further proved to be a Schism; or, A Survey of Dr Owens Review of his Tract of Schism.* London.

Cellini, G. 1966a. Gli indivisibili nel pensiero matematico e filosofico di Bonaventura Cavalieri. *Periodico di matematiche* 44:1–21.

————. 1966b. Le dimonstrazioni di Cavalieri del suo principio. *Periodico di matematiche* 44:85–105.

Child, Arthur. 1953. Making and Knowing in Hobbes, Vico, and Dewey. *University of California Publications in Philosophy* 16:271–310.

Clavelin, Maurice. 1974. *The Natural Philosophy of Galileo: Essay on the Origins and Formation of Classical Mechanics*. Trans. A. J. Pomerans. Cambridge: MIT Press.

Clavius, Christopher. 1612. *Christophori Clavii Bambergensis E Societate Jesu Opera Mathematica V Tomis distributa.* 5 vols. Mainz: Reinhard Eltz.

Cleary, John Joseph. 1982. Aristotle's Theory of Abstraction: A Problem about the Mode of Being of Mathematical Objects. Ph.D. diss., Boston University.

Costabel, Pierre. 1985. Descartes et la mathématique de l'infini. *Historia Scientiarum* 29:37–49.

———. 1992. Pierre de Carcavy et ses relations italiennes. In Bucciantini and Torrini 1992, 35–49.

Costabel, Pierre, and Monette Martinet. 1986. *Quelques savants et amateurs de science au XVIIe siècle*. Cahiers d'Histoire et de Philosophie des Sciences, n.s., no. 14. Paris: Société Française d'Histoire des Sciences et des Techniques.

Crombie, A. C. 1977. Mathematics and Platonism in the Sixteenth-Century Italian Universities and in Jesuit Educational Policy. In *PRISMATA: Naturwissenschaftgeschichtliche Studien*, ed. Y. Maeyama and W. G. Saltzer, 63–94. Wiesbaden: Franz Steiner Verlag.

Curley, Edwin M. 1992. "I Durst Not Write So Boldly" or, How to Read Hobbes's Theological-Political Treatise. In *Hobbes e Spinoza, Atti del Gonvegno Internazionale, Urbino 14–17 ottobre, 1988*, ed. Daniela Bostrengi, 497–593. Naples: Bibliopolis.

———. 1995. Hobbes versus Descartes. In Ariew and Grene, 97–109.

———. 1996a. Calvin and Hobbes, or, Hobbes as an Orthodox Christian. *Journal of the History of Philosophy* 34:257–72.

———. 1996b. Reply to Professor Martinich. *Journal of the History of Philosophy* 34:285–87.

Dal Pra, Mario. 1962. Note sulla logica di Hobbes. *Rivista critica di storia della filosofia* 17:411–33.

Damerow, Peter, Gideon Freudenthal, Peter McLaughlin, and Jürgen Renn. 1992. *Exploring the Limits of Preclassical Mechanics, A Study of Conceptual Development in Early Modern Science: Free Fall and Compounded Motion in the Work of Descartes, Galileo, and Beeckman*. New York and Berlin: Springer-Verlag.

De Gandt, François, ed. 1987a. *L'oeuvre de Torricelli: Science galiléenne et nouvelle géométrie*. Publications de la Faculté des Lettres et Sciences Humaines de Nice, no. 32. Paris: Les Belles Lettres.

———. 1987b. Les indivisibles de Torricelli. In De Gandt 1987a, 147–206.

———. 1991. Cavalieri's Indivisibles and Euclid's Canons. In *Revolution and Continuity: Essays in the History and Philosophy of Early Modern Science*, ed. Peter Barker and Roger Ariew, 157–82. Washington: Catholic University of America Press.

———. 1992. L'evolution de la theorie des indivisibles et l'apport de Torricelli. In Bucciantini and Torrini 1992, 103–18.

De Jong, Willem R. 1986. Hobbes's Logic: Language and Scientific Method. *History and Philosophy of Logic* 7:123–42.

Dear, Peter. 1988. *Mersenne and the Learning of the Schools.* Cornell History of Science Series, ed. Pearce Williams. Ithaca: Cornell University Press.

———. 1995a. *Discipline and Experience: The Mathematical Way in the Scientific Revolution.* Science and Its Conceptual Foundations, ed. David L. Hull. Chicago: University of Chicago Press.

———. 1995b. Mersenne's Suggestion: Cartesian Meditation and the Mathematical Model of Knowledge in the Seventeenth Century. In Ariew and Grene, 44–62.

Debus, Allen G. 1970. *Science and Education in Seventeenth-Century England: The Webster-Ward Debate.* London: Macdonald; New York: American Elsevier.

Dell, William. 1816. *The Works of William Dell, Minister of the Gospel and Master of Gonville and Caius College, in Cambridge.* New York: John Sarpless.

Des Chene, Dennis. 1996. *Physiologia: Natural Philosophy in Late Aristotelian and Cartesian Thought.* Ithaca: Cornell University Press.

Descartes, René. 1964–76. *Oeuvres de Descartes.* Ed. Charles Adam and Paul Tannery. Rev. ed. Paris: Vrin.

Digby, Kenelme. 1644. *Two Treatises: in the one of which, The Nature of Bodies; in the other, The Nature of Mans Soule; is looked into: in way of discovery, of the Immortality of Reasonable Soules.* Paris: Gilles Blaizot.

Dijksterhuis, E. J. 1931–32. John Pell in zijn strijd over de rectificatie van den cirkel. *Euclides* 8:286–96.

———. [1956] 1987. *Archimedes.* Trans. C. Dikshoorn. With a bibliographic essay by Wilbur R. Knorr. Princeton: Princeton University Press.

D'Israeli, Isaac. [1814] 1970. *Quarrels of Authors; or, Some Memoirs of our Literary History.* 3 vols. Reprint, New York: Johnson Reprint.

Engfer, Hans-Jürgen. 1982. *Philosophie als Analysis: Studien zur Entwicklung philosophischer Analysiskonzeptionen unter dem Einfluß mathematischer Methodenmodelle im 17. und frühen 18. Jahrhundert.* Stuttgart–Bad Cannstatt: Frommann-Holzboog.

Euclid. [1925] 1956. *The Thirteen Books of Euclid's "Elements" Translated from the Text of Heiberg.* Ed. and trans. Thomas L. Heath. 3 vols. Cambridge: Cambridge University Press. Reprint, New York: Dover.

Favaro, Antonio. 1921. Galileo Galilei e Tommasso Hobbes. *Atti e memorie della R. Accademia di Scienze Lettere ed Arti in Padova* 36:22–25.

Federspiel, Michel. 1991. Sur la définition euclidienne de la droite. In Rashed 1991b, 115–30.

Feingold, Mordechai. 1984. *The Mathematicians' Apprenticeship: Science, Universities, and Society in England, 1560–1640.* Cambridge: Cambridge University Press.

———. 1985. A Friend of Hobbes and an Early Translator of Galileo: Robert Payne of Oxford. In *The Light of Nature: Essays in the History and Philosophy of Science Presented to A. C. Crombie,* ed. J. D. North and J. J. Roche, 265–80. The Hague: Martinus Nijhoff.

Festa, Egidio. 1990. La querelle de l'atomisme: Galilée, Cavalieri et les jésuites. *La recherche* 224:1038–47.

———. 1992. Quelques aspects de la controverse sur les indivisibles. In Bucciantini and Torrini 1992, 193–208.

Fiebig, Hans. 1973. *Erkenntnis und technische Erzeugung: Hobbes' operationale Philosophie der Wissenschaft.* Monographien zur Naturphilosophie, vol. 14. Meisenheim am Glan: Anton Hain.

Fletcher, J. M. J. 1940. Seth Ward, Bishop of Salisbury, 1667–1689. *The Wiltshire Archaeological and Natural History Magazine.* 49:1–16.

Flew, Antony. 1986. A Strong Programme for the Sociology of Belief. *Inquiry* 25:365–85.

Fontialis, Jacobus. 1740. *Opera Posthuma.* Namur, Belgium.

Fowler, David. 1992. Logistic and Fractions in Early Greek Mathematics: A New Interpretation. In Benoit, Chemla, and Ritter 1992, 133–47.

Friedman, Michael. 1998. On the Sociology of Scientific Knowledge and Its Philosophical Agenda. *Studies in History and Philosophy of Science* 29: 239–71.

Fritz, Kurt von. 1959. Gleichheit, Kongruenz und Ähnlichkeit in der antiken Mathematik bis auf Euklid. *Archiv für Begriffsgeschichte* 4:7–81.

Galileo Galilei. 1973. *Les nouvelles pensées de Galilée.* Trans. Marin Mersenne. Critical edition with introduction and notes by Pierre Costabel and Michel-Pierre Lerner. Paris: Vrin.

———. 1974. *Two New Sciences, Including Centers of Gravity & Force of Percussion.* Ed. and trans. Stillman Drake. Madison: University of Wisconsin Press.

Garber, Daniel. 1992. *Descartes' Metaphysical Physics.* Science and Its Conceptual Foundations, ed. David L. Hull. Chicago: University of Chicago Press.

———. 1995. J.-B. Morin and the *Second Objections.* In Ariew and Grene, 63–83.

Garcia, Alfred. 1986. *Thomas Hobbes: Bibliographie internationelle de 1620 à 1986.* Bibliothèque de Philosophie politique et juridique, Textes et Documents. Caen: Centre de philosophie politique et juridique, Université de Caen.

Gargani, Aldo G. 1971. *Hobbes e la scienza.* Biblioteca di cultura filosofica, no. 36. Turin: Giulio Einaudi.

Gassendi, Pierre. [1658] 1964. *Opera Omnia.* 6 vols. Reprint, Stuttgart: Frommann-Holzboog.

Gaukroger, Stephen. 1992. The Nature of Abstract Reasoning: Philosophical Aspects of Descartes' Work in Algebra. In *The Cambridge Companion to Descartes,* ed. John Cottingham, 91–114. Cambridge: Cambridge University Press.

Giancotti, Emilia. 1990. La funzione dell'idea di Dio nel sistema naturale e politico di Hobbes. In Borrelli 1990, 15–34.

Giorello, Giulio. 1990. Pratica geometrica e immagine della matematica in Thomas Hobbes. In Napoli and Canziani 1990, 214–44.

Giusti, Enrico. 1980. *Bonaventura Cavalieri and the Theory of Indivisibles.* Milan: Edizioni Cremonese.

———. 1993. *Euclides reformatus: La teoria delle proporzioni nella scuola galileiana*. Turin: Bollati Boringhieri.

Grant, Hardy. 1990. Geometry and Politics: Mathematics in the Thought of Thomas Hobbes. *Mathematics Magazine* 63:147–54.

———. 1996. Hobbes and Mathematics. In Sorell 1996a, 108–29.

Grattan-Guinness, Ivor. 1996. Numbers, Magnitudes, Ratios, and Proportions in Euclid's *Elements:* How Did He Handle Them? *Historia Mathematica* 23:355–75.

Greaser, Andreas, ed. 1987. *Mathematics and Metaphysics in Aristotle*. Bern and Stuttgart: Paul Haupt.

Green, Ian. 1994. Anglicanism in Stuart and Hannoverian England. In *A History of Religion in Britain: Practice and Belief from Pre-Roman Times to the Present*, ed. Sheridan Gilley and W. J. Shields, 168–87. Oxford: Blackwell.

Green, V. H. H. 1964. *Religion at Oxford and Cambridge*. London: SCM Press.

Grosholz, Emily. 1987. Some Uses of Proportion in Newton's *Principia*, Book I: A Case Study in Applied Mathematics. *Studies in History and Philosophy of Science* 18:209–20.

———. 1991. *Cartesian Method and the Problem of Reduction*. Oxford: Oxford University Press, Clarendon Press.

Guenancia, Pierre. 1990. Hobbes-Descartes: le nom et la chose. In Zarka and Bernhardt 1990, 67–79.

Guitian, Georges. 1996. Les exilés royalistes anglais en France 1640–1660. Ph.D. diss., Université Paris VIII.

Guldin, Paul. 1635–41. *Centrobaryca, seu de centro gravitatis dissertatio*. . . . 4 vols. Vienna.

Halliwell, James Orchard, ed. 1841. *A Collection of Letters Illustrative of the Progress of Science in England from the Reign of Queen Elizabeth to that of Charles the Second*. London: Historical Society of Science.

Hankinson, R. J. 1995. Philosophy of Science. In *The Cambridge Companion to Aristotle*, ed. Jonathan Barnes, 109–39. Cambridge: Cambridge University Press.

Hanson, Donald W. 1990. The Meaning of "Demonstration" in Hobbes's Science. *History of Political Thought* 9:587–626.

Heath, Thomas L. [1921] 1981. *A History of Greek Mathematics*. 2 vols. Oxford: Oxford University Press, Clarendon Press. Reprint, New York: Dover.

Henry, John. 1982. Atomism and Eschatology: Catholicism and Natural Philosophy in the Interregnum. *British Journal for the History of Science* 15:211–40.

Herbert, Gary B. 1989. *Thomas Hobbes: The Unity of Scientific and Moral Wisdom*. Vancouver: University of British Columbia Press.

Hervey, Helen. 1952. Hobbes and Descartes in the Light of Some Unpublished Letters of the Correspondence between Sir Charles Cavendish and Dr. John Pell. *Osiris* 10:67–90.

Hill, Christopher. 1975. The Radical Critics of Oxford and Cambridge in the 1650s. In *Change and Continuity in Seventeenth-Century England*, 127–48. Cambridge: Harvard University Press.

Hill, Katherine. 1996. Neither Ancient nor Modern: Wallis and Barrow on the

Composition of Continua. Part One: Mathematical Styles and the Composition of Continua. *Notes and Records of the Royal Society of London* 50:165–78.

———. 1997. Neither Ancient nor Modern: Wallis and Barrow on the Composition of Continua. Part Two: The Seventeenth-Century Context: The Struggle between Ancient and Modern. *Notes and Records of the Royal Society of London* 51:13–22.

Hintikka, Jaakko, and Unto Remes. 1974. *The Method of Analysis: Its Geometric Origin and General Significance.* Boston: D. Reidel.

Hobbes, Thomas. 1650. *Humane Nature: Or, The fundamental Elements of Policie. Being a Discoverie Of the Faculties, Acts, and Passions of the Soul of Man, from their original causes; According to such Philosophical Principles as are not commonly known or asserted.* London: T. Newcomb for Francis Bowman.

———. 1651. *Leviathan, Or The Matter, Forme, & Power of a Common-Wealth Ecclesiasticall and Civill.* London: Andrew Crooke.

———. 1655. *Elementorum Philosophiae Sectio Prima De Corpore.* London: Andrew Crooke.

———. 1656a. *Elements of Philosophy, the First Section, Concerning Body. Written in Latine by Thomas Hobbes of Malmesbury and now translated in to English. To which are added Six Lessons to the Professors of Mathematicks of the Institution of Sr. Henry Saville, in the University of Oxford.* London: R. & W. Leybourn for Andrew Crooke.

———. 1656b. *Six Lessons to the Professors of the Mathematiques, one of Geometry, the other of Astronomy: In the Chaires set up by the Noble and Learned Sir Henrey Savile, in the University of Oxford.* London: Andrew Crooke.

———. 1657. Στίγμαι Αγεομετρίας, Αγροικίας, Αντιπολιτείας, Αμαθείας; or, *Markes of the Absurd Geometry, Rural Language, Scottish Church-Politicks, And Barbarismes of John Wallis Professor of Geometry and Doctor of Divinity.* London: Andrew Crooke.

———. 1658. *Elementorum Philosophiae Sectio Secunda De Homine.* London: Andrew Crook.

———. 1660. *Examinatio et Emendatio Mathematicae Hodiernae. Qualis explicatur in libris Johannis Wallisii Geometriae Professoris Saviliani in Academia Oxoniensi. Distributa in sex Dialogos.* London: Andrew Crooke.

[———]. 1661a. *La Duplication du Cube par V.A.Q.R.* Paris: n.p.

———. 1661b. *Dialogus Physicus, sive de Natura Aeris Conjectura sumpta ab Experimentis nuper Londini habitis in Collegio Greshamensi Item De Duplicatio Cubi.* London: Andrew Crooke.

———. 1662a. *Problemata Physica . . . Adjunctae sunt etiam Propositiones duae de Duplicatione Cubi, & Dimensione Circuli.* London: Andrew Crooke.

[———]. 1662b. *Mr Hobbes Considered in his Loyalty, Religion, Reputation, and Manners. By way of Letter to Dr. Wallis.* London: Andrew Crooke.

———. 1666. *De Principiis et Ratiocinatione Geometrarum. Ubi ostenditur incertitudinem falsitatemque non minorem inesse scriptis eorum, quam scriptis Physicorum & Ethicorum. Contra fastum Professorum Geometriae.* London: Andrew Crooke.

———. 1668. *Thomae Hobbes Malmesburiensis Opera Philosophica, Quae Latinè scripsit Omnia.* Amsterdam: J. Blaeu.

———. 1669a. *Quadratura Circuli, Cubatio Sphaerae, Duplicatio Cubi, Breviter demonstrata.* London: Andrew Crooke.

———. 1669b. *Quadratura Circuli, Cubatio Sphaerae, Duplicatio Cubi, Una cum Responsione ad objectiones Geometriae Professoris Saviliani Oxoniae editas.* London: Andrew Crooke.

———. 1671a. *Three Papers Presented to the Royal Society Against Dr. Wallis, Together with Considerations on Dr. Wallis his Answer to them.* London: for the Author.

[———]. 1671b. *Rosetum Geometricum, Sive Propositiones Aliquot Frustra antehac tentatae, Cum Censura brevi Doctrinae Wallisianae de Motu.* London: John Crooke for William Crooke.

[———]. 1672. *Lux Mathematica. Excussa Collisionibus Johannis Wallisii Theol. Doctoris, Geometriae in celeberrima Academia Oxoniensi Professoris Publici, et Thomae Hobbesii Malmesburiensis.* London: John Crooke for William Crooke.

———. 1674. *Principia et Problemata Aliquot Geometrica Antè Desperata, Nunc breviter Explicata and Demonstrata.* London: William Crooke.

———. 1678. *Decameron Physiologicum: Or, Ten Dialogues of Natural Philosophy. By Thomas Hobbes of Malmesbury. To which is added the Proportion of a straight Line to half the Arc of a Quadrant.* London: John Crooke for William Crooke.

———. 1680. *Considerations upon the Reputation, Loyalty, Manners, & Religion of Thomas Hobbes of Malmesbury. Written by himself, By Way of Letter to a Learned Person.* London: William Crooke.

———. 1682. *Seven Philosophical Problems, And Two Propositions of Geometry, By Thomas Hobbes of Malmesbury, With an Apology for Himself, and his Writings.* London: William Crooke.

———. [1839–45] 1966a. *Thomae Hobbes Malmesburiensis Opera Philosophica Quae Latine Scripsit Omnia in Unum Corpus Nunc Primum Collecta.* Ed. William Molesworth. 5 vols. Reprint, Aalen, Germany: Scientia Verlag.

———. [1839–45] 1966b. *The English Works of Thomas Hobbes of Malmesbury, now First Collected and Edited by Sir William Molesworth.* Ed. William Molesworth. 11 vols. Reprint, Aalen, Germany: Scientia Verlag.

———. [1651] 1968. *Leviathan, or the Matter, Forme, & Power of a Common-Wealth Ecclesiastical and Civill.* Ed. C. B. Macpherson. Hammondsworth: Penguin.

———. [1889] 1969a. *The Elements of Law Natural and Politic.* Ed. Ferdinand Tönnies. Reprint, with an introduction by M. M. Goldsmith, London: Frank Cass.

———. [1889] 1969b. *Behemoth; or, The Long Parliament.* Reprint, with an introduction by M. M. Goldsmith, London: Frank Cass.

————. 1976. *Thomas White's "De Mundo" Examined*. Ed. and trans. Harold Whitmore Jones. London: Bradford University Press.

————. 1994. *The Correspondence*. Ed. Noel Malcolm. *The Clarendon Edition of the Works of Thomas Hobbes*, vols. 6–7. Oxford: Oxford University Press, Clarendon Press.

Hobson, E. W., H. P. Hudson, A. N. Singh, and A. B. Kempke. 1953. *Squaring the Circle and Other Monographs*. Reprint, New York: Chelsea Publishing Company.

Hoffman, Joseph E. 1957. *Geschichte der Mathematik*. 4 vols. bound as one. Berlin: Walter de Gruyter.

————. 1966. Über die Kreismessung von Chr. Huygens, ihre Vorgeschichte, ihren Inhalt, ihre Bedeutung und ihr Nackwirken. *Archive for History of the Exact Sciences* 3:102–36.

Holder, William. 1678. *A supplement to the Philosophical Transactions of July 1670, With some Reflexions on Dr. John Wallis, his Letter there Inserted*. London: Henry Brome.

Hostetler, Gordon. 1945. Linguistic Theories of Thomas Hobbes and George Campbell. *ETC: Review of General Semantics* 2:170–80.

Hübener, Wolfgang. 1977. Ist Thomas Hobbes Ultranominalist gewesen? *Studia Leibnitiana* 9:77–100.

Hungerland, I., and G. R. Vick 1973. Hobbes's Theory of Signification. *Journal of the History of Philosophy* 11:459–82.

Hunter, Michael. 1979. The Debate over Science. In *The Restored Monarchy: 1660–1688*, ed. J. R. Jones, 176–95. Totowa, N.J.: Rowman and Littlefield.

————. 1981. *Science and Society in Restoration England*. Cambridge: Cambridge University Press.

————. 1982. *The Royal Society and Its Fellows, 1660–1700: The Morphology of an Early Scientific Institution*. London: British Society for the History of Science.

————. 1990. Science and Heterodoxy: An Early Modern Problem Reconsidered. In *Reappraisals of the Scientific Revolution*, ed. Robert S. Westman and David C. Lindberg, 437–60. Cambridge: Cambridge University Press.

Hussey, Edward. 1991. Aristotle on Mathematical Objects. *Apeiron: A Journal for Ancient Philosophy and Science*. 24:105–33.

Hutton, Ronald. 1985. *The Restoration: A Political and Religious History of England and Wales, 1658–1667*. Oxford: Oxford University Press, Clarendon Press.

Huygens, Christiaan. 1888–1950. *Les oeuvres complètes de Christiaan Huygens*. Ed. La Société Hollandaise des Sciences. 22 vols. The Hague: Martinus Nijhoff.

Huygens, Loedwijk. 1982. *The English Journal 1651–52*. Ed. and trans. A. G. H. Bachrach and R. G. Collmer. Publications of the Sir Thomas Browne Institute. Leiden: E. J. Brill/Leiden University Press.

Inhetveen, Rudiger. 1981. Können die Gegenstände der Geometrie bewegt werden? In *Vernunft, Handlung, und Erfahrung: Über die Grundlagen und*

Ziele der Wissenschaft, ed. Oswald Schwemmer, 64–68. Munich: C. H. Beck.

Isermann, Michael. 1991. *Die Sprachtheorie im Werk von Thomas Hobbes.* Studium Sprachwissenschaft, ed. Helmut Gipper and Peter Schmitter, no. 15. Münster: Nodus.

Jacob, James R. 1983. *Henry Stubbe, Radical Protestantism, and the Early Enlightenment.* Cambridge: Cambridge University Press.

Jacoli, Ferdinand. 1869. Notizia sconosciuta relativa a Bonaventura Cavalieri. *Bullettino di bibliografia e di storia delle scienze matematiche e fisiche* 2:299–312.

Jacquot, Jean. 1949. Un amateur de science, ami de Hobbes et Descartes: Sir Charles Cavendish. *Thales* 6:81–88.

———. 1952a. Un document inedit: Les notes de Charles Cavendish sur la première version du «De Corpore» de Hobbes. *Thales* 8:26–86.

———. 1952b. Notes on an Unpublished Work of Thomas Hobbes. *Notes and Records of the Royal Society of London* 9:188–95.

———. 1952c. Sir Charles Cavendish and His Learned Friends: A Contribution to the History of Scientific Relations between England and the Continent in the Earlier Part of the 17th Century. *Annals of Science* 8:13–27; 175–91.

Jardine, Nicholas. 1988. Epistemology of the Sciences. In *The Cambridge History of Renaissance Philosophy,* ed. Charles B. Schmitt, Quentin Skinner, Eckhard Kessler, and Jill Kraye, 685–712. Cambridge: Cambridge University Press.

Jennings, Richard C. 1984. Truth, Rationality, and the Sociology of Science. *British Journal for the Philosophy of Science* 35:201–11.

Jesseph, Douglas M. 1989. Philosophical Theory and Mathematical Practice in the Seventeenth Century. *Studies in History and Philosophy of Science* 20:215–44.

———. 1993a. Hobbes and Mathematical Method. *Perspectives on Science* 1:306–41.

———. 1993b. Of Analytics and Indivisibles: Hobbes on the Methods of Modern Mathematics. *Revue d'histoire des sciences* 46:153–93.

———. 1993c. *Berkeley's Philosophy of Mathematics.* Science and Its Conceptual Foundations, ed. David Hull. Chicago: University of Chicago Press.

———. 1996. Hobbes and the Method of Natural Science. In Sorell 1996a, 86–107.

———. 1999a. Leibniz on the Foundations of the Calculus: The Question of the Reality of Infinitesimal Magnitudes. *Perspectives on Science* 6:7–41.

———. 1999b. The Decline and Fall of Hobbesian Geometry. *Studies in History and Philosophy of Science* 30:1–29.

Johnston, David. 1986. *The Rhetoric of "Leviathan": Thomas Hobbes and the Politics of Cultural Transformation.* Studies in Moral, Political, and Legal Philosophy, ed. Marshall Cohen. Princeton: Princeton University Press.

Jones, J. F. 1983. Intelligible Matter and Geometry in Aristotle. *Apeiron* 17: 94–102.

Kästner, Abraham G. 1796–1800. *Geschichte der Mathematik seit der Wied-*

erherstellung der Wissenschaften bis an das Ende des achtzehnten Jahrhunderts. Göttingen: J. G. Rosenbusch.

Keller, Eve. 1992. In the Service of "Truth" and "Victory": Geometry and Rhetoric in the Political Works of Thomas Hobbes. *Prose Studies* 15: 129–52.

Kersting, Wolfgang. 1988. Erkenntnis und Methode in Thomas Hobbes' Philosophie. *Studia Leibnitiana* 20:126–39.

Klein, Jacob. 1968. *Greek Mathematical Thought and the Origin of Algebra.* Trans. Eva Brann. Cambridge: MIT Press.

Knorr, Wilbur Richard. 1986. *The Ancient Tradition of Geometric Problems.* Basel, Boston, and Berlin: Birkhäuser Verlag.

Köhler, Max. 1902. Die Naturphilosophie des Th. Hobbes in ihrer Abhängigkeit von Bacon. *Archiv für Geschichte der Philosophie* 15:371–99.

———. 1903. Studien zur Naturphilosophie des Th. Hobbes. *Archiv für Geschichte der Philosophie* 16:58–96.

Krämer, Sybille. 1991. *Berechenbare Vernunft: Kalkül und Rationalismus im 17. Jahrhundert.* Quellen und Studien zur Philosophie, ed. Günter Patzig, Erhard Scheibe, and Wolfgang Wieland, vol. 29. Berlin: Walter de Gruyter.

Kripke, Saul A. 1982. *Wittgenstein on Rules and Private Language: An Elementary Exposition.* Cambridge: Harvard University Press.

Krook, Dorothea. 1956. Thomas Hobbes's Doctrine of Meaning and Truth. *Philosophy: The Journal of the Royal Institute of Philosophy* 31:3–22.

Kuhn, Thomas S. 1970. *The Structure of Scientific Revolutions.* 2d ed. Chicago: University of Chicago Press.

———. 1992. *The Trouble with the Historical Philosophy of Science.* Robert and Maurine Rothschild Distinguished Lecture. Cambridge: Department of the History of Science, Harvard University.

Lachterman, David. 1989. *The Ethics of Geometry: A Genealogy of Modernity.* New York: Routledge.

Lalovera, Antonius [Antoine de La Loubère] 1651. *Quadratura circuli et hyperbolae segmentorum.* . . . Toulouse: P. Bosc.

Largeault, J. 1971. *Enquête sur le nominalisme.* Publications de la faculté des lettres et sciences humaines de Paris-Sorbonne, Série "Recherches", vol. 65. Louvain-Paris: Nauwelaerts.

Lasswitz, Kurd. 1890. *Die Geschichte der Atomistik von Mittelalter bis Newton.* 2 vols. Hamburg: Leopold Voss.

Lattis, James M. 1994. *Between Copernicus and Galileo: Christoph Clavius and the Collapse of Ptolemaic Cosmology.* Science and Its Conceptual Foundations, ed. David L. Hull. Chicago: University of Chicago Press.

Lear, Jonathan. 1982. Aristotle's Philosophy of Mathematics. *Philosophical Review* 91:161–92.

Leibniz, Gottfried Wilhelm. 1923–. *Sämtliche Schriften und Briefe,* ed. Prussian (later German) Academy of Sciences in Berlin. 7 series, 30 vols. to date. Darmstadt, Leipzig, and Berlin: Akademie Verlag.

Leijenhorst, Cornelis H. 1998. *Hobbes and the Aristotelians: The Aristotelian Setting of Thomas Hobbes's Natural Philosophy.* Quaestiones Infinitae, vol. 25. Utrecht: Zeno Institute for Philosophy.

Lindberg, David C., ed. 1978. *Science in the Middle Ages*. The Chicago History of Science and Medicine, ed. Allen G. Debus. Chicago: University of Chicago Press.

Lloyd, S. A. 1992. *Ideals as Interests in Hobbes's "Leviathan": The Power of Mind over Matter*. Cambridge: Cambridge University Press.

Longomontanus [Christian Severin Longborg]. 1612. *Cyclometria ex lunulis reciproce demonstrata . . . inventore Christiano S. Longomontano*. Copenhagen: H. Waldkirch.

———. 1627. *Cyclometria verè et absolutè . . . inventa. Cui accesit Introductio ad canonem trigonometriae sub initium et finem quadrantis circuli instaurandum*. Hamburg: n.p.

———. 1634. *Christiani Severini Longomontani . . . Inventio quadraturae circuli*. Copenhagen: n.p.

———. 1643. *Christiani Severini Longomontani, Problema quod, tam aequationibus in numeris, quam comparatione ad alia, diversimode conspirantia, absolutam circuli mensuram praestat*. Copenhagen: S. Sartorii.

———. 1644a. *Christiani Severini Longomontani . . . Rotundi in Plano, seu circuli absoluta mensura, duobus libellis comprehensa*. Amsterdam: J. Blaeu. Includes Pell 1644 as an addendum.

———. 1644b. Ἐλέγξεος *Joannis Pellii contra Christianum S. Longomontanum De Mensura Circuli* Ἀνασκευὴ. Copenhagen: n.p.

Loux, Michael J. 1974. The Ontology of William of Ockham. In *Ockham's Theory of Terms: Part 1 of the "Summa Logicae."* Ed. and trans. Michael J. Loux, 1–22. Notre Dame: Notre Dame University Press.

Lynch, William T. 1991. Politics in Hobbes' Mechanics: The Social as Enabling. *Studies in History and Philosophy of Science* 22:295–320.

Maanen, Jan A. van. 1986. The Refutation of Longomontanus' Quadrature by John Pell. *Annals of Science* 43:315–42.

Macdonald, Hugh, and Mary Hargreaves. 1952. *Thomas Hobbes: A Bibliography*. London: Bibliographical Society.

Macpherson, Crawford. B. 1968. Introduction to *Leviathan,* by Thomas Hobbes, 9–63. Harmondsworth: Penguin.

Mahoney, Michael S. 1978. Mathematics. In Lindberg 1978, 145–78.

———. 1980. The Beginnings of Algebraic Thought in the Seventeenth Century. In *Descartes: Philosophy, Mathematics, and Physics,* ed. Stephen Gaukroger, 141–55. Brighton, Sussex: Harvester Press.

———. 1990. Barrow's Mathematics: Between Ancients and Moderns. In *Before Newton: The Life and Times of Isaac Barrow,* ed. Mordechai Feingold, 179–249. Cambridge: Cambridge University Press.

———. 1994. *The Mathematical Career of Pierre de Fermat 1601–1665*. 2d ed. Princeton: Princeton University Press.

Maierù, Luigi. 1984. Il "meraviglioso problema" in Oronce Finé, Girolamo Cardano, e Jacques Peletier. *Bollettino di storia delle scienze matematiche* 4:141–70.

———. 1990. ". . . in Christophorum Clavium de Contactu Linearum Apologia": Considerazioni attorno alla polemica fra Peletier e Clavio circa l'angolo di contatto (1577–1589). *Archive for History of Exact Sciences* 41: 115–37.

———. 1991. John Wallis: Lettura della Polemica fra Peletier e Clavio circa l'angolo di contatto. *Giornate di Storia della Matematica* 4:317–64.

———. 1994. *Fra Descartes e Newton: Isaac Barrow e John Wallis.* Soveria Manelli: Rubbettino.

Malcolm, Noel. 1988. Hobbes and the Royal Society. In Rogers and Ryan 1988, 43–66.

———. 1990. Hobbes's Science of Politics and His Theory of Science. In Napoli and Canziani, 145–57.

Malet, Antoni. 1997a. Barrow, Wallis, and the Remaking of Seventeenth-Century Indivisibles. *Centaurus* 39:67–92.

———. 1997b. Isaac Barrow on the Mathematization of Nature: Theological Voluntarism and the Rise of Geometrical Optics. *Journal of the History of Ideas* 58:265–87.

Malherbe, Michel. 1990. Le religion materialiste de Thomas Hobbes. In Borrelli 1990, 51–70.

Mancosu, Paolo. 1992a. Aristotelian Logic and Euclidean Mathematics: Seventeenth-Century Developments of the *Quaestio de Certitudine Mathematicarum*. *Studies in History and Philosophy of Science* 23:241–65.

———. 1992b. Descartes's *Géométrie* and Revolutions in Mathematics. In *Revolutions in Mathematics*, ed. Donald Gillies, 83–116. Oxford: Oxford University Press, Clarendon Press.

———. 1996. *Philosophy of Mathematics and Mathematical Practice in the Seventeenth Century.* New York: Oxford University Press.

Mancosu, Paolo, and Ezio Vailati. 1991. Torricelli's Infinitely Long Solid and Its Philosophical Reception in the Seventeenth Century. *Isis* 82:50–70.

Mandey, Venterus. 1682. *Mellificium Mensionis: or, the Marrow of Measuring.* London: for the author.

Martinich, A. P. 1992. *The Two Gods of "Leviathan": Thomas Hobbes on Religion and Politics.* Cambridge: Cambridge University Press.

———. 1995. *A Hobbes Dictionary.* The Blackwell Philosophers Dictionaries. Oxford: Blackwell.

———. 1996. On the Proper Interpretation of Hobbes's Philosophy. *Journal of the History of Philosophy* 34:273–84.

———. 1997. *Thomas Hobbes.* British History in Perspective, ed. Jeremy Black. New York: St. Martin's Press.

McGregor, J. F., and B. Reay, eds. 1984. *Radical Religion in the English Revolution.* Oxford: Oxford University Press.

Medina, José. 1985. Les mathématiques chez Spinoza et Hobbes. *Revue philosophique* 2:177–88.

Meibom, Marcus. 1655. *De Proportionibus Dialogus.* Copenhagen: Melchior Martzanus.

Mersenne, Marin. 1644. *F. Marini Mersenni Minimi Cogitata Physico-Mathematica. In quibus tam naturae quam artis effectus admirandi certissimis demonstrationibus explicantur.* Paris: Antonii Bertier.

————. 1933–1988. *Correspondance du P. Marin Mersenne, religieux minime.* Ed. C. de Waard, R. Pintard, R. Lenoble, B. Rochot, and A. Beaulieu. 17 vols. Paris: CNRS.

Metzger, Hans-Dieter. 1991. *Thomas Hobbes und die englische Revolution.* Stuttgart-Bad Canstatt: Frommann-Holzboog.

Meyer, Manfred. 1992. *Leiblichkeit und Konvention: Struktur und Aporien der Wissenschaftsbegründung bei Hobbes und Poincaré.* Freiburg and Munich: Karl Alber.

Mintz, Samuel I. 1952. Galileo, Hobbes, and the Circle of Perfection. *Isis* 43:98–100.

————. 1962. *The Hunting of Leviathan: Seventeenth-Century Reactions to the Materialism and Moral Philosophy of Thomas Hobbes.* Cambridge: Cambridge University Press.

Mitchell, Alex F., and John Struthers, eds. 1874. *Minutes of the Sessions of the Westminster Assembly of Divines, while engaged in preparing their Directory for Church Government, Confession of Faith, and Catechisms (November 1644 to March 1649).* Edinburgh: William Blackwood and Sons.

Mitchell, Joshua. 1991. Luther and Hobbes on the Question: Who Was Moses, Who Was Christ? *Journal of Politics* 53:676–700.

Mittelstrass, Jürgen. 1979. The Philosopher's Conception of Mathesis Universalis from Descartes to Leibniz. *Annals of Science* 36:593–610.

Molland, George. 1976. Shifting the Foundations: Descartes' Transformation of Ancient Geometry. *Historia Mathematica* 3:21–49.

Moranus, G. 1655. *Animadversiones in Elementorum Philosophiae Sectionem I. De Corpore editam a Thoma Hobbes Anglo Malmesburiensi.* Brussels: n.p.

Moreau, Pierre-François. 1989. *Hobbes: Philosophie, science, religion.* Paris: Presses Universitaires de France.

Moser, H. 1923. *Thomas Hobbes: Seine logische Problematik und ihre erkenntnistheoretischen Voraussetzungen.* Berlin: Verlag Hellersberg.

Mueller, Ian. 1970. Aristotle on Geometrical Objects. *Archiv für Geschichte der Philosophie* 52:156–71.

————. 1981. *Philosophy of Mathematics and Deductive Structure in Euclid's "Elements."* Cambridge: MIT Press.

————. 1991. Sur les principes des mathématiques chez Aristote et Euclide. In Rashed 1991b, 102–13.

Mullan, David George. 1986. *Episcopacy in Scotland: The History of an Idea, 1560–1638.* Edinburgh: John Donald.

Mullett, Michael. 1994. Radical Sects and Dissenting Churches, 1600–1750. In *A History of Religion in Britain: Practice and Belief from Pre-Roman Times to the Present,* ed. Sheridan Gilley and W. J. Shields, 188–210. Oxford: Blackwell.

Murdoch, John. 1968. The Medieval Euclid: Salient Aspects of the Translations of the *Elements* by Adelard of Bath and Campanus of Novara. *Revue de synthèse* 49:67–94.

Napoli, Andrea, and Guido Canziani, eds. 1990. *Hobbes Oggi.* Proceedings of the international conference directed by Arrigo Pacchi (Milan-Locarno, 18–21 May 1988). Milan: Franco Angeli.

Naux, Charles. 1983. Le père Christophore Clavius: Sa vie et son oeuvre. *Revue des questions scientifiques.* 154:55–67; 181–93; 325–47.

Newton, Isaac. 1967–81. *The Mathematical Papers of Isaac Newton.* Ed. D. T. Whiteside and M. A. Hoskins. 8 vols. Cambridge: Cambridge University Press.

North, J. D. 1983. Finite and Otherwise: Aristotle and Some Seventeenth-Century Views. In *Nature Mathematized,* ed. William R. Shea, 113–48. Dordrecht and Boston: Reidel.

Ockham, William. 1974. *Ockham's Theory of Terms: Part 1 of the "Summa Logicae."* Ed. and trans. Michael J. Loux. Notre Dame: Notre Dame University Press.

Oldenburg, Henry, 1965–77. *The Correspondence of Henry Oldenburg.* Ed. and trans. A. Rupert Hall and Marie Boas Hall. 11 vols. Madison: University of Wisconsin Press; London: Mansell.

Oughtred, William. 1693. *Clavis Mathematicæae Denuo Limata, Sive potius Fabricata.* 5th ed. Oxford: Litchfield.

Owen, John. 1970. *The Correspondence of John Owen (1616–1683), With an account of his life and work.* Ed. Peter Toon. Cambridge and London: James Clarke.

Pacchi, Arrigo. 1965. *Convenzione e ipotesi nella formazione della filosofia naturale di Thomas Hobbes.* Pubblicazioni dell'Istituto di Storia Della Filosofia dell'Università degli Studi di Milano, no. 7. Florence: La Nuova Italia Editrice.

———. 1968. Una "biblioteca ideale" di Thomas Hobbes: Il ms E2 dell'archivo di Chatsworth. *Acme* 21:5–42.

Pappus of Alexandria. [1875] 1965. *Pappi Alexandrini Collectionis quae Supersunt, E Libris Manu Scripts Edidit.* 3 vols. Ed. and trans. Friedrich Hultsch. Berlin. Reprint, Amsterdam: Adolf Hakkert.

Parsons, Robert. 1680. *A Sermon Preached at the Funeral of the Rt. Honorable John, Earl of Rochester.* Oxford: Richard Davis and Thomas Bowman.

Pedersen, Kristi Møller. 1970. Roberval's Comparison of the Arclength of a Spiral and a Parabola. *Centaurus* 15:26–43.

Peletier, Jacques. 1557. *In Euclidis Elementa Geometrica Demonstrationum Libri sex.* . . . Lyon: Tornaes.

Pell, John. 1644. I. Pellius contra Ch. S. Longomontanum. Amsterdam: J. Blaeu. Printed as an addendum to Longomontanus 1644a.

———. 1647. *Controversiae de verâ circuli mensurâ anno* MDCXLIV *exortae, inter Christianum Severini, Longomontanum . . . et Iohannes Pellium . . . Prima pars.* Amsterdam: J. Blaeu.

Pereira, Benedictus. 1576. *De communibus omnium rerum naturalium principiis et affectionibus libri quindecim.* . . . Lyon.

Perez-Ramos, Antonio. 1988. *Francis Bacon's Idea of Science and the Maker's Knowledge Tradition.* Oxford: Oxford University Press, Clarendon Press.

Peters, Richard. 1956. *Hobbes.* Harmondsworth: Penguin.

Pope, Walter. 1697. *The Life of the Right Reverend Father in God, Seth, Lord Bishop of Salisbury, and Chancellor of the Most Noble Order of the Garter.*

With a Brief Account of Bishop Wilkins, Mr. Lawrence Rooke, Dr. Isaac Barrow, Dr. Turberville, and others. London: n.p.

Prag, Adolf. 1931. John Wallis (1616–1703). Zur Ideengeschichte der Mathematik im 17. Jahrhundert. *Quellen und Studien zur Geschichte der Mathematik, Astronomie und Physik,* pt. B, 1:381–411.

Prall, Stuart E., ed. 1968. *The Puritan Revolution: A Documentary History.* New York: Doubleday.

Prins, Jan. 1988. De Oorsprong en Betekenis van Hobbes' Geometrische Methodenideaal. *Tijdschrift voor Philosophie* 50:248–71.

———. 1990. Hobbes and the School of Padua: Two Incompatible Approaches of Science. *Archiv für Geschichte der Philosophie* 72:26–46.

———. 1993. Ward's Polemic with Hobbes on the Sources of His Optical Theories. *Revue d'histoire des sciences* 46:195–224.

Probst, Siegmund. 1993. Infinity and Creation: The Origin of the Controversy between Thomas Hobbes and the Savilian Professors Seth Ward and John Wallis. *British Journal for the History of Science* 26:271–79.

———. 1997. Die mathematische Kontroverse zwischen Thomas Hobbes und John Wallis. Ph.D. diss., Universität Regensburg.

Proclus. 1970. *A Commentary on the First Book of Euclid's "Elements."* Trans. and ed. Glen R. Morrow. Princeton: Princeton University Press.

Purver, Margery. 1967. *The Royal Society: Concept and Creation.* Introduction by H. R. Trevor-Roper. London: Routledge and Kegan Paul.

Pycior, Helena M. 1987. Mathematics and Philosophy: Wallis, Hobbes, Barrow, and Berkeley. *Journal of the History of Ideas* 48:265–86.

———. 1997. *Symbols, Impossible Numbers, and Geometric Entanglements: British Algebra through the Commentaries on Newton's "Universal Arithmetick."* Cambridge: Cambridge University Press.

Quine, W. V. 1990. Elementary Proof that Some Angles Cannot be Trisected by Ruler and Compass. *Mathematics Magazine* 63:95–105.

Rashed, Roshdi. 1991a. L'analyse et la synthèse selon Ibn Al-Haytham. In Rashed 1991b, 131–62.

———, ed. 1991b. *Mathématiques et philosophie de l'antiquité à l'âge classique: Hommage à Jules Vuillemin.* Paris: CNRS.

Rigaud, Stephen Jordan, ed. [1841] 1965. *Correspondence of scientific men of the seventeenth century.* Foreword by J. E. Hoffman. Oxford. Reprint, Hildesheim, Germany: Georg Olms.

Robbe, Martin. 1960. Zu Problemen der Sprachphilosophie bei Thomas Hobbes. *Deutsche Zeitschrift für Philosophie* 8:433–50.

Robertson, G. C. [1886] 1910. *Hobbes.* Edinburgh and London: William Blackwood.

Roberval, Gilles Personne de. 1693. Ouvrages de mathématique de M. Roberval. *Divers ouvrages de Mathématique et de Physique par MMrs. de l'Academie Royale des Sciences* 1:69–305.

Robinet, André. 1979. Pensée et langage chez Hobbes: Physique de la parole et *translatio. Revue internationale de philosophie.* 33:452–83.

———. 1990. Hobbes: structure et nature du *conatus.* In Zarka and Bernhardt 1990, 139–51.

Röd, Wolfgang. 1970. *Geometrischer Geist und Naturrecht: Methodenge-schichtliche Untersuchungen zur Staatsphilosphie im 17. und 18. Jahrhundert.* Bayerishe Akademie der Wissenschaften, Philosophische-Historische Klasse, Abhandlungen, n.s., vol. 70. Munich: Bavarian Academy of Sciences.

Rogers, G. A. J. 1988. Hobbes's Hidden Influence. In Rogers and Ryan, 189–206.

———. 1990. Religion and the Explanation of Action in the Thought of Thomas Hobbes. In Borrelli 1990, 35–50.

Rogers, G. A. J., and Alan Ryan, eds. 1988. *Perspectives on Thomas Hobbes.* Mind Association Occasional Series. Oxford: Oxford University Press, Clarendon Press.

Roncaglia, Gino. 1995. Smiglecius on *entia rationis*. *Vivarium* 31:27–49.

Roux, Louis. 1990. Modèles théologiques et modèles scientifiques dans la pensée de Thomas Hobbes. In Borrelli 1990, 83–94.

Sacksteder, William. 1980. Hobbes: The Art of the Geometricians. *Journal of the History of Philosophy* 18:131–46.

———. 1981a. Hobbes: Geometrical Objects. *Philosophy of Science* 48: 573–90.

———. 1981b. Some Ways of Doing Language Philosophy: Nominalism, Hobbes, and the Linguistic Turn. *Review of Metaphysics* 34:459–85.

———. 1992. Three Diverse Science in Hobbes: First Philosophy, Geometry, and Physics. *Review of Metaphysics* 45:739–72.

Saito, Ken. 1986. Compounded Ratio in Euclid and Apollonius. *Historia Scientiarum* 31:25–59.

———. 1993. Duplicate Ratio in Book VI of Euclid's *Elements. Historia Scientiarum* 2d ser., 3:115–35.

———. 1995. Doubling the Cube: A New Interpretation of Its Significance for Early Greek Geometry. *Historia Mathematica* 22:119–37.

Sasaki, Chikara. 1985. The Acceptance of the Theory of Proportion in the Sixteenth and Seventeenth Centuries: Barrow's Reaction to the Analytic Mathematics. *Historia Scientiarum* 29:83–116.

Savile, Henry. 1621. *Praelectiones Tresdecim in Principium Elementorum Euclidis, Oxonii Habitae* MDCXX. Oxford: John Litchfield for Jacob Short.

Schaffer, Simon. 1988. Wallifaction: Thomas Hobbes on School Divinity and Experimental Pneumatics. *Studies in History and Philosophy of Science* 19:275–98.

Schumann, Karl. 1984. Francis Bacon und Hobbes' Widmungsbrief zu De Cive. *Zeitschrift für philosophische Forschung* 38:165–90.

———. 1985. Geometrie und Philosophie bei Thomas Hobbes. *Philosophisches Jahrbuch* 92:161–77.

———. 1995. Hobbes dans les Publications de Mersenne en 1644. *Archives de philosophie* 58:2–7.

———. 1997. Introduction to *Elemente der Philosophie, Erste Abteilung, Der Körper,* by Thomas Hobbes. Ed. and trans. Karl Schumann. Philosophische Bibliothek, vol. 501. Hamburg: Felix Meiner Verlag.

Scott, J. F. 1938. *The Mathematical Work of John Wallis, D. D., F.R.S. (1616–1703).* London: Taylor and Francis.

Scriba, Christoph J. 1966. *Studien zur Mathematik des John Wallis (1616–1703): Winkelteilungen, Kombinationslehere und zahlentheoretische Probleme.* Wiesbaden: Franz Steiner Verlag.

———. 1967. A Tentative Index of the Correspondence of John Wallis, F.R.S. *Notes and Records of the Royal Society* 22:58–93.

———. 1970. The Autobiography of John Wallis, F.R.S. *Notes and Records of the Royal Society* 25:17–46.

Shapin, Steven. 1982. History of Science and Its Sociological Reconstructions. *History of Science* 20:157–211.

———. 1992. Discipline and Bounding: The History and Sociology of Science as Seen through the Externalism-Internalism Debate. *History of Science* 30:333–69.

Shapin, Steven, and Simon Schaffer. 1985. *Leviathan and the Air-Pump: Hobbes, Boyle, and the Experimental Life.* Princeton: Princeton University Press.

Shapiro, Barbara J. 1969. *John Wilkins, 1614–1672: An Intellectual Biography.* Berkeley and Los Angeles: University of California Press.

Sherry, David. 1997. On Mathematical Error. *Studies in History and Philosophy of Science* 28:393–416.

Shields, W. J. 1994. Reformed Religion in England, 1520–1640. In *A History of Religion in Britain: Practice and Belief from Pre-Roman Times to the Present,* ed. Sheridan Gilley and W. J. Shields, 151–67. Oxford: Blackwell.

Silver, Victoria. 1996. Hobbes on Rhetoric. In Sorell 1996a, 329–45.

Siorvanes, Lucas. 1996. *Proclus: Neo-Platonic Philosophy and Science.* New Haven: Yale University Press.

Skinner, Quentin. 1965. History and Ideology in the English Revolution. *Historical Journal* 8:151–78.

———. 1966. The Ideological Context of Hobbes's Political Thought. *Historical Journal* 9:286–317.

———. 1969. Thomas Hobbes and the Nature of the Early Royal Society. *Historical Journal* 12:217–39.

———. 1972. Conquest and Consent: Thomas Hobbes and the Engagement Controversy. In *The Interregnum: The Quest for Settlement 1646–1660,* ed. Gerald E. Aylmer, 79–98. Problems in Focus. London: Macmillan.

———. 1996. *Reason and Rhetoric in the Philosophy of Thomas Hobbes.* Cambridge: Cambridge University Press.

Smigleckius, Martin. 1634. *Logica Martini Smigleckii Societatis Jesu S. Theologiae Doctoris, Selectis disputationibus & quaestionibus illustrata.* 2 vols. Oxford: H. Crypps, E. Forrest, and H. Curteyne.

Snell, Willebrord. 1621. *Cyclometricus, De circuli dimensione secundum Logistarum abacos, & ad Mechanicem accuratissima; atque omnium parabilissima.* Leyden: Elzevir.

Sommerville, Johann P. 1992. *Thomas Hobbes: Political Ideas in Historical Context.* London: Macmillan.

Sorbière, Samuel. 1664. *Relation d'un voyage en Angleterre*. Paris: Louis Billaine.

Sorell, Tom. 1986. *Hobbes*. Arguments of the Philosophers, ed. Ted Honderich. London: Routledge & Kegan Paul.

———. 1988. Descartes, Hobbes, and the Body of Natural Science. *Monist* 71:515–25.

———. 1995. Hobbes's Objections and Hobbes's System. In Ariew and Grene, 83–96.

———, ed. 1996a. *The Cambridge Companion to Hobbes*. Cambridge: Cambridge University Press.

———. 1996b. Hobbes's Scheme of the Sciences. In Sorell 1996a, 45–61.

Spragens, Thomas A. 1973. *The Politics of Motion: The World of Thomas Hobbes*. Lexington: University Press of Kentucky.

Springborg, Patricia. 1996. Hobbes on Religion. In Sorell 1996a, 346–80.

Stein, Howard. 1990. Eudoxos and Dedekind: On Ancient Greek Theory of Ratios and Its Relation to Modern Mathematics. *Synthese* 84:163–211.

Strong, Tracy. 1993. How to Write Scripture: Words, Authority, and Politics in Thomas Hobbes. *Critical Inquiry* 20:128–59.

Stubbe, Henry. 1657. *Clamor, Rixa, Joci, Mendacia, Furta, Cachiny; or, a Severe Enquiry into the late Oneirocritica Published by John Wallis, Grammar-Reader in Oxon*. London: n.p.

———. 1658. *The Savilian Professours Case Stated. Together with the severall reasons urged against his capacity of standing for the publique Office of Antiquary in the University of Oxford. . . .* London: J. T. for Andrew Crooke.

———. 1670a. *The Plus Ultra Reduced to a Nonplus*. London.

———. 1670b. *A Censure upon certain Passages contained in the History of the Royal Society, as being destructive to the Established Religion of the Church of England*. Oxford: Richard Davis.

———. 1670c. *Legends no Histories; or, A Specimen of some Animadverseions upon the History of the Royal Society*. London: n.p.

———. 1671. *A reply unto the Letter written to Mr H. S. in defense of the History of the Royal Society*. Oxford: Richard Davis.

———. 1973. *Lettere di Henry Stubbe a Thomas Hobbes (8 luglio 1656–6 maggio 1657)*. Ed. Onofrio Nicastro. Siena: Università degli Studi, Facoltà de Lettere e Filosofia.

Sylla, Edith. 1984. Compounding Ratios: Bradwardine, Oresme, and the First Edition of Newton's *Principia*. In *Transformation and Tradition in the Sciences*, ed. Everet Mendelsohn, 11–43. Cambridge: Cambridge University Press.

Tacquet, Andreas. 1669. *Opera Mathematica*. 6 vols. Antwerp: Jacob Meurs.

Talaska, Richard A. 1988. Analytic and Synthetic Method According to Hobbes. *Journal of the History of Philosophy* 26:207–37.

Tenison, Thomas. 1670. *The Creed of Mr. Hobbes Examined; in a feigned Conference Between Him and a Student in Divinity*. London: Francis Tyton.

Thomas, Keith. 1987. Numeracy in Early Modern England (The Prothero Lecture). *Transactions of the Royal Historical Society.* 37:103–32.

Thomason, S. K. 1982. Euclidean Infinitesimals. *Pacific Philosophical Quarterly* 63:168–85.

Tönnies, Ferdinand. 1925. *Thomas Hobbes: Leben und Lehre.* 3d ed. Frommanns Klassiker der Philosophie, ed. Georg Mehlis, no. 2. Stuttgart: Frommann.

———. 1975. *Studien zur Philosophie und Gesellschaftslehre im 17. Jahrhundert.* Ed. E. G. Jacoby. Stuttgart-Bad Cannstatt: Frommann-Holzboog.

Tricaud, François. 1990. La doctrine du salut dans le *Léviathan.* In Borrelli 1990, 3–14.

Triplett, Tim. 1986. Relativism and the Sociology of Mathematics: Remarks on Bloor, Flew, and Frege. *Inquiry* 29:439–50.

Tuck, Richard. 1988a. Hobbes and Descartes. In Rogers and Ryan 1988, 11–42.

———. 1988b. Optics and Sceptics: The Philosophical Foundations of Hobbes's Political Thought. In *Conscience and Casuistry in Early Modern Europe,* ed. Edmund Leites, 235–63. Cambridge: Cambridge University Press; Paris: Editions de la Maison des Sciences de l'Homme.

———. 1989. *Hobbes.* Past Masters. Oxford: Oxford University Press, Clarendon Press.

Vaughn, Robert. 1838. *The Protectorate of Oliver Cromwell, and the State of Europe During the Early Part of the Reign of Louis XIV.* 2 vols. London: Henry Colburn.

Vavilov, Valery. 1992. Calculating π: The Contribution of Christiaan Huygens. *Quantum* 2:45–50.

Viéte, François. 1646. *Francisci Vietæae Opera Mathematica, in unum Volumen congesta, ac recognita, Operâ atque studio Francisci à Schooten Leyudensis, Matheseos Professoris.* Leyden: Elzevir.

Vita, Vincenzo. 1973. Gli indivisibili di Roberval. *Archimede* 25:38–46.

Vitrac, Bernard. 1992. Logistique et fractions dans le monde hellénistique. In Benoit, Chemla, and Ritter 1992, 149–72.

Walker, Eric C. 1970. *William Dell: Master Puritan.* Cambridge: W. Heffer and Sons.

Walker, Evelyn. 1932. *A Study of the "Traité des Indivisibles" of Gilles Persone de Roberval.* New York: Teachers' College, Columbia University.

Wallis, John. 1642. *Truth Tried: or, Animadversions On a Treatise published by the Right Honorable Robert Lord Brook, Entituled "The Nature of Truth, Its Union and Unity with the Soule," Which (saith he) is One in its Essence, Faculties, Acts; One with Truth.* London: Richard Bishop for Samuel Gellibrand.

———. 1655. *Elenchus Geometriae Hobbianae, sive, Geometricorum, quae in "Elementis Philosophiae," à Thoma Hobbes Malmesburiensi proferuntur, Refutatio.* Oxford: H. Hall for John Crooke.

———. 1656a. *Due Correction for Mr. Hobbes; or Schoole Discipline, for not saying his Lessons right. In Answer To His Six Lesons, directed to the Professors of Mathematics.* Oxford: L. Litchfield for Thomas Robinson.

————. 1656b. . . . *Operum Mathematicorum Pars Altera: qua continentur De Angulo Contactus et Semicirculi Disquisitio Geometrica. De Sectionibus Conicis Tractatus. Arithmetica Infinitorum. Eclipsis Solaris Observatio.* Oxford: L. Litchfield for T. Robinson.

————. 1657a. *Mens Sobria Seriò Commendata.* Oxford: Leonard Litchfield for Thomas Robinson.

————. 1657b. *Hobbiani Puncti Dispunctio; or The Undoing of Mr Hobs's Points: In Answer to M. Hobs's* STIGMAI, *Id est, Stigmata Hobbii.* Oxford: Leonard Litchfield for Thomas Robinson.

————. 1657c. . . . *Operum Mathematicorum Pars Prima: qua continentur Oratio Inauguralis. Mathesis Universalis; sive, Arithemeticum Opus Integrum, tum Numerosam Arithmeticam tum Speciosam complectens. Adversus Meibomii, de Proportionibus Dialogum, Tractatus Elencticus.* Oxford: L. Litchfield for T. Robinson.

————. 1658. *Commercium Epistolicum de Quaestionibus quibusdam Mathematicis nuper habitum.* Oxford: A. Litchfield for Thomas Robinson.

————. 1662. *Hobbius Heauton-timorumenos; or A Consideration of Mr. Hobbes his Dialogues in An Epistolary Discourse Addressed to the Honourable Robert Boyle, Esq.* Oxford: A. & L. Litchfield for Samuel Thomson.

————. 1666. Review of Hobbes, *De Principiis et Ratiocinatione Geometrarum* [in a letter addressed to Henry Oldenburg]. *Philosophical Transactions of the Royal Society* 16:289–94.

————. 1669a. *Thomae Hobbes Quadratura Circuli, Cubatio Sphaerae, Duplicatio Cubi; Confutata.* Oxford: A. Litchfield for Thomas Gilbert.

————. 1669b. *Thomae Hobbes Quadratura Circuli, Cubatio Sphæræ, Duplicatio Cubi; (Secundò Edita) Denuo Refutata.* Oxford: A. Litchfield for Thomas Gilbert.

————. 1670. *Mechanica; sive, De Motu, Tractatus Geometricus.* London: M. Pitt.

————. 1671. An Answer to Four Papers of Mr. Hobs, lately Published in the Months of August, and this present September, 1671. *Philosophical Transactions of the Royal Society* 6:2241–50.

————. 1678. *A Defence of the Royal Society, and the Philosophical Transactions, particularly those of July, 1670. In Answer to the Cavils of Dr. William Holder.* London: T. Moore.

————. 1684. *A Defense of the Treatise of the Angle of Contact.* London: John Playford for Richard Davis.

————. 1685. *A Treatise of Algebra, Both Historical and Practical. Shewing, The Original, Progress, and Advancement thereof, from time to time; and by what Steps it hath attained to the Height at which it now is.* London: John Playford for Richard Davis.

————. 1692a. *A Defense of the Christian Sabbath. In answer to a Treatise of Mr. Tho. Bampfield Pleading for Saturday-Sabbath.* London: L. Litchfield for C. Comingsby.

————. 1692b. *Theological discourses; containing viii letters and iii sermons concerning the Blessed Trinity.* London: Tho. Parkhurst.

————. 1693–99. *Johannis Wallis S.T.D . . . Opera Mathematica*. 3 vols. Oxford: at the Sheldonian Theater.

————. 1696. *An answer to Dr. Sherlock's Examination of the Oxford decree; in a letter from a member of that university, to his friend in London*. London: M. Whitlock.

————. 1697. *A defense of infant-baptism. In answer to A letter (here recited) from an anti-paedo-Baptist*. Oxford: L. Litchfield for H. Clements.

————. 1791. *Sermons; Now First Printed from The Original Manuscripts of John Wallis, D. D. Some Time Savilian Professor of Geometry in the University of Oxford, Keeper of the Archives, Member of the Royal Society, and Chaplain Ordinary to King Charles II. To Which are Prefixed, Memoirs of the Author, With Some Original Anecdotes; and A Recommendatory Introduction, by the Rev. C. E. De Coetlogon, M. A.* Ed. W. Wallis. London: J. Nichols for G.G.J. and J. Robinson.

————. 1970. The Autobiography of John Wallis, F.R.S. Ed. Christoph J. Scriba. *Notes and Records of the Royal Society* 25:17–46.

Wallner, C. R. 1903. Die Wandlungen des Indivisibilienbegriffs von Cavalieri bis Wallis. *Bibliotheca Mathematica*, 3d ser. 4:28–47.

Ward, John. [1740] 1967. *The Lives of the Professors of Gresham College, to which is prefixed the Life of the founder sir Thomas Gresham*. Reprint, Sources of Science, no. 71, ed. Harry Woolf. New York: Johnson Reprint.

Ward, Seth. 1652. *A Philosophicall Essay Towards an Eviction of The Being and Attributes of God, The Immortality of the Souls of Men, The Truth and Authority of Scripture. Together with an Index of the Heads of every Particular Part*. Oxford: L. Litchfield for John Adams and Edward Forrest.

————. 1656. *In Thomae Hobbii Philosophiam Exercitatio Epistolica*. Oxford: H. Hall for Richard Davis.

————. 1670. *The Christians Victory over Death: A Sermon at the Funeral of the Most Honourable George, Duke of Albemarle, &c. In the Collegiate Church of S. Peter Westminster on the XXXth of April, M.DC.LXX*. London: James Collins.

[Ward, Seth]. 1654. *Vindicae Academiarum; Containing, Some briefe Animadversions upon Mr Webster's Book, Stiled the Examination of Academies. Together with an Appendix concerning what M. Hobbs and M. Dell have published on this Argument*. Oxford: L. Litchfield for Thomas Robinson.

Watkins, J. W. N. 1965. *Hobbes's System of Ideas: A Study in the Political Significance of Scientific Theories*. London: Hutchinson.

Webster, Charles. 1973. William Dell and the Idea of University. In *Changing Perspectives in the History of Science*, ed. Mikulas Teich and Robert Young, 110–26. London: Heinemann.

Webster, John. 1654. *Academiarum Examen; or the Examination of the Academies*. London: Giles Calvert.

Weinreich, Hermann. 1911. *Über die Bedeutung des Hobbes für das naturwissenschaftliche und mathematische Denken*. Leipzig: Robert Noske.

Weiss, Ulrich. 1978. Hobbes' Rationalismus: Aspekte der neueren deutschen Hobbes-Rezeption. *Philosophisches Jahrbuch* 85:167–96.

————. 1980. *Das philosophische System von Thomas Hobbes*. Stuttgart–Bad Cannstatt: Frommann-Holzboog.

————. 1983a. Wissenschaft und menschliches Handeln. Zu Thomas Hobbes' anthropologischen Fundierung von Wissenschaft. *Zeitschrift für philosophische Forschung* 37:37–55.

————. 1983b. Der mos geometricus als Paradigma von Wissenschaft bei Hobbes und Descartes. *Deutsche Zeitschrift für Philosophie*. 40:1295–1302.

Westfall, Richard S. 1958. *Science and Religion in Seventeenth-Century England*. New Haven: Yale University Press.

Whitehall, J. 1680. *Behemoth Arraign'd: Or, A Vindication of Property Against a Fanatical Pamphlet Stiled Behemoth: or, the History of the Civil Wars of England, from 1640 to 1660. Subscribed by Tho. Hobbes of Malmesbury*. London: Thomas Fox.

Whiteside, Derek T. 1960–62. Patterns of Mathematical Thought in the Later Seventeenth Century. *Archive for History of the Exact Sciences* 1:179–338.

Willms, Bernard. 1990. One Head, One Sword, One Crozier: The Significance of Theology in Hobbes's *Leviathan*. In Borrelli 1990, 71–82.

Wittgenstein, Ludwig. 1956. *Remarks on the Foundations of Mathematics*. Oxford: Blackwell.

Wood, Anthony à. [1813] 1967. *Athenae Oxonienses; An Exact History of All the Writers and Bishops Who Have Had Their Education in the University of Oxford, to Which are Added the Fasti, or Annals of the Said University*. 4 vols. London. Reprint, Sources of Science, no. 55. New York: Johnson Reprint.

Yoder, Joella G. 1988. *Unrolling Time: Christiaan Huygens and the Mathematization of Nature*. Cambridge: Cambridge University Press.

Zarka, Yves-Charles, 1985. Empirisme, nominalisme et matérialisme chez Hobbes. *Archives de philosophie* 48:177–233.

————. 1987. *La décision métaphysique de Hobbes: Conditions de la politique*. Paris: Vrin.

————. 1996. First Philosophy and the Foundation of Knowledge. In Sorell 1996a, 62–85.

————, ed. 1992. *Hobbes et son vocabulaire: Études de lexicographie philosophique*. Bibliothèque d'Histoire de la Philosophie, n.s. Paris: Vrin.

Zarka, Yves-Charles, and Jean Bernhardt, eds. 1990. *Thomas Hobbes: Philosophie première, théorie de la science et politique*. Colloquium for the four hundredth birthday of Thomas Hobbes. Paris: Presses Universitaires de France.

Index